Advances in

Nutritional Research

Volume 4

Advances in

Nutritional Research

A Continuation Order Plan is available for this series. A continuation order will bring delivery of each new volume immediately upon publication. Volumes are billed only upon actual shipment. For further information please contact the publisher.

Advances in Nutritional Research

Volume 4

Edited by Harold H. Draper
University of Guelph
Guelph, Ontario, Canada

Plenum Press · New York and London

The Library of Congress cataloged the first volume of this title as follows:

Advances in nutritional research. v. 1–
New York, Plenum Press, c1977–
1 v. ill. 24 cm.
Key title: Advances in nutritional research, ISSN 0149-9483

1. Nutrition–Yearbooks.
QP141.A1A3 613.2'05 78-640645

DOI 10.1007/978-1-4613-9934-6

MyCopy version of the original edition 1982
A Division of Plenum Publishing Corporation
233 Spring Street, New York, N.Y. 10013

Contributors

William J. Bettger, Department of Molecular, Cellular and Developmental Biology, University of Colorado, Boulder, Colorado 80309. Present address: Department of Biochemistry, St. Louis University School of Medicine, St. Louis, Missouri 63104

Peggy R. Borum, Division of Nutrition, Department of Biochemistry, Vanderbilt University, Nashville, Tennessee 37232

Harry P. Broquist, Division of Nutrition, Department of Biochemistry, Vanderbilt University, Nashville, Tennessee 37232

John T. Brosnan, Department of Biochemistry, Memorial University of Newfoundland, St. John's, Newfoundland A1B 3X9, Canada

Margaret E. Brosnan, Department of Biochemistry, Memorial University of Newfoundland, St. John's, Newfoundland A1B 3X9, Canada

R. H. Dadd, Division of Entomology and Parasitology, University of California, Berkeley, California 94720

Krishnamurti Dakshinamurti, Department of Biochemistry, Faculty of Medicine, University of Manitoba, Winnipeg, Manitoba R3E 0W3, Canada

W. G. Friend, Department of Zoology, University of Toronto, Toronto, Ontario M5S 1A1, Canada

John G. Haddad, Jr., Endocrine Division, University of Pennsylvania Medical School, Philadelphia, Pennsylvania 19104

Richard G. Ham, Department of Molecular, Cellular and Developmental Biology, University of Colorado, Boulder, Colorado 80309

Bruce W. Hollis, Department of Medicine, Case Western Reserve University;

and Department of Medicine, Division of Endocrinology and Mineral Metabolism, Veterans Administration Medical Center, Cleveland, Ohio 44106

Bruce J. Holub, Department of Nutrition, College of Biological Science, University of Guelph, Guelph, Ontario N1G 2W1, Canada

Phillip W. Lambert, Department of Medicine, Case Western Reserve University; and Department of Medicine, Division of Endocrinology and Mineral Metabolism, Veterans Administration Medical Center, Cleveland, Ohio 44106

S. Harvey Mudd, Laboratory of General and Comparative Biochemistry, National Institute of Mental Health, Bethesda, Maryland 20205

Bernard A. Roos, Department of Medicine, Case Western Reserve University; and Department of Medicine, Division of Endocrinology and Mineral Metabolism, Veterans Administration Medical Center, Cleveland, Ohio 44106

Raymond Clifford Noble, Department of Biochemistry, The Hannah Research Institute, Ayr KA6 5HL, Scotland

John Herbert Shand, Department of Biochemistry, The Hannah Research Institute, Ayr KA6 5HL, Scotland

Preface

Volume 4 of *Advances in Nutritional Research* reflects the increased importance that recently has been attached to nutrition in many fields of clinical medicine. This heightened interest in nutrition stems from the demonstration that the intake of specific nutrients may have far-reaching consequences, not only for normal metabolism, but also for metabolic processes affecting clinical or subclinical disease. Conversely, many disease states have been shown to have previously unrecognized effects on nutrient function and metabolism.

In addition to topics of obvious relevance to human clinical nutrition, this volume contains chapters dealing with the nutrition of cells grown in culture and of species that may provide insights into nutritional disorders of man. Together with its predecessors, Volume 4 provides graduate students and established investigators with authoritative accounts of the status of research on a range of topics of current interest in experimental and clinical nutrition.

Contents

Chapter 1

Vitamin-Responsive Genetic Abnormalities

S. Harvey Mudd

1. Introduction

This review will focus on certain aspects of vitamin-responsive genetic disorders. To be included, a condition must be gentically determined, and its characteristic chemical and/or biochemical manifestations must be alleviated by larger than physiological amounts of a particular vitamin or by use of an unusual route of administration of that vitamin. Such conditions have recently attracted much attention from human geneticists and from those concerned clinically and biochemically with inborn errors of metabolism. Previous reviews provide adequate coverage of the clinical features of these conditions and of many details concerning the history, structure, and the biochemical role of the particular vitamins to be discussed (Frimpter *et al.*, 1969; Mudd, 1971, 1974a, 1977; Scriver, 1973; Rosenberg, 1976). In general, these matters will be beyond the scope of this chapter. Here, emphasis will be on the general properties that characterize vitamin-responsive conditions, the mechanism or mechanisms underlying vitamin responsiveness, and the implications of our present understanding of these matters both for those concerned with individual patients with

Abbreviations used: PLP, pyridoxal 5′-phosphate; OH-Cbl, hydroxocobalamin; CN-Cbl, cyanocobalamin; AdoCbl, adenosylcobalamin; MeCbl, methylcobalamin; TC II, transcobalamin II.

S. Harvey Mudd • Laboratory of General and Comparative Biochemistry, National Institute of Mental Health, Bethesda, Maryland 20205.

metabolic diseases and for those concerned with more general aspects of human nutrition.

2. Historical Perspective

In 1937, Albright and his colleagues demonstrated that a 16-year-old boy with intractable rickets and hypophosphatemia was clinically benefited by use of vitamin D but only if this compound were given at almost 1000 times its usual requirement (Albright *et al.*, 1937). Subsequently, it was shown that such "vitamin D-resistant rickets" was familial (Christensen, 1940–1941), following what has been established as an X-linked dominant mode of transmission (Rasmussen and Anast, 1978). Together, these findings provided the first clear demonstration of the phenomenon that is now called vitamin responsiveness in genetic disease.

Some years later, a second example of vitamin responsiveness was furnished when Hunt and his associates (1954) demonstrated that severe seizures in a newborn girl could be controlled by ten times the usual requirement of pyridoxine. It is now known that this disorder is inherited as an autosomal recessive trait (Scriver and Whelan, 1969). Within little more than a decade of the discovery of this first B_6-responsive condition, a number of additional genetic disorders were found to be responsive to pyridoxine. In one of these disorders, cystathioninuria, the specific underlying enzyme defect was pinpointed for the first time in a condition known to be vitamin responsive by the demonstration that γ-cystathionase activity was deficient (Frimpter, 1965; Finkelstein *et al.*, 1966). Moreover, the mutant enzymes from two patients displayed abnormal interactions with the cofactor, pyridoxal 5′-phosphate (PLP) (Frimpter, 1965). At the time, precedents for such abnormal interactions were known from microbial genetics, and the implications of such abnormalities for vitamin-responsive conditions in humans had already been suggested by Scriver (1964). Abnormal interaction between mutant apoenzymes and their cofactors has emerged as an important general cause of vitamin responsiveness, although much remains to be learned about the specifics of such abnormalities. These questions will be discussed more extensively in ensuing sections.

About a decade ago, a somewhat different group of vitamin-responsive conditions came into sharper focus with the discovery that certain patients with methylmalonic aciduria or with methylmalonic aciduria and homocystinuria had defects in their abilities to metabolically convert the vitamin forms of vitamin B_{12}, hydroxocobalamin (OH-Cbl) or cyanocobalamin (CN-Cbl), to the coenzymatically active cobalamin derivatives, adenosylcobalamin (AdoCbl) or methylcobalamin (MeCbl) (Mudd *et al.*, 1969; Rosenberg *et al.*, 1969; Mahoney *et al.*, 1971). Together with known genetically determined abnormalities affecting the absorption or cellular uptake of cobalamin (Imerslund,

1960; Gräsbeck *et al.*, 1960; Lampkin and Maurer, 1967; Hakami *et al.*, 1971) or of folic acid (Luhby *et al.*, 1961; Lanzkowsky, 1970), these conditions exemplify abnormalities in the processing of vitamins from their orally ingested forms to suitable tissue cofactor derivatives. The study of such processing defects has been an important area of progress in the recent study of vitamin-responsive conditions and will be discussed more thoroughly in the following sections.

3. Genetic Abnormalities Currently Known to Be Vitamin Responsive

With increasing recognition of the sorts of situations in which vitamin responsiveness may be expected to occur and with the rapid recognition of many inborn errors of metabolism in recent years, the list of vitamin-responsive genetic disorders has expanded rapidly. The currently recognized disorders are listed in Table I, in which are included only those situations in which the evidence suggesting responsiveness is relatively conclusive. Classically, responsiveness to a vitamin has been assessed by relief of the clinical manifestations of a genetic condition. Alleviation of intractable infantile convulsions by high doses of pyridoxine or of severe rickets by high doses of vitamin D are early examples. More recent instances would include the relief of severe ketoacidosis in certain methylmalonic aciduric patients by vitamin B_{12} or of life-threatening megaloblastic anemia in transcobalamin II (TC II)-deficient patients by the same vitamin. When such conditions are accompanied by recognized biochemical abnormalities, response may also be judged by diminuition of these biochemical manifestations. Thus, the methylmalonic aciduria of the patients mentioned above has been markedly decreased as their ketoaciduria has been alleviated by vitamin B_{12} therapy. When a genetic condition is characterized only by chronic, slowly developing, and relatively irreversible clinical difficulties (e.g., cystathionine β-synthase deficiency) or by no established clinical abnormalities (e.g., γ-cystathionase deficiency), vitamin responsiveness may best be judged by alleviation of the characteristic biochemical abnormality (homocystinuria and cystathioninuria, respectively, in these two conditions). In such situations, it may be difficult to judge whether alleviation of a biochemical abnormality by a vitamin has been accompanied by true clinical benefit.

As is shown in Table I, seven pyridoxine-responsive conditions have currently been identified. Most of these conditions have been recognized for some time, but a recent addition is gyrate atrophy of the retina caused by ornithine : 2-oxoacid aminotransferase deficiency. It now appears that at least some patients with this disease may be clinically and biochemically responsive to B_6. In addition to these B_6-responsive conditions, there are genetic disorders now known to be responsive to vitamin B_{12}, folic acid, thiamine, lipoic acid, and biotin. The list is not limited to diseases responsive to vitamins of the B group, as

Table I. Vitamin-Responsive Genetic Conditions

Vitamin bringing about response	Condition	Underlying defect	Demonstration of vitamin response	Reviews or recent citations
Pyridoxine	Infantile convulsions	? Glutamic acid decarboxylase (Yoshida *et al.*, 1971)	Hunt *et al.*, 1954	Frimpter *et al.*, 1969
Pyridoxine	B_6-responsive anemia	? δ-Aminolaevulinic acid synthetase (Aoki *et al.*, 1973)	Harris *et al.*, 1956	Frimpter *et al.*, 1969
Pyridoxine	Xanthurenic aciduria	Kynureninase (Tada *et al.*, 1967)	Knapp, 1960	Frimpter *et al.*, 1969
Pyridoxine	Cystathioninuria	γ-Cystathionase (Frimpter, 1965; Finkelstein *et al.*, 1966)	Frimpter *et al.*, 1963	Frimpter *et al.*, 1969; Mudd and Levy, 1978
Pyridoxine	Homocystinuria	Cystathionine β-synthase (Mudd *et al.*, 1964)	Barber and Spaeth, 1967	Mudd and Levy, 1978
Pyridoxine	Hyperoxaluria	Soluble α-ketoglutarate : glyoxylate carboligase (Koch *et al.*, 1967)	Smith and Williams, 1967; Gibbs and Watts, 1970	Williams and Smith, 1978
Pyridoxine	Gyrate atrophy of choroid and retina	Ornithine : ketoacid transaminase (Takki, 1974)	Weleber *et al.*, 1978	Shih *et al.*, 1978; O'Donnell *et al.*, 1978
Cobalamin	Megaloblastic anemia	Absent or abnormal intrinsic factor (Katz *et al.*, 1974)		Lampkin and Maurer, 1967; McIntyre *et al.*, 1965; Mudd, 1977
Cobalamin	Megaloblastic anemia	Abnormal ileal uptake (Imerslund, 1960; Gräsbeck *et al.*, 1960)		Mackenzie *et al.*, 1972; Mudd, 1977

Cobalamin	Megaloblastic anemia	Transcobalamin II (Hakami *et al.*, 1971; Hitzig *et al.*, 1974)	Hakami *et al.*, 1971	Mudd, 1977; Gimpert *et al.*, 1975
Cobalamin	Methylmalonic aciduria	Adenosylcobalamin formation: *cbl* A (Mahoney *et al.*, 1975) *cbl* B (? Cbl^{II} reductase or adenosyltransferase) (Mahoney *et al.*, 1975)	Rosenberg *et al.*, 1968	Rosenberg, 1976; Mudd, 1977; Rosenberg, 1978
Cobalamin	Methylmalonic aciduria and homocystinuria	Adenosylcobalamin and methylcobalamin formation: *cbl* C (Mahoney *et al.*, 1971) *cbl* D (Willard *et al.*, 1978)	Goodman *et al.*, 1972; Anthony and Mc-Leay, 1976	Rosenberg, 1978; Mellman *et al.*, 1979; Willard and Rosenberg, 1979a
Folic acid	Megaloblastic anemia	Transport of folate (Luhby *et al.*, 1961)	Luhby *et al.*, 1961	Lanzkowsky, 1977; Erbe, 1979
Folic acid	Homocystinuria	Methylenetetrahydrofolate reductase (Mudd *et al.*, 1972)	Freeman *et al.*, 1975	Rosenblatt *et al.*, 1979
Thiamine	Megaloblastic anemia	Unknown (Rogers *et al.*, 1969)	Rogers *et al.*, 1969	
Thiamine	Maple syrup urine disease (branched-chain ketoaciduria)	Branched chain α-ketoacid decarboxylase (Scriver *et al.*, 1971)	Scriver *et al.*, 1971	Danner *et al.*, 1978
Thiamine	Hyperpyruvicacidemia	Pyruvate decarboxylase (Lonsdale *et al.*, 1969; Blass *et al.*, 1970)	Lonsdale *et al.*, 1969	
Thiamine	Lactic acidosis	Pyruvate carboxylase (low-K_m form) (Brunette *et al.*, 1972)	Brunette *et al.*, 1972	Rosenberg, 1976

(*continued*)

Table 1. (*Continued*)

Vitamin bringing about response	Condition	Underlying defect	Demonstration of vitamin response	Reviews or recent citations
Lipoic acid	Lactic acidosis	Pyruvate carboxylase (Clayton *et al.*, 1967; Hommes *et al.*, 1968)	Clayton *et al.*, 1967; Hommes *et al.*, 1968	
Biotin	Methylcrotonylglycinuria and excretion of propionate metabolites	Holocarboxylase synthetase (Bartlett and Gompertz, 1976; Weyler *et al.*, 1977)	Gompertz *et al.*, 1971	Weyler *et al.*, 1977; Bartlett and Gompertz, 1978
Biotin	Propionic acidemia	Propionyl-CoA carboxylase (Gompertz *et al.*, 1970)	Barnes *et al.*, 1970	Hillman *et al.*, 1978
Ascorbic acid	Ehlers–Danlos syndrome (Type VI)	Collagen lysyl hydroxylase (Krane *et al.*, 1972)	Elsas *et al.*, 1978	
Vitamin D	X-linked familial hypophosphatemic rickets	Uncertain	Albright *et al.*, 1937	Rosenberg, 1976; Rasmussen and Anast, 1978
Vitamin D	Vitamin D-dependent rickets	Possible 25-hydroxycholecalciferol 1-α-hydroxylase (Fraser *et al.*, 1973), or other (Balsan *et al.*, 1975)	Fraser and Salter, 1958	Rosenberg, 1976; Rasmussen and Anast, 1978

shown by the presence in Table I not only of two vitamin-D-responsive conditions but also of an ascorbate-responsive form of the Ehlers-Danlos syndrome. Altogether, this list now comprises more than 20 disorders responsive to a wide variety of vitamins. There is no reason to suppose the list is final or that additional conditions responsive to these or other vitamins may not be found.

4. Mechanisms Underlying Vitamin Responsiveness

In a consideration of the mechanisms underlying vitamin responsiveness, it is useful to recall the obvious fact that provision of a normal dietary supply of a vitamin to an individual in general is merely the first step in insuring that that individual is able to carry out at a normal rate the biochemical reactions that ultimately depend on cofactor forms of that vitamin. An orally administered vitamin must be absorbed from the gastrointestinal tract, transported through the blood to the tissues where it is required, taken up by the tissues, and often converted chemically to a cofactor derivative which, finally, must interact in a complicated and specific manner with an apoenzyme protein. Even the early steps in this progression, at least for some vitamins, involve specific interactions with special proteins. An overall view of vitamin-responsive disease leads one to conclude that in this area, as in many others, if something can go wrong, it will go wrong. In biochemical terms, for any reaction in which a specific protein is involved, there is likely to be a corresponding genetic disease caused by malfunction or lack of that protein.

4.1. Defects in the Processing of a Vitamin Prior to Its Interaction in Cofactor Form with a Particular Apoenzyme

The vitamin B_{12}-responsive conditions illustrate well the variety of defects that may arise in processing of a vitamin. These situations will, therefore, be discussed in some detail.

4.1.1. Abnormalities of Intrinsic Factor

Ingested cobalamin complexes with intrinsic factor, a glycoprotein secreted in the stomach. Such binding is required for normal absorption, a process which occurs in the distal ileum. More than 30 patients with specific genetically determined abnormalities of intrinsic factor have been described over the past 25 years (Table I and McIntyre *et al.*, 1965; McNicholl and Egan, 1968; Katz *et al.*, 1972). These cases are characterized by the onset of megaloblastic anemia early in life. Serum B_{12} is below normal, and intrinsic factor is either absent or altered. The methods used to study many of these cases were inadequate to permit a final distinction between the latter two possibilities (Katz *et al.*, 1974).

In a recent study, Katz and co-workers (1974) succeeded in demonstrating an abnormal, structurally altered intrinsic factor in a patient with congenital B_{12} malabsorption. This abnormal intrinsic factor was isolated from the patient's gastric juice by affinity chromatography and appeared normal in B_{12} binding, molecular weight, total amino acid and carbohydrate composition, and immunodiffusion. The probable basis for the malabsorption was demonstrated by the failure of the patient's intrinsic factor-B_{12} complex to bind normally to human ileal mucosal homogenates. The association constant was 60-fold lower than that for the normal intrinsic factor–B_{12} complex. These results demonstrate that the B_{12} and the ileal binding sites of the intrinsic factor molecule reside in different regions of the protein and that structural abnormalities may affect one function without affecting the other. Further investigations of patients with intrinsic factor abnormalities may possibly reveal a variety of functional changes in intrinsic factor, each of which might lead to malabsorption (Mudd, 1977).

4.1.2. Ileal Abnormalities Leading to Abnormal Uptake

More than 80 patients with abnormal ileal uptake of a normal intrinsic factor–B_{12} complex have now been described (Table I; Gräsbeck and Kvist, 1967; Ben-Bassat *et al.*, 1969). Again, the clinical picture is characterized by development of early megaloblastic anemia. However, these patients possess normal intrinsic factor. Serum B_{12}-binding proteins are also normal. There is no generalized malabsorption, although nonspecific malabsorption may occur secondarily when the patients are B_{12} deficient. The specificity of the absorptive defect is thus best demonstrated after appropriate treatment with parenteral cobalamin. It is assumed that in each of these patients a specific defect is present in the intestinal mucosa that prevents the normal absorption of cobalamin. Possibilities for such defects are many (Mudd, 1977). However, in no single patient has a decision been made among these possibilities. Biopsies of the ileal mucosa have been normal by light and electron microscopy (Gräsbeck and Kvist, 1967; MacKenzie *et al.*, 1972). Attempts to demonstrate abnormalities in the release of cobalamin from its complex with intrinsic factor have been inconclusive (Gräsbeck and Kvist, 1967). In a recent study of one family, it was shown that attachment of the intrinsic factor–B_{12} complex to ileal mucosal homogenates was not altered (MacKenzie *et al.*, 1972). Further investigations of these types are needed and may be expected to reveal genetic heterogeneity among different families with this sort of cobalamin malabsorption as well as to throw light on the normal processing of cobalamin during ileal uptake.

Both abnormalities of intrinsic factor and of ileal uptake are responsive to vitamin B_{12} in the special sense that administration of a normal amount of vitamin is therapeutically effective if the vitamin is given by an abnormal route (that is a parenteral one which avoids the need for cobalamin to traverse the defective step). Since such therapy is so effective, there has been relatively little exploration of the likely possibility that administration of massive amounts of cobalamin orally would also be therapeutically effective.

4.1.3. Abnormalities of Transcobalamin II

Newly absorbed cobalamin emerging from ileal cells into the portal blood becomes bound to a serum β-globulin, now termed transcobalamin II (TC II). Portions may also bind to other serum binding proteins, for example, transcobalamin I. The TC II-cobalamin complex is cleared from the blood into tissues much more rapidly than are the other complexes (Ellenbogen, 1975). There is ample evidence that TC II, but not TC I, facilitates the uptake of cobalamin by HeLa cells, liver, reticulocytes, erythrocytes, and fibroblasts (Ellenbogen, 1975; Mahoney and Rosenberg, 1975). In 1971, Hakami and co-workers reported the first cases of TC II deficiency in two siblings who presented at 3 and 5 weeks of age with failure to thrive, vomiting and diarrhea, progressive pancytopenia, and megaloblastic bone marrow changes. Levels of serum vitamin B_{12} were normal, but no TC II was detected in serum (Hakami *et al.*, 1971).

An additional case has been reported from Switzerland (Hitzig *et al.*, 1974). Early in life, this child manifested severe malabsorption because of atrophy of the small intestinal mucosa. He was unable to form specific antibodies and plasma cells and had pancytopenia resulting from bone marrow insufficiency. Subsequent investigations revealed that there was virtually no serum protein capable of binding cobalamin and migrating with TC II during chromatography or polyacrylamide gel electrophoresis. Further studies (Gimpert *et al.*, 1975) failed to detect the presence in the serum of any substance that was immunologically cross-reactive with antibody to normal TC II, so by this criterion, TC II appeared to be absent rather than present in an altered form incapable of binding B_{12}.

Very recently, a patient with an abnormal, rather than absent, TC II was described (Haurani *et al.*, 1979). A 34-year-old woman with megaloblastic anemia from childhood had abnormally high serum concentrations of both cobalamin and TC II. Her TC II behaved normally during gel filtration, bound cobalamin, and shared immunologic properties with normal TC II, but failed to facilitate the uptake of cobalamin by the cells studied (HeLa cells and stimulated lymphocytes). These findings indicate a functional distinction between the domain of TC II responsible for cobalamin binding and that responsible for facilitation of cellular uptake. The observation that patients with absent or abnormal TC II during times of relapse when they have megaloblastic anemia are neither homocystinuric, cystathioninuric, nor methylmalonicaciduric (Scott *et al.*, 1972; Haurani *et al.*, 1979) suggests that bone marrow, and perhaps other rapidly proliferating tissues, are more sensitive to the need for normally functional TC II than is liver.

In each of these cases, the progressive downhill clinical course was relieved by large doses of vitamin B_{12}. The mechanism whereby B_{12} is taken up by the target cells in the absence of TC II remains to be clarified in detail.

4.1.4. Intracellular Processing of Cobalamin

Following its cellular uptake, the TC II–cobalamin complex comes to lie in the lysosomal compartment where it undergoes degradation leading to the release of free cobalamin into the cystosol (Pletsch and Coffey, 1971; Youngdahl-Turner *et al.*, 1978, 1979). The vitamin is then metabolized alternatively either to AdoCbl or to MeCbl, the two coenzymatically active cobalamin derivatives. Conversion to AdoCbl involves reduction of the cobalt moiety of cobalamin from its 3+ valence state (Cbl^{III}), successively to the 2+ (Cbl^{II}), then to the 1+ (Cbl^{I}) state. In bacteria, two separable enzymes catalyze these two reductive steps (Walker *et al.*, 1969). Transfer of the adenosyl moiety from ATP to Cbl^{I} by a specific adenosyltransferase completes the synthesis of AdoCbl. The last two enzymic steps in this sequence are found in mammalian mitochondria (Mahoney *et al.*, 1975; Fenton and Rosenberg, 1978). Formation of MeCbl seems likely to occur in the cystosol and, by analogy with AdoCbl formation and with chemical alkylation of cobalamin, very probably involves prior reduction of Cbl^{III} to Cbl^{I}.

Methylcobalamin functions as a cofactor in the enzymic conversion of homocysteine to methionine. Lack of this enzymic reaction leads to a form of homocystinuria (Mudd, 1974b). AdoCbl functions as a cofactor in the enzymic conversion of the coenzyme-A derivative of methylmalonic acid to the corresponding derivative of succinic acid. Lack of this activity leads to a form of methylmalonic aciduria (Rosenberg, 1978).

Among the reactions involved in the processing of OH-Cbl to the cofactor forms, MeCbl and AdoCbl, four distinct genetic lesions are presently known. Two of these produce similar phenotypes: affected patients are both homocystinuric and methylmalonic aciduric. A variety of evidence shows that cells and tissues of these patients fail to accumulate normal amounts of either AdoCbl or MeCbl (Mudd, 1974b; Rosenberg, 1978).

Such patients have been further separated into two distinguishable genetic complementation classes (*cbl* C and *cbl* D) by study of fused cultured fibroblasts (Willard *et al.*, 1978). Presumably some early step (or steps) in cobalamin processing common to the formation of both AdoCbl and MeCbl is affected. Cultured fibroblasts of the *cbl* C type have been shown to bind, internalize, and degrade the TC II–cobalamin complex normally (Youngdahl-Turner *et al.*, 1978). The same is apparently true for fibroblasts of the *cbl* D type (Mellman *et al.*, 1979). The cobalamin resulting from degradation of the TC II–cobalamin complex is, however, not converted at normal rates to either AdoCbl or MeCbl and is not retained normally within the cells (Rosenberg, 1978; Mellman *et al.*, 1979). For *cbl* C cells, these abnormalities are accentuated if the cobalamin is administered initially in the form of CN-Cbl rather than OH-Cbl. In contrast, *cbl*

D cells do not show such marked differences in their relative abilities to process CN-Cbl and OH-Cbl, being roughly intermediate between *cbl* C cells and normal cells in this respect (Mellman *et al.*, 1979).

Together, these data are strongly suggestive that *cbl* C cells are defective in a Cbl^{III} reductase activity needed normally to produce Cbl^{II} for both MeCbl and AdoCbl formation. A less likely possibility is a defect at an earlier stage such as release of cobalamin from the lysosomes following degradation of the TC II–cobalamin complex. Whether *cbl* D mutations complement *cbl* C mutations by intra- or interallelic interactions, and whether the *cbl* D mutation is merely a more "leaky" form of the *cbl* C mutation or affects another metabolic step remain to be established (Willard *et al.*, 1978; Mellman *et al.*, 1979).

Two additional groups of patients have been shown to have errors in the intracellular processing of cobalamin. Such patients are methylmalonic aciduric but not homocystinuric. They form MeCbl normally but fail to accumulate AdoCbl. Heterogeneity among such patients has been shown by the demonstration that broken cell preparations from certain of them (*cbl* B type) are unable to convert Cbl^{II} to AdoCbl in the presence of ATP and appropriate reductants, whereas broken cell preparations from others (*cbl* A type) are able to carry out this conversion normally. The lesion in *cbl* B patients is thought to be in either Cbl^{II} reductase or ATP:Cbl^{I} adenosyltransferase, whereas the lesion in the *cbl* A group remains uncertain (Mahoney *et al.*, 1975). The distinction between the *cbl* A and *cbl* B groups has been confirmed by demonstration that cells from the two groups complement one another when fused (Gravel *et al.*, 1975).

With respect to vitamin responsiveness, at last some patients from the *cbl* C and *cbl* D groups have been shown to decrease homocystinuria and methylmalonic aciduria (Goodman *et al.*, 1972; Anthony and McLeay, 1976), and some from the *cbl* A and *cbl* B groups have been shown to decrease methylmalonic aciduria when given massive doses of cobalamin. Such chemical responses may be accompanied by alleviation of severe ketoacidosis in acutely ill patients (Rosenberg *et al.*, 1968; Rosenberg, 1978).

Growth of fibroblasts in the presence of high concentrations of cobalamin increases the AdoCbl and MeCbl content of some lines of *cbl* C and *cbl* D cells (Linnell *et al.*, 1976) and the AdoCbl content of some lines of *cbl* A cells (Rosenberg *et al.*, 1969). Increments in AdoCbl content are accompanied by increases in ability to metabolize methylmalonyl-CoA as reflected either indirectly by propionate oxidation (Rosenberg *et al.*, 1969; Mudd *et al.*, 1979a; Willard and Rosenberg, 1979a) or by methylmalonyl-CoA mutase holoenzyme content (Willard and Rosenberg 1979a; Mellman *et al.*, 1979).

Similarly, increments in MeCbl are accompanied by increases in the ability to methylate homocysteine (Mudd *et al.*, 1970a) and by increased N^5-methyltetrahydrofolate:homocysteine methyltransferase holoenzyme activity (Mudd *et al.*, 1970a; Goodman *et al.*, 1970; Mellman *et al.*, 1979). Among *cbl* B cell lines, some have responded to growth in the presence of high concentrations of cobalamin with increases in holomutase activity (Willard and Rosenberg, 1979a), but other lines have shown neither an increase in AdoCbl content (Linnell *et al.*, 1976) nor an increase in holomutase activity (Willard and Rosenberg, 1979a). The molecular bases for these variable responses of *cbl* B lines in culture are not yet clear, nor indeed are the bases for the responses of the *cbl* A, *cbl* C, and *cbl* D lines. Even in the latter lines, the accumulation of AdoCbl may not vary in a straightforward manner with the holomutase activity, as emphasized recently by Willard and Rosenberg (1979a).

Furthermore, as these same authors have pointed out, to date there have been few rigorous studies of clinical responsiveness of patients with the various errors of cobalamin processing. Among the few instances in which data are available on both the clinical response of the patient to cobalamin therapy and the response of his cells to growth in the presence of high concentrations of cobalamin, clinical response, or lack thereof, often correlates with *in vitro* increases in coenzyme and holoenzyme concentrations, but exceptions have also been reported: the cells of a clinically nonresponsive patient showed unequivocal increases in holomutase; the cells of another patient, clinically responsive to cobalamin, did not show the expected holomutase increase (Willard and Rosenberg, 1979a). Clearly, a complete understanding of these complex situations will be possible only when direct studies of the defective enzymes and the kinetics of their interactions with their cobalamin substrates have been undertaken.

4.1.5. Possible Toxic Effects of Vitamins

Studies of fibroblasts grown in the presence of high concentrations of cobalamins have revealed another effect of possible theoretical and therapeutic importance (Table II) (Willard and Rosenberg, 1979b). When total methylmalonyl-CoA mutase activity was measured in normal cells and in cells with defects in their capacity to form AdoCbl, growth with very high OH-Cbl in

Table II. Effect of Cobalamin Supplementation in Culture on Methylmalonyl-CoA Mutase Activity

	Total mutase activity after growth in[a]		
Cells	Basal medium	1 μg/ml OH-Cbl	1 μg/ml AdoCbl
Control	1.4	1.5	1.3
cbl A	2.3	0.3	1.8
cbl B	2.3	1.5	1.8
cbl B	2.9	1.2	1.8

[a]Cells were grown for 4 days in basal medium containing <50 pg/ml cobalamin or in the indicated supplemented medium. Data are expressed as nmol succinate formed/min/mg protein. OH-Cbl, hydroxycobalamin; AdoCbl, adenosylcobalamin. Data adapted from Willard and Rosenberg (1979b).

the medium was found to depress the total amount of enzyme in the mutant cells but not in the control cells. It was also shown that OH-Cbl can compete with AdoCbl for the mutase active site in cell extracts and that the inhibition by OH-Cbl increased as the AdoCbl concentration decreased (Willard and Rosenberg, 1979b). It appears that in mutant cells, in which the intracellular AdoCbl concentration may be very low, sufficient OH-Cbl or one of its reduced forms may accumulate under supplemented growth conditions to compete for binding to mutase apoenzyme and thus inhibit mutase activity. Thus, the effectiveness of cobalamin therapy in patients defective in AdoCbl synthesis may not only be a matter of how much AdoCbl can be produced but rather may reflect a delicate balance between formation of holomutase because of accumulated AdoCbl and inhibition of mutase activity by high concentrations of OH-Cbl or its metabolites. Mahoney *et al.* (1976) reported an example of possible OH-Cbl toxicity in a patient with a *cbl* B defect. Following an initial favorable response to cobalamin therapy, this patient relapsed during continued cobalamin administration. Administration of lower doses of OH-Cbl or of more appropriate forms of cobalamin (possibly AdoCbl itself) may be preferable for patients such as this (Willard and Rosenberg, 1979b).

The possibilities of similar toxic effects in other vitamin-responsive conditions certainly are real. Perhaps those dealing with such conditions will have to give these possibilities more serious consideration than has generally been the case heretofore.

4.1.6. Processing Errors for Other Vitamins

Beyond the realm of cobalamin biochemistry, there are only a few diseases known to be caused by defects in processing of vitamins. One interesting example has been provided very recently with the identification of a child with deficient activity of two biotin-dependent carboxylases: 3-methylcrotonyl-CoA carboxylase and propionyl-CoA carboxylase (Bartlett and Gompertz, 1976; Weyler *et al.*, 1977). The chemical abnormalities in the patient resulting from each of these deficiencies were responsive to large doses of biotin. The activities of both carboxylases in extracts of fibroblasts cultured from the patient were abnormally dependent on the concentration of biotin in the growth medium (Bartlett and Gompertz, 1976; Weyler *et al.*, 1977). For example, the activities in the patient's fibroblast extracts were less than 0.3% of control values when the cells had been grown in medium with a minimal biotin content (5×10^{-9} M). A 20-fold increase in the biotin content of the growth medium had little effect on the activities of the two carboxylases in control cells, but restored these activities in the mutant cells to normal. The carboxylases in the mutant cells grown with high biotin had normal K_m's for the substrates to be carboxylated (Weyler *et al.*, 1977).

The results of genetic complementation studies were also consistent with the presence of normal apocarboxylase in these mutant cells when they were grown with relatively low biotin (Wolf, 1980). These findings support the hypothesis that the underlying defect in this biotin-responsive patient may be in a holocarboxylase synthetase that would normally serve to attach biotin to each of the two apocarboxylases in question. The reaction catalyzed by this holocarboxylase synthetase proceeds in two partial reactions:

$$\text{biotin} + \text{ATP} \xrightarrow{Mg^{2+}} \text{biotinyl-5'-adenylate} + \text{PPi} \quad (1)$$

$$\text{biotinyl-5'-adenylate} + \text{apocarboxylase} \rightarrow \text{holocarboxylase} + \text{AMP} \quad (2)$$

$$\text{Sum: biotin} + \text{ATP} + \text{apocarboxylase} \rightarrow \text{holocarboxylase} + \text{AMP} + \text{PPi} \quad (3)$$

[The properties of biotin holocarboxylase synthetases purified from various sources are described by Murthy and Mistry (1974) and references cited there.] Which partial reaction is defective in the affected patient has not been demonstrated. Acetyl-CoA carboxylase activity in these mutant fibroblasts was normal when cells were grown at a minimal biotin concentration. Apparently, the defect in holocarboxylase synthetase activity affecting the attachment of biotin to 3-methylcrotonyl-CoA carboxylase and to propionyl-CoA carboxylase did not affect the attachment of biotin to acetyl-CoA carboxylase.

For vitamin B_6, no genetic disease has yet been shown unequivocally to be caused by a defect in the processing of this vitamin. A few individuals with sideroblastic anemia have been reported to respond to PLP but not to pyridoxine (Gehrmann, 1965; Mason and Emerson, 1973). It seems possible that these patients have tissue-specific defects in the formation of phosphorylated B_6 derivatives, for example in the bone marrow, but enzyme assays bearing on this possibility have not been reported, nor is it clear that the condition is genetically determined (Cartwright and Deiss, 1975).

Another possible exception is the single infant studied by Scriver and Hutchison (1963). This child had repeated convulsions, excreted abnormal amounts of cystathionine, and had an elevated output of xanthurenic acid in response to a tryptophan load. All of these manifestations of possible PLP deficiency responded favorably to an increase in the daily oral intake of pyridoxine from normal up to 2.5 mg per day—four to five times the normal dose for an infant. The indications that several PLP-dependent enzyme activities were simultaneously deficient in this girl and responded to increased vitamin intake suggest a defect in vitamin processing. Unfortunately, more direct investigations of this possibility have not been reported.

Two conditions are known that affect the processing of folic acid or its derivatives. Genetically determined defective absorption of folate has been recognized in a few patients (Luhby *et al.*, 1961; Lanzkowsky *et al.*, 1969; Lanskowsky, 1970, 1977; Santiago-Borrero *et al.*, 1973). There is evidence that transport of folate into the cerebrospinal fluid is also defective in these patients (Erbe, 1979). The molecular basis of this disease remains unknown, although presumably, some protein normally involved in intestinal transport of folate and in selective uptake of 5-methyltetrahydrofolate into the cerebrospinal fluid is absent or impaired. Massive oral doses of folate (5–40 mg daily) have usually been capable of controlling the severe megaloblastic anemia that accompanies this abnormality and, to a lesser extent, the neurological abnormalities (Erbe, 1979). However, in one patient, even 100 mg folate given each day by the oral route was ineffective, and remission could be maintained only with 15-mg doses of folic acid given intramuscularly each month (Santiago-Borrero *et al.*, 1973). Whether a residual capability of the defective system is able to transport sufficient folate when presented with very high concentrations or whether transport occurs via an additional mechanism remains to be clarified in these patients.

A number of patients have now been identified who are unable to convert 5,10-methylenetetrahydrofolate to 5-methyltetrahydrofolate because of deficient activity of 5,10-methylenetetrahydrofolate reductase (Mudd *et al.*, 1972; Rosenblatt *et al.*, 1979; Erbe, 1979). 5-Methyltetrahydrofolate serves as a methyl donor in the methylation of homocysteine to methionine, and patients with this defect are homocystinuric. A few of these patients apparently responded to large doses of folic acid with decreases in the excretion of homocystine (Shih *et al.*, 1972; Freeman *et al.*, 1975), whereas others did not (Wong *et al.*, 1977; Baumgartner *et al.*, 1977). Since even in those patients who responded biochemically, clinical benefit was absent (Shih *et al.*, 1972) or inconstant (Erbe, 1979), folate therapy requires more evaluation before being accepted. Cultured fibroblasts from 5,10-methylenetetrahydrofolate reductase-deficient patients often have residual activities of this enzyme, ranging from 4–25% of mean control values, and the amount of the residual activity generally correlates inversely with the clinical severity of the disease (Rosenblatt *et al.*, 1979). If folate administration leads to increased 5,10-methylenetetrahydrofolate concentrations, it seems possible that the action of such residual reductases could then produce sufficient 5-methyltetrahydrofolate to prevent excessive homocystinuria. However, in the few cases reported, folate responsiveness did not correlate with the amount of the residual activity of the defective reductase (Mudd *et al.*, 1972; Shih *et al.*, 1972; Freeman *et al.*, 1975; Wong *et al.*, 1977), so the molecular details determining folate-responsiveness remain obscure. Inhibitory effects of folates other than 5-methyltetrahydrofolate which may accumulate during folate administration are possible complicating factors (Erbe, 1979).

A final instance of a defect in vitamin processing concerns vitamin D. Current evidence suggests that some individuals with vitamin D-dependent rickets suffer from an inability to convert 25-hydroxyvitamin D_3 to 1,25-dihydroxyvitamin D_3 (Fraser *et al.*, 1973), the compound that appears to be the most important biologically active form of vitamin D_3 (Rassmussen and Anast, 1978). However, direct evidence to support this defect in 1-vitamin D_3 hydroxylase has not yet been provided, and it is possible that other abnormalities are present, at least in some patients with vitamin D-dependent rickets (Balsan *et al.*, 1975; Rasmussen and Anast, 1978). Any consideration of the molecular basis of vitamin D responsiveness in such patients must await further elucidation of the etiology of the disease.

In summary, relatively few diseases have been identified in which a genetic defect affects vitamin processing. Most of the known diseases of this type affect cobalamin metabolism. This may be attributed chiefly to at least two facts. First, the processes for the absorption, plasma transport, and tissue uptake of cobalamin may be unusually complicated and therefore excessively subject to genetic errors. Second, and probably more important, as far as is presently known, cobalamin is involved in only two enzymatic reactions in mammals. Both can be severely defective without the defect being lethal *in utero*. Thus, patients with defects involving cobalamin processing may survive long enough to come to medical attention. Such may not necessarily be the case for individuals with severe genetic defects affecting the metabolism of vitamins the cofactor forms of which participate in a great number of reactions.

Many patients with diseases affecting vitamin processing have been responsive to large doses of the vitamin in question. However, other patients with grossly similar defects have not been responsive. The factors critical to such distinctions have not been clearly identified in any single instance.

4.2. Defects in Particular Apoenzymes that Interact with the Cofactor Forms of Vitamins

A second general category of vitamin-responsive conditions includes those in which a genetic change affects the structure of a particular apoenzyme so that its interaction with its cofactor is modified. To establish that a genetic defect should be included in this group, it is generally necessary to demonstrate that there is no abnormality in any of the steps in the processing of the vitamin to its cofactor form. This is usually done by demonstrating that other reactions that depend on the same cofactor are normally functional. This has been accomplished, for example, for many patients with B_6-responsive cystathioninuria, homocystinuria, or xanthurenic aciduria. In general, in these patients only a single enzyme activity can be considered to be defective. For any disease in this group, it would be desirable to answer two key questions. First, is the effect of the vitamin on the defective enzyme itself, or does the effect of the vitamin occur through stimulation of some alternative metabolic pathway that may permit removal of excess metabolite accumulated because of the primary block? This is clearly an important question in judging the therapeutic potential of vitamin administration. Second, what is the molecular mechanism of the vitamin effect?

4.2.1. Cystathionine β-Synthase Deficiency

To illustrate some of the difficulties and complexities that arise in an attempt to answer these questions, the disease homocystinuria caused by cystathionine β-synthase deficiency will be discussed in some detail. This disease has a number of advantages as a model case for investigation of vitamin responsiveness. It is relatively common for an inborn error of metabolism, and several hundred patients with this condition have been identified throughout the world. Almost all have been given clinical trials of pyridoxine. Roughly half respond little if at all; about half do respond with decreases in homocystinuria and hypermethioninemia (the second chemical abnormality characteristic of this disease). Although there may be intermediate degrees of response, it has usually been possible to classify a given patient quite clearly as either responsive or nonresponsive. Thus, there is adequate clinical material for study. Further, cystathionine β-synthase deficiency is manifest in fibroblasts cultured from deficient patients, a fact that greatly facilitates studies of the properties of the mutant enzyme (Mudd and Levy, 1978).

The metabolic pathway in which cystathionine β-synthase participates is summarized by equations (4)-(7):

$$\text{methionine} \rightarrow\rightarrow\rightarrow \text{homocysteine} \tag{4}$$

$$\text{homocysteine} + \text{serine} \rightarrow \text{cystathionine} \tag{5}$$

$$\text{cystathionine} \rightarrow \text{cysteine} + \alpha\text{-ketobutyrate} \tag{6}$$

$$\text{cysteine} \rightarrow\rightarrow\rightarrow SO_4^{=} \tag{7}$$

Cystathionine β-synthase is the PLP-dependent enzyme that catalyzes reaction 5, the condensation of homocysteine with serine to form the thioether cystathionine. In mammals, homocysteine arises entirely from methionine, and cystathionine is cleaved only to cysteine. Methionine is considered to be an essential amino acid for humans, whereas cysteine is not because it can be formed by this pathway. Cysteine, in turn, is catabolized, so that approximately 80% of its sulfur ends in inorganic sulfate.

Initial attempts to investigate experimentally the decrease in homocystinuria brought about by pyridoxine treatment in cystathionine β-synthase deficient patients were based on the working hypothesis that this situation might be explained by the "K_m mutation" model. In this model, it is supposed that a particular patient has a mutant enzyme altered solely so that its K_m for cofactor is higher than normal. It is further supposed that to satisfy the physiological requirement for this enzyme, somewhat less than 100% of normal maximal velocity is adequate. If the cofactor concentration is low relative to the K_m of the mutant enzyme, there will then be a relatively large difference between the reaction velocity of the normal enzyme and that of the mutant enzyme. The former may satisfy the physiological requirement; the latter may not. As cofactor concentration is raised, the relative activity of the mutant enzyme will increase more than that of the normal enzyme until, eventually, the mutant enzyme can also satisfy the physiological requirement. Under conditions of very high cofactor concentration, there will be little difference between the activities of the normal and mutant enzymes. The latter property provides a means to test whether this model is adequate to explain the vitamin responsiveness of any particular patient. If it is, the activity of the mutant enzyme should approach that of the normal enzyme as cofactor concentration is raised.

To test this model, hepatic cystathionine β-synthase activity was assayed in two patients with deficiencies of the enzyme. Initially, on normal B_6 intakes, the activities were 1% and 2% of the normal mean control specific activities. Increases of B_6 intakes to 100–200 mg daily brought about dramatic responses as judged by decreases in homocystine excretions and plasma methionine concentrations. Surprisingly, repeat assays showed that under these circumstances hepatic cystathionine β-synthase activities remained very low: 3% and 4% of normal. In each instance, the assays during response were performed in the presence of the relatively high concentration of 1 mM PLP added *in vitro*. Clearly, the data do not fit the K_m model just described, for the mutant enzyme activities remained very far from normal in spite of both *in vivo* and *in vitro* administration of B_6 and its cofactor derivative. In fact, these findings raised very serious doubts as to whether the vitamin response in these patients was indeed brought about by an effect on cystathionine β-synthase activity. The possibility that pyridoxine was stimulating some alternative pathway for homocysteine removal had to be seriously entertained (Mudd *et al.*, 1970b).

In the intervening years, many lines of evidence have been uncovered that together offer convincing evidence that the B_6 response in most homocystinuric patients is mediated by enhancement of cystathionine β-synthase activity rather than enhancement of an alternative metabolic pathway. In summary, these lines of evidence are the following.

1. Changes in concentrations of metabolites during B_6 response are compatible with this hypothesis. All compounds accumulated proximal to the block in cystathionine β-synthase decrease during B_6 therapy, whereas plasma cyst(e)ine and urinary sulfate, the measurable metabolites distal to the block, increase (Mudd *et al.*, 1970b). No increase of any compound suggestive of the functioning of an alternative pathway has been reported.

2. As increasing numbers of cystathionine β-synthase deficient patients have been classified as to their pyridoxine responsiveness, it has become apparent that pyridoxine responsiveness is constant within sibships: all cystathionine β-synthase-deficient members of a particular sibship are either responsive to B_6, or none are responsive. A recent survey of the literature (Mudd and Levy, 1978)

disclosed reports of at least 24 sibships with a total of at least 54 patients in which constancy of responsiveness held true. No well-documented exceptions were discovered. This finding not only indicates that the capacity to respond to B_6 is genetically determined, but it also suggests that the genetic determinant governing responsiveness is closely linked with, or identical to, that determining cystathionine β-synthase deficiency itself. If the two genetic factors were unlinked and segregated separately, variation of responsiveness within cystathionine β-synthase deficient members of the same sibship might be expected. Thus, the simplest interpretation of the constancy of responsiveness within sibships is that the same mutation that makes an individual cystathionine β-synthase deficient also determines whether he will be B_6 responsive. A single alteration of apoenzyme structure could well determine both properties.

3. A third indication that B_6 responsiveness may occur through an effect on cystathionine β-synthase itself comes from studies of the activities of this enzyme in fibroblasts cultured from deficient patients. Table III summarizes the results of two large series of such studies (Uhlendorf *et al.*, 1973; Fowler *et al.*, 1978). In this table, the results from both studies have been combined and grouped into sibships, since, again, the results with respect to residual enzyme activities were always constant within sibships. Among sibships responsive to B_6, 22 had residual activities of cystathionine β-synthase in cultured fibroblasts; only three lacked such activity. The residual activities ranged from about 10% of control down to 0.25 to 0.1% of control in the two experimental series. In contrast, among sibships not responsive to B_6, only four had detected residual activity of cystathione β-synthase, whereas 13 had no detected activity. Thus, there are good correlations between the presence of a small residual activity of cystathionine β-synthase and B_6 responsiveness on the one hand, and between the lack of residual activity and B_6 nonresponsiveness on the other.

With the advent of the use of the fibroblast system, liver biopsy for study of cystathionine β-synthase deficient patients has largely ceased. Nevertheless, the results available from assays of hepatic cystathionine β-synthase done by adequately sensitive methods confirm the correlation between B_6 responsiveness and the presence of residual cystathionine β-synthase activity. From all reported

Table III. Clinical Responsiveness to Pyridoxine and Presence of Detected Cystathionine β-Synthase Activity (Number of Sibships)

Enzyme activity[a]	Clinical: Responsive	Clinical: Not responsive
Detected	22	4
Not detected	3	13

[a]Assayed without added pyridoxal 5′-phosphate. Sensitivity for detection of activity: 0.1–0.25% of mean normal. Data combined from the results reported by Uhlendorf *et al.* (1973) and Fowler *et al.* (1978).

results, residual activities have been detected in liver extracts from each of seven B_6-responsive patients but not detected in extracts from any of three B_6 nonresponders (Mudd and Levy, 1978). Among responders, hepatic activities have invariably been higher during B_6 response than on basal diets (1.5- to 4-fold), but during response, even when assayed at high concentrations of PLP, cystathionine β-synthase activities have remained very much below normal. The simplest interpretation of these findings is that most cystathionine β-synthase-deficient patients who respond to B_6 do so by virtue of possession of at least a small residual activity of cystathionine β-synthase. Were the response to B_6 mediated by enhancement of some other activity, there is no obvious reason why the correlation in question should have been found. However, both the liver and the fibroblast studies suggest that during B_6 response, cystathionine β-synthase activities are increased only severalfold above the basal activities present on a normal B_6 intake and that during response to B_6, cystathionine β-synthase activities remain far below normal.

4. Several findings make these quantitative relationships understandable. Measurements of sulfate excretion after acute loads of methionine (Mudd *et al.*, 1970b) have shown that normal humans possess the capacity to metabolize daily at least 60 mmole of homocysteine through the cystathionine β-synthase step, a capacity which is large relative to the need imposed by the normal dietary intake of about 10 mmole of methionine. Cystathionine β-synthase-deficient patients on basal diets, however, could convert only about 3 mmole of methionine sulfur to sulfate, an amount insufficient to metabolize the normal daily intake. On high B_6 intakes, the capacity of these same patients rose modestly, yet remained far below normal, in keeping with the observed increases in hepatic cystathionine β-synthase activities. Now, however, the capacity to metabolize methionine became just about sufficient to handle a normal daily load of 10 mmole.

In another series of experiments, studies were made of nitrogen and sulfur balances of B_6-responsive and -nonresponsive cystathionine β-synthase-deficient patients on synthetic diets containing varying amounts of cystine (Poole *et al.*, 1975). The results indicated that cystathionine β-synthase-deficient, B_6-responsive patients convert about 2.5–4.0 mmole of methionine to cysteine each day while on normal B_6 intakes. This result confirms and extends the conclusions drawn on the basis of sulfate excretions. It appears that even patients with only a few percent residual activity of cystathionine β-synthase can metabolize 25–40% of a normal methionine intake through this enzymatic step. The relatively small enhancements of residual activities that occur during B_6 treatment may then confer on the patients the metabolic capacities to just handle the normal load.

There may well be a very important general lesson in this situation. Experience with many other inborn errors of metabolism suggests that very often normal humans have capacities of many enzyme activities large relative to those required under normal physiological conditions (for discussion of several specific examples see Mudd, 1974a). Relatively modest enhancement of a small residual activity of such an enzyme may be of extreme benefit to a deficient individual.

Given the conclusion that the B_6 response is usually mediated by a modest enhancement of residual mutant cystathionine β-synthase activity, what can be said about the molecular mechanism of this effect? The clearest conclusion seems to be a negative one: the effect is not a straightforward K_m effect of the type discussed above. This conclusion is supported by the fact that in no reported assay of hepatic enzyme in a B_6 responder has cystathionine β-synthase activity been restored to near normal by administration of pyridoxine to the patient or PLP to the *in vitro* assay system. Nor has cystathionine β-synthase activity in fibroblasts cultured from responsive patients been brought close to normal by *in vitro* addition of PLP. Further, as shown in Table IV, residual cystathionine β-synthase activities in extracts of fibroblasts from clinically responsive patients have not in general been more stimulated by *in vitro* addition of PLP than is cystathionine β-synthase in extracts of fibroblasts from normal subjects or from patients clinically not responsive to B_6.

In Table IV, the ratio of activity with added PLP to that without added PLP is shown for all cystathionine β-synthase-deficient sibships for whom activity was detected in fibroblast extracts. Enzymes from 16 responsive sibships were stimulated twofold or less. Greater stimulations were unusual among B_6 responders. Among the relatively few B_6 nonresponders with detected residual activities, the stimulations did not cluster at a low ratio but were more dispersed and tended to be higher. Finally, in studies of a much more limited series of mutant enzymes resolved from firmly bound endogenous PLP, Fowler and co-workers (1978) did not detect any consistent distinction between the dissociation constants for PLP of enzymes derived from clinically responsive patients and those of clinically nonresponsive patients.

Very recently, Lipson *et al.* (1980) pointed out that all reported studies of cystathionine β-synthase in cultured fibroblasts have been carried out with cells grown in media containing a high concentration (1000 ng/ml) of pyridoxal. By

Table IV. Clinical Responsiveness to Pyridoxine and *in Vitro* Stimulation by Pyridoxal 5′-Phosphate (Numbers of Sibships)

Activity (+ PLP)/ activity (no added PLP)[a]	Clinical	
	Responsive	Not responsive
1-2	16	1
2-3	3	1
3-8	3	1
8-13	—	1

[a]Only extracts with detected activity in the absence of added pyridoxal 5′-phosphate (PLP) are shown. Data are combined from the results reported by Uhlendorf *et al.* (1973) and Fowler *et al.* (1978).

growing fibroblasts in B_6-depleted medium, these workers were able to decrease intracellular PLP by more than 95% and the portion of cystathionine β-synthase present as holoenzyme to less than 10% of the total (compared to 70% in control cells before B_6 depletion). Maximal saturation of apocystathionine β-synthase with PLP was attained when such depleted cells from normal subjects were grown with a concentration of 10 ng/ml pyridoxal added to the medium. With cells from a clinically responsive patient, maximal saturation of residual apocystathionine β-synthase was attained only at a medium concentration of 25–50 mg/ml pyridoxal, and with cells from a clinically nonresponsive patient, saturation was reached at 100 ng/ml. The apparent K_ms of cystathionine β-synthases from depleted cells from four responsive patients were two to four times the control values, and in two lines from nonresponsive patients were 16- and 63-fold normal.

These results were interpreted as suggesting that clinical responsiveness of a cystathionine β-synthase-deficient patient to B_6 may depend not only on the presence of at least a small residual activity of this enzyme but also on the mutant enzyme having an affinity for PLP which is no more than moderately reduced. The latter condition is important because the capacity of cells to increase their PLP concentrations when exposed to increases in B_6 may be limited. A meaningful increase in cystathionine β-synthase activity must be achieved at an attainable PLP concentration. Clinical nonresponsiveness to pyridoxine may be the result of an almost complete lack of residual activity of cystathionine β-synthase (probably the most common situation; see Table III) or the result of a residual mutant activity having such a reduced affinity for PLP that insufficient increase in this activity is achieved at a concentration of cofactor attainable by the cells (Lipson *et al.*, 1980).

These studies of Lipson and colleagues provide a useful experimental tool (growth of fibroblasts in a B_6-depleted medium) and a promising formulation of the molecular factors that determine clinical responsiveness or nonresponsiveness. These results also emphasize the need to focus on the amount and distribution of PLP in tissues as affected by B_6 supply. In liver of rats on B_6-sufficient diets, PLP is localized chiefly in the cytosol. Deficient B_6 intake decreases total hepatic PLP, with the predominant decrease occurring in the cytosol (Bosron *et al.*, 1978). However, increases in B_6 from two to 200 times the daily dietary allowance recommended by the National Research Council lead to no further increases in total hepatic PLP (Li *et al.*, 1974) or may even result in decreased total PLP (Cohen *et al.*, 1973). This strict upper limit on the hepatic PLP content is set largely by hydrolysis of unbound PLP by a phosphatase activity associated with the plasma membrane (Li *et al.*, 1974; Lumeng and Li, 1975).

The interaction between apocystathionine β-synthase and PLP may bring about results beyond the immediate effects on enzyme activity. Cystathionine β-synthase from normal human liver (Mudd *et al.*, 1970b) or fibroblasts (Kim and Rosenberg, 1974; Fleisher *et al.*, 1978) is stabilized to *in vitro* heat inactivation by high concentrations of PLP. Mutant cystathionine β-synthase activities from different patients have had either the same (Mudd *et al.*, 1970b), increased (Kim and Rosenberg, 1974; Fleisher *et al.*, 1978), or decreased (Fleisher *et al.*, 1978) thermostability, and the thermostability has either been increased (Mudd *et al.*, 1970b; Kim and Rosenberg, 1974; Fleisher *et al.*, 1978), or decreased (Fleisher *et al.*, 1978) by high concentrations of PLP. The limited studies reported to date have revealed no striking correlation between these parameters and the clinical responsiveness or nonresponsiveness to B_6 of the patients from whom the mutant enzymes derived.

A number of data in the literature are compatible with the possibility that coenzyme dissociation may be a rate-limiting step in the degradation of PLP-dependent enzymes (Litwak and Rosenfield, 1973). Kominami and Katunuma and co-workers have demonstrated and characterized a group of proteases that degrade specifically the apoenzyme, but not the holoenzyme, forms of several PLP-dependent enzymes. Members of this class of proteases (referred to as group-specific proteases for pyridoxal enzymes) are found in a variety of tissues of the rat, including liver, in which the group-specific protease is located chiefly in mitochondria (Katunuma *et al.*, 1975; Kominami *et al.*, 1972, 1975; Banno *et al.*, 1975; Aoki *et al.*, 1975; Kominami and Katunuma, 1976). The physiological role of these proteases has not yet been fully clarified (Kominami and Katunuma, 1976), although it has been shown that both serine dehydratase (Khairallah and Pitot, 1968) and ornithine aminotransferase (Kominami and Katunuma, 1976), each of which is a good substrate for pyridoxal-specific proteases (Katunuma *et al.*, 1975), have shorter half-lives in B_6-depleted than in normal rats, whereas tyrosine aminotransferase, which is not a substrate (Katunuma *et al.*, 1975), has the same half-life in B_6-depleted and in normal rats (Lee *et al.*, 1977; but see also Snape *et al.*, 1980). The relevance of these proteases to the turnover of cystathionine β-synthase is subject to further uncertainty, since it has not yet been reported whether the synthase is degraded by these enzymes. Nevertheless, it seems likely that a full understanding of the mechanisms by which changes in pyridoxine intake affect normal and mutant cystathionine β-synthases will be attained only when the role of PLP in the turnover of this enzyme has been clarified.

4.2.2. Other Abnormalities

The failure of the K_m model to account for clinical vitamin responsiveness in cystathionine β-synthase deficiency does not appear to be an isolated instance. Among methylmalonic aciduric patients with defects in methylmalonyl-CoA mutase, at least six have now been identified in whom the apoenzymes have K_ms for adenosylcobalamin 250- to 5000-fold greater than normal. The methylmalonic aciduria of only one of these patients was clinically responsive to cobalamin administration, and the response in even that case was reported as being not unequivocal (Willard and Rosenberg, 1979a; Morrow *et al.*, 1978). In these patients, possession of an abnormally high K_m was not sufficient to insure clinical responsiveness. In each of these cases, the V_{max}s of the mutant mutase enzymes, measured in the presence of optimal adenosylcobalamin, were below normal, ranging from 1-50% of control values. Perhaps "pure" K_m mutations are rare events.

In cystathioninuria resulting from γ-cystathionase deficiency, among the few cases studied, clinical responsiveness to B_6 has not correlated with restoration of a near normal activity of mutant γ-cystathionase by *in vitro* PLP but has correlated with the presence or absence of detected residual activity of mutant enzyme (Mudd and Levy, 1978). The situation seems to be similar to that which exists in homocystinuria caused by cystathionine β-synthase deficiency.

Danner and associates studied partially purified branched-chain α-ketoacid dehydrogenases using mitochondrial inner membranes prepared from control fibroblasts and from fibroblasts cultured from two patients with thiamine-responsive forms of maple syrup urine disease resulting from deficient activities of this enzyme. The branched chain α-ketoacid dehydrogenases in both mutant cell lines had apparent K_ms for thiamine pyrophosphate identical to the control enzyme. Heat inactivation profiles and stabilization by thiamine pyrophosphate were also similar. Enzyme activity in peripheral leucocytes of one patient was 2-5% of control (depending on whether valine, leucine, or isoleucine decarboxylation was assayed) while the subject was receiving the Recommended Dietary Allowance of thiamine. The activity slowly rose to 17-26% of control after 3 weeks during which the patient received doses of thiamine as high as 150 mg daily (Danner *et al.*, 1978).

It was also observed that branched chain α-ketoacid dehydrogenase activity in normal liver increased 1.5- to 2.0-fold after human subjects had taken thiamine, 100 mg daily, for at least 18 days (Danner *et al.*, 1975). It was suggested that the increases in enzyme-specific activities in both normal and mutant tissues were caused by decreased turnover brought about by increased available thiamine pyrophosphate (Danner *et al.*, 1978). Extensive data on patients with maple syrup urine disease not responsive to thiamine are not available. However, it appears that some (Wong *et al.*, 1972; Elsas *et al.*, 1972), but not all (Schulman *et al.*, 1970), such patients have a virtually complete lack of detected residual activity of branched chain α-ketoacid dehydrogenase. It would be instructive to determine whether the residual activities in those thiamine-nonresponsive patients who have such activities fail to rise in response to thiamine administration.

An overall view of the observations relevant to vitamin responsiveness in those diseases in which a cofactor interacts with a single structurally altered apoenzyme suggests that the detailed molecular mechanisms underlying vitamin responsiveness may be much more complicated than had been anticipated during early studies. The presence of a mutant enzyme with an abnormally elevated K_m

for its cofactor more often than not turns out to be neither a necessary nor a sufficient condition to confer vitamin responsiveness on a patient. The presence of some residual activity of the mutant enzyme is likely to be necessary for vitamin responsiveness, but other molecular properties of the mutant enzyme, as yet poorly defined, are likely also to be necessary. Proof that a particular mechanism is the operative one may require extensive studies of the enzymes from both responsive and nonresponsive patients to pinpoint the structural factors crucial in insuring responsiveness. A more positive conclusion from these results is that, since the situation is so complicated, each patient with one of these diseases merits a clinical trial to determine his vitamin responsiveness. Failure of some patients to respond positively cannot be extrapolated beyond those particular patients.

4.3. Vitamin-Responsive Genetic Diseases in Which the Effect of the Vitamin Is on a Metabolic Step or Process Other Than the Primary Abnormality in the Disease

A third general mechanism for vitamin responsiveness is found in situations in which the vitamin affects a metabolic reaction or process different from the one primarily responsible for the genetic disease. The vitamin responsiveness of several diseases listed in Table I appears to reflect such effects. For example, Brunette and co-workers (1972) showed that a patient with lactic acidosis resulting from a defect in pyruvate carboxylase was responsive to thiamine. Thiamine pyrophosphate is not a cofactor for the defective enzyme but is a cofactor for another enzyme that disposes of pyruvate, pyruvate dehydrogenase. In assays with intact leucocytes from both control subjects and the patient with pyruvate carboxylase deficiency, addition of thiamine to the medium enhanced the flux through the pyruvate dehydrogenase step. Presumably, such an effect accounted for the clinical relief of lactic acidosis by thiamine.

Other examples are furnished by the response of additional cases of lactic acidosis caused by pyruvate carboxylase deficiency to lipoic acid (Clayton *et al.*, 1967; Hommes *et al.*, 1968) and by the response to pyridoxine of some cases of primary oxaluria caused by deficient activity of soluble 2-oxo-glutarate: glyoxylate carboligase (Williams and Smith, 1978). Patients with homocystinuria caused by cystathionine β-synthase deficiency develop an increased requirement for folic acid, and their homocystinuria may sometimes be relieved by above-normal doses of folate (Carey *et al.*, 1968; Morrow and Barness, 1972). This is because of enhanced removal of homocysteine via its N^5-methyltetrahydrofolate-dependent methylation to methionine. Betaine (Komrower and Sardharwalla, 1971) or its metabolic precursor choline (Perry *et al.*, 1968) may also alleviate the homocystinuria of cystathionine β-synthase-deficient patients, presumably through accelerated removal of homocysteine via an alternative methylation reaction dependent on betaine. Either betaine or choline led to striking rises in plasma concentrations of methionine to abnormal heights as homocystinemia and homocystinuria diminished. The therapeutic effect of betaine or choline would thus be determined by the as yet unresolved question of whether methionine or homocyst(e)ine accumulation is more dangerous for such patients.

Similar phenomena may be expected in other attempts to use vitamin treatment to affect reactions other than the one that is primarily defective in a genetic disorder, and they underline the need for especially careful scrutiny of the effects of such treatments. A somewhat less equivocal situation is the relief of homocystinuria caused by defects in the methylation of homocysteine via the N^5-methyltetrahydrofolate-dependent pathway through use of high doses of betaine (S. I. Goodman and S. H. Mudd, unpublished observations). In this instance, betaine stimulates a reaction which in effect parallels the defective metabolic step: the conversion of homocysteine to methionine.

Final examples of this category of vitamin response are provided by recent studies of Schulman and associates which have shown that vitamin E, 30–35 IU/kg body weight, may improve erythrocyte survival and/or polymorphonuclear leucocyte function in patients with any of a variety of disorders affecting the production of adequate amounts of reduced glutathione (Schulman *et al.*, 1980a). These diseases include glutathione synthetase deficiency (Spielberg *et al.*, 1979; J. D. Schulman, personal communication, 1980), an occasional patient with chronic hemolyzing glucose-6-phosphate dehydrogenase deficiency (Spielberg *et al.*, 1979; J. D. Schulman, personal communication, 1980), and patients with the Mediterranean form of glucose-6-phosphate dehydrogenase deficiency (Schulman *et al.*, 1980b). In all these situations, the antioxidant action of vitamin E is thought to prevent damage that would otherwise occur because of lack of normal amounts of reduced glutathione.

5. Significance of Vitamin-Responsive Genetic Conditions

Most of the genetic conditions discussed in this chapter are rare, and one may understandably wonder whether they have meaning for more everyday situations. In the opinion of the author, there are two major reasons why nutritionists may wish to be aware of these conditions. First, as with all of nature's genetic experiments, these diseases furnish unique tools to illuminate normal metabolic processes. Historically, the first indication or the most definitive proof for the existence of many steps in normal processing of vitamins has come from the studies of inborn errors of metabolism. Examples discussed here might include the evidence provided for the role of specific proteins in cellular uptake of cobalamin and in transport of folate. Many details of the intracellular processes for cobalamin metabolism will surely be further elucidated through study of patients with methylmalonic aciduria. An understanding of B_6 responsiveness in cystathionine β-synthase deficiency demands further clarification of the means by which mammalian protein turnover is regulated and of the role of cofactor–apoenzyme interactions in such turnover. Mutant cells from B_6-responsive and -nonresponsive patients furnish one means to gain such understanding.

Second, the conditions discussed furnish excellent illustrations of the great amount of genetic heterogeneity that may exist in the human population with respect to the needs of different individuals for specific vitamins. Taking into account the many other findings that currently indicate that genetic heterogeneity within populations may be more the rule than the exception (Lewontin, 1974; Harris, 1975, 1976; but see also McConkey *et al.*, 1979; Walton *et al.*, 1979; Edwards and Hopkinson, 1980), one may at least raise the question of whether there is, properly speaking, a "normal" requirement for a vitamin or whether the patients discussed here do not represent merely the extremes on a broad distribution curve of the vitamin requirements of individuals within the population.

Individuals with genetically determined modifications of transketolase leading to higher than normal K_ms for thiamine pyrophosphate tend to develop the Wernicke–Korsakoff syndrome when receiving a marginal thiamine intake (Blass and Gibson, 1977). This situation furnishes an interesting example of the generalization that a given change in a protein may be benign or deleterious depending on the environment, including the nutritional circumstances in which the possessor of that protein is placed. Thus, rather than there being a single "ideal" human with a "normal" requirement for a vitamin, about which actual persons cluster as imperfect approximations, there may well exist an array of individuals in any population with genetically determined differences in their vitamin requirements spread over a wide range. If this is so, determination of nutritional standards for vitamins takes on an increased complexity to which nutritionists may wish to give a good deal of thought in the years to come.

References

Albright, F., Butler, A. M., and Bloomberg, E., 1937, Rickets resistant to vitamin D therapy, *Am. J. Dis. Child.* **54**:529.

Anthony, M., and McLeay, A. C., 1976, A unique case of derangement of vitamin B_{12} metabolism, *Proc. Aust. Assoc. Neurol.* **13**:61.

Aoki, Y., Urata, G., and Takaku, F., 1973, δ-Aminolevulinic acid synthetase in erythroblasts of patients with primary siderblastic anemia, *Acta Haematol. Jpn.* **86**:74.

Aoki, Y., Urata, G., Takaku, F., and Katunuma, N., 1975, A new protease inactivating δ-aminolevulinic acid synthetase in mitochondria of human bone marrow cells, *Biochem. Biophys. Res. Commun.* **65**:567.

Balsan, S., Garabedian, M., Sorgniard, R., Holick, M. F., and DeLuca, H. F., 1975, 1,25-Dihydroxyvitamin D_3 and 1,α-hydroxyvitamin D_3 in children: Biologic and therapeutic effects in nutritional rickets and different types of vitamin D resistance, *Pediatr. Res.* **9**:586.

Banno, Y., Shiotani, T., Towatari, T., Yoshikawa, D., Katsunuma, T., Afting, E.-G., and Katunuma, N., 1975, Studies on new intracellular proteases in various organs of rat. 3. Control of group-specific protease under physiological conditions, *Eur. J. Biochem.* **52**:59.

Barber, G. W., and Spaeth, G. L., 1967, Pyridoxine therapy in homocystinuria, *Lancet* **2**:337.
Bartlett, K., and Gompertz, D., 1976, Combined carboxylase defect: Biotin responsiveness in cultured fibroblasts, *Lancet* **2**:804.
Bartlett, K., and Gompertz, D., 1978, Biotin activation of carboxylase activity in cultured fibroblasts from a child with a combined carboxylase defect, *Clin. Chim. Acta* **84**:399.
Barnes, N. D., Hull, D., Balgobin, L., and Gompertz, D., 1970, Biotin-responsive propionicacidaemia, *Lancet* **2**:244.
Baumgartner, E. R., Schweizer, K., and Wick, H., 1977, Different congenital forms of defective remethylation in homocystinuria: Clinical, biochemical and morphologic studies, *Pediatr. Res.* **11**:1015.
Ben-Bassat, I., Feinstein, A., and Ramot, B., 1969, Selective vitamin B_{12} malabsorption with proteinuria in Israel, *Isr. J. Med. Sci.* **5**:62.
Blass, J. P., and Gibson, G. E., 1977, Abnormality of a thiamine-requiring enzyme in patients with Wernicke–Korsakoff syndrome, *N. Engl. J. Med.* **297**:1367.
Blass, J. P., Avigan, J., and Uhlendorf, B. W., 1970, A defect in pyruvate decarboxylase in a child with an intermittent movement disorder, *J. Clin. Invest.* **49**:423.
Bosron, W. F., Veitch, R. L., Lumeng, L., and Li, T.-K., 1978, Subcellular localization and identification of pyridoxal 5′-phosphate-binding proteins in rat liver, *J. Biol. Chem.* **253**:1488.
Brunette, M. G., Delvin, E., Hazel, B., and Scriver, C. R., 1972, Thiamine-responsive lactic acidosis in a patient with deficient low-K_m pyruvate carboxylase activity in liver, *Pediatrics* **50**:702.
Carey, M. C., Fennelly, J. J., and FitzGerald, O., 1968, Homocystinuria. II. Subnormal serum folate levels, increased folate clearance and effects of folic acid therapy, *Am. J. Med.* **45**:26.
Cartwright, G. E., and Deiss, A., 1975, Sideroblasts, siderocytes and sideroblastic anemia, *N. Engl. J. Med.* **292**:185.
Christensen, J. R., 1940–1941, Three familial cases of atypical late rickets, *Acta Paediatr. Scand.* **28**:247.
Clayton, B. E., Dobbs, R. H., and Patrick, A. D., 1967, Leigh's subacute necrotizing encephalopathy: Clinical and biochemical study, with special reference to therapy with lipoate, *Arch. Dis. Child.* **42**:467.
Cohen, P. A., Schneidman, K., Ginsberg-Fellner, F., Sturman, J. A., Knittle, J., and Gaull, G. E., 1973, High pyridoxine diet in the rat: Possible implications for megavitamin therapy, *J. Nutr.* **103**:143.
Danner, D. J., Davidson, E. D., and Elsas, L. J. II, 1975, Thiamine increases the specific activity of human liver branched chain α-ketoacid dehydrogenase, *Nature* **254**:529.
Danner, D. J., Wheeler, F. B., Lemmon, S. K., and Elsas, L. J. II, 1978, *In vivo* and *in vitro* response of human branched chain α-ketoacid dehydrogenase to thiamine and thiamine pyrophosphate, *Pediatr. Res.* **12**:235.
Edwards, Y., and Hopkinson, D. A., 1980, Are abundant proteins less variable? *Nature* **284**:511.
Ellenbogen, L., 1975, Absorption and transport of cobalamin. Intrinsic factor and the transcobalamins, in: *Cobalamin, Biochemistry and Pathophysiology* (B. M. Babior, ed.), pp. 215–286, John Wiley & Sons, New York.
Elsas, L. J., Pask, B. A., Wheeler, F. B., Perl, D. P., and Trusler, S., 1972, Classical maple syrup urine disease: Cofactor resistance, *Metabolism* **21**:929.
Elsas, L. J. II, Miller, R. L., and Pinnell, S. R., 1978, Inherited human collagen lysyl hydroxylase deficiency: Ascorbic acid response, *J. Pediatr.* **92**:378.
Erbe, R. W., 1979, Genetic aspects of folate metabolism, *Adv. Hum. Genet.* **9**:293.
Fenton, W. A., and Rosenberg, L. E., 1978, Mitochondrial metabolism of hydroxocobalamin: Synthesis of adenosylcobalamin by intact rat liver mitochondria, *Arch. Biochem. Biophys.* **189**:441.
Finkelstein, J. D., Mudd, S. H., Irreverre, F., and Laster, L., 1966, Deficiencies of cystathionase

and homoserine dehydratase activities in cystathionuria, *Proc. Natl. Acad. Sci. U.S.A.* **55**:865.

Fleisher, L. D., Longhi, R. C., Tallan, H. H., and Gaull, G. E., 1978, Cystathionine β-synthase deficiency: Differences in thermostability between normal and abnormal enzyme from cultured human cells, *Pediatr. Res.* **12**:293.

Fowler, B., Kraus, J., Packman, S., and Rosenberg, L. E., 1978, Homocystinuria. Evidence for three distinct classes of cystathionine β-synthase mutants in cultured fibroblasts, *J. Clin. Invest.* **61**:645.

Fraser, D., and Salter, R. B., 1958, The diagnosis and management of the various types of rickets, *Pediatr. Clin. North Am.* **5**:417.

Fraser, D., Kooh, S. W., Kind, H. P., Holick, M. F., Tanaka, Y., and DeLuca, H. F., 1973, Pathogenesis of hereditary vitamin D-dependent rickets. An inborn error of vitamin D metabolism involving defective conversion of 25-hydroxyvitamin D to 1α,25-dihydroxyvitamin D, *N. Engl. J. Med.* **289**:817.

Freeman, J. M., Finkelstein, J. D., and Mudd, S. H., 1975, Folate-responsive homocystinuria and "schizophrenia." A defect in methylation due to deficient 5,10-methylenetetrahydrofolate reductase activity, *N. Engl. J. Med.* **292**:491.

Frimpter, G. W., 1965, Cystathioninuria: Nature of the defect, *Science* **149**:1095.

Frimpter, G. W., Haymovitz, A., and Horwith, M., 1963, Cystathionuria, *N. Engl. J. Med.* **268**:333.

Frimpter, G. W., Andelman, R. J., and George, W. F., 1969, Vitamin B_6-dependency syndromes. New horizons in nutrition, *Am. J. Clin. Nutr.* **22**:794.

Gehrmann, G., 1965, Pyridoxine-responsive anaemias, *Br. J. Haematol.* **11**:86.

Gibbs, D. A., and Watts, R. W. E., 1970, The action of pyridoxine in primary hyperoxaluria, *Clin. Sci.* **38**:277.

Gimpert, E., Jakob, M., and Hitzig, W. H., 1975, Vitamin B_{12} transport in blood. I. Congenital deficiency of transcobalamin II, *Blood* **45**:71.

Gompertz, D., Storrs, C. N., Bau, D. C. K., Peters, T. J., and Hughes, E. A., 1970, Localization of enzyme defect in propionicacidemia, *Lancet* **1**:1140.

Gompertz, D., Draffan, G. H., Watts, J. L., and Hull, D., 1971, Biotin-responsive β-methylcrotonylglycinuria, *Lancet* **2**:22.

Goodman, S. I., Moe, P. G., Hammond, K. B., Mudd, S. H., and Uhlendorf, B. W., 1970, Homocystinuria with methylmalonic aciduria: Two cases in a sibship, *Biochem. Med.* **4**:500.

Goodman, S. I., Keyser, A. J., Mudd, S. H., Schulman, J. D., Turse, H., and Lewy, J., 1972, Responsiveness of congenital methylmalonic aciduria to derivatives of vitamin B_{12}, *Pediatr. Res.* **6**:138.

Gräsbeck, R., and Kvist, G., 1967, Defecto congenito y selectivo de absorcion de la vitamina B_{12} con proteinuria, *Rev. Clin. Esp.* **106**:448.

Gräsbeck, R., Gordin, R., Kantero, I., and Kuhlback, B., 1960, Selective vitamin B_{12} malabsorption and proteinuria in young people, *Acta Med. Scand.* **167**:289.

Gravel, R. A., Mahoney, M. J., Ruddle, F. H., and Rosenberg, L. E., 1975, Genetic complementation in heterokaryons of human fibroblasts defective in cobalamin metabolism, *Proc. Natl. Acad. Sci. U.S.A.* **72**:3181.

Hakami, N., Neiman, P. E., Canellos, G. P., and Lazerson, J., 1971, Neonatal megaloblastic anemia due to inherited transcobalamin II deficiency in two siblings, *N. Engl. J. Med.* **285**:1163.

Harris, H., 1975, *The Principles of Human Biochemical Genetics,* North-Holland, Amsterdam.

Harris, H., 1976, Molecular evolution: The neutralist–selectionist controversy, *Fed. Proc.* **35**:2079.

Harris, J. W., Whittington, R. M., Weisman, R., Jr., and Horrigan, D. L., 1956, Pyridoxine responsive anemia in the human adult, *Proc. Soc. Exp. Biol. Med.* **91**:427.

Haurani, F. I., Hall, C. A., and Rubin, R., 1979, Megaloblastic anemia as a result of an abnormal transcobalamin II (Cardeza), *J. Clin, Invest.* **64**:1253.

Hillman, R. E., Keating, J. P., and Williams, J. C., 1978, Biotin-responsive proprionic acidemia presenting as the rumination syndrome, *J. Pediatr.* **92**:439.

Hitzig, W. H., Dohmann, U., Pluss, H. J., and Vischer, D., 1974, Hereditary transcobalamin II deficiency: Clinical findings in a new family, *J. Pediatr.* **85**:622.

Hommes, F. A., Polman, H. A., and Reerink, J. D., 1968, Leigh's encephalomyelopathy: An inborn error of gluconeogenesis, *Arch. Dis. Child.* **43**:423.

Hunt, A. D., Jr., Stokes, J., Jr., McCrory, W. W., and Stroud, H. H., 1954, Pyridoxine dependency: Report of a case of intractable convulsions in an infant controlled by pyridoxine, *Pediatrics* **13**:140.

Imerslund, O., 1960, Idiopathic chronic megaloblastic anemia in children, *Acta Paediatr. Scand.* [*Suppl.*] **119**:1.

Katunuma, N., Kominami, E., Kobayashi, K., Banno, Y., Suzuki, K., Chichibu, K., Hamaguchi, Y., and Katsunuma, T., 1975, Studies on new intracellular proteases in various organs of rat. 1. Purification and comparison of their properties, *Eur. J. Biochem.* **52**:37.

Katz, M., Lee, S. K., and Cooper, B. A., 1972, Vitamin B_{12} malabsorption due to biologically inert intrinsic factor, *N. Engl. J. Med.* **287**:425.

Katz, M., Mahlman, C. S., and Allen, R. H., 1974, Isolation and characterization of an abnormal human intrinsic factor, *J. Clin. Invest.* **53**:1274.

Khairallah, E. A., and Pitot, H. C., 1968, Studies on the turnover of serine dehydrase: Amino acid induction, glucose repression, and pyridoxine stabilization, in: *Symposium on Pyridoxal Enzymes* (K. Yamada, N. Katunuma, and H. Wada, eds.), pp. 159–164, Maruzen, Tokyo.

Kim, Y. J., and Rosenberg, L. E., 1974, On the mechanism of pyridoxine responsive homocystinuria. II. Properties of normal and mutant cystathionine β-synthase from cultured fibroblasts, *Proc. Natl. Acad. Sci. U.S.A.* **71**:4821.

Knapp, A., 1960, Uber eine neue hereditare von vitamin B_6 abhangige Storung im Tryptophanstoffwechsel, *Clin. Chim. Acta* **5**:6.

Koch, J., Stokstad, E. L. R., Williams, H. E., and Smith, L. H., Jr., 1967, Deficiency of 2-oxoglutarate:glyoxylate carboligase activity in primary hyperaluria, *Proc. Natl. Acad. Sci. U.S.A.* **57**:1123.

Kominami, E., and Katunuma, N., 1976, Studies on new intracellular proteases in various organs of rats. Participation of proteases in degradation of ornithine aminotransferase *in vitro* and *in vivo, Eur. J. Biochem.* **62**:425.

Kominami, E., Kobayashi, K., Kominami, S., and Katunuma, N., 1972, Properties of a specific protease for pyridoxal enzymes and its biological role, *J. Biol. Chem.* **247**:6848.

Kominami, E., Banno, Y., Chichibu, K., Shiotani, T., Hamaguchi, Y., and Katunuma, N., 1975, Studies on new intracellular proteases in various organs of rat. 2. Mode of limited proteolysis, *Eur. J. Biochem.* **52**:51.

Komrower, G. M., and Sardharwalla, I. B., 1971, The dietary treatment of homocystinuria, in: *Inherited Disorders of Sulphur Metabolism* (N. A. J. Carson and D. N. Raine, eds.), pp. 254–263, Churchill Livingstone, London.

Krane, S. M., Pinnell, S. R., and Erbe, R. W., 1972, Lysyl-protocollagen hydroxylase deficiency in fibroblasts from siblings with hydroxylysine-deficient collagen, *Proc. Natl. Acad. Sci. U.S.A.* **69**:2899.

Lampkin, B. C., and Maurer, A. M., 1967, Congenital pernicious anemia with coexistent transitory malabsorption of vitamin B_{12}, *Blood* **30**:495.

Lanzkowsky, P., 1970, Congenital malabsorption of folate, *Am. J. Med.* **48**:580.

Lanzkowsky, P., 1977, Congenital malabsorption of folate associated with mental retardation, in: *Nutritional Deficiency Secondary to Inborn Errors of Metabolism Its Relation to Physical and Mental Development* (N. Shimazono and T. Arakawa, eds.), pp. 213–226, Japanese Malnutrition Panel of the U.S.–Japan Cooperative Medical Science Program, c/o Wayo Women's University, Ichikawa, Chiba Prefecture, 272 Japan.

Lanzkowsky, P., Erlandson, M. E., and Bezan, A. I., 1969, Isolated defect of folic acid absorption associated with mental retardation and cerebral calcification, *Blood* **34**:452.

Lee, K.-L., Darke, P. L., and Kenney, F. T., 1977, Role of coenzyme in aminotransferase turnover, *J. Biol. Chem.* **252**:4958.

Lewontin, R. C., 1974, *The Genetic Basis of Evolutionary Change*, Columbia University Press, New York.

Li, T.-K., Lumeng, L., and Veitch, R. L., 1974, Regulation of pyridoxal 5′-phosphate metabolism in liver, *Biochem. Biophys. Res. Commun.* **61**:677.

Linnell, J. C., Matthews, D. M., Mudd, S. H., Uhlendorf, B. W., and Wise, I. J., 1976, Cobalamins in fibroblasts cultured from normal control subjects and patients with methylmalonic aciduria, *Pediatr. Res.* **10**:179.

Lipson, M. H., Kraus, J., and Rosenberg, L. E., 1980, Affinity of cystathionine β-synthase for pyridoxal 5′-phosphate in cultured cells: A mechanism for pyridoxine responsive homocystinuria, *J. Clin. Invest.* **66**:188.

Litwak, G., and Rosenfield, S., 1973, Coenzyme dissociation, a possible determinant of short half-life of inducible enzymes in mammalian liver, *Biochem. Biophys. Res. Commun.* **52**:181.

Lonsdale, D., Faulkner, W. R., Price, J. M., and Smeby, R. R., 1969, Intermittent cerebellar ataxia associated with hyperpyruvic acidemia, hyperphenylalaninemia and hyperalaninuria, *Pediatrics* **43**:1025.

Luhby, A. L., Eagle, F. J., Roth, E., and Cooperman, J. M., 1961, Relapsing megaloblastic anemia in an infant due to a specific defect in gastrointestinal absorption of folic acid, *Am. J. Dis. Child.* **102**:482.

Lumeng, L., and Li, T.-K., 1975, Characterization of the pyridoxal 5′-phosphate and pyridoxamine 5′-phosphate hydrolase activity in rat liver, *J. Biol. Chem.* **250**:8126.

MacKenzie, I. L., Donaldson, R. M., Jr., Trier, J. S., and Mathan, V. I., 1972, Ileal mucosa in familial selective vitamin B_{12} malabsorption, *N. Engl. J. Med.* **286**:1021.

Mahoney, M. J., and Rosenberg, L. E., 1975, Inborn errors of cobalamin metabolism, in: *Cobalamin, Biochemistry and Pathophysiology* (B. M. Babior, ed.), pp. 369–402, John Wiley & Sons, New York.

Mahoney, M. J., Rosenberg, L. E., Mudd, S. H., and Uhlendorf, B. W., 1971, Defective metabolism of vitamin B_{12} in fibroblasts from patients with methylmalonicaciduria, *Biochem. Biophys. Res. Commun.* **44**:375.

Mahoney, M. J., Hart, A. C., Steen, V. D., and Rosenberg, L. E., 1975, Methylmalonicacidemia: Biochemical heterogeneity in defects of 5′-deoxyadenosylcobalamin synthesis, *Proc. Natl. Acad. Sci. U.S.A.* **72**:2799.

Mahoney, M. J., Nicholson, J. F., Hart, A. C., Rosenberg, L. E., and Challop, R., 1976, Cobalamin (vitamin B_{12}) toxicity in methylmalonicacidemia, *Pediatr. Res.* **10**:368.

Mason, D. Y., and Emerson, P. M., 1973, Primary acquired sideroblastic anaemia: Response to treatment with pyridoxal-5-phosphate, *Br. Med. J.* **1**:389.

McConkey, E. H., Taylor, B. J., and Phan, D., 1979, Human heterozygosity: A new estimate, *Proc. Natl. Acad. Sci. U.S.A.* **76**:6500.

McIntyre, O. R., Sullivan, L. W., Jeffries, G. H., and Silver, R. H., 1965, Pernicious anemia in childhood, *N. Engl. J. Med.* **272**:981.

McNicholl, B., and Egan, B., 1968, Congenital pernicious anemia: Effects on growth, brain, and absorption of B_{12}, *Pediatrics* **42**:149.

Mellman, I., Huntington, F. W., Youngdahl-Turner, P., and Rosenberg, L. E., 1979, Cobalamin coenzyme synthesis in normal and mutant human fibroblasts. Evidence for a processing enzyme activity deficient in *cbl* C cells, *J. Biol. Chem.* **254**:11847.

Morrow, G. III, and Barness, L. A., 1972, Combined vitamin responsiveness in homocystinuria, *J. Pediatr.* **81**:946.

Morrow, G. III, Resvin, B., Clark, R., Lebowitz, J., and Whelan, D. T., 1978, A new variant of methylmalonic acidemia—defective coenzyme–apoenzyme binding in cultured fibroblasts, *Clin. Chim. Acta* **85**:67.

Mudd, S. H., 1971, Pyridoxine-responsive genetic disease, *Fed. Proc.* **30**:970.

Mudd, S. H., 1974a, Inborn errors of metabolism. Vitamin-responsive genetic disease, *J. Clin. Pathol.* [*Suppl.*] **8**:38.

Mudd, S. H., 1974b, Homocystinuria and homocysteine metabolism: Selected aspects, in: *Heritable Disorders of Amino Acid Metabolism* (W. L. Nyhan, ed.), pp. 429–451, John Wiley & Sons, New York.

Mudd, S. H., 1977, Cobalamin-responsive genetic disorders, in: *Nutritional Deficiency Secondary to Inborn Errors of Metabolism Its Relation to Physical and Mental Development* (N. Shimazono and T. Arakawa, eds.), pp. 251–265, Japanese Malnutrition Panel of the U.S.-Japan Cooperative Medical Science Program, c/o Wayo Women's University, Ichikawa, Chiba Prefecture, 272 Japan.

Mudd, S. H., and Levy, H. L., 1978, Disorders of transsulfuration, in: *The Metabolic Basis of Inherited Disease* (J. B. Stanbury, J. B. Wyngaarden, and D. S. Fredrickson, eds.), pp. 458–503, McGraw-Hill, New York.

Mudd, S. H., Finkelstein, J. D., Irreverre, F., and Laster, L., 1964, Homocystinuria: An enzymatic defect, *Science* **143**:1443.

Mudd, S. H., Levy, H. L., and Abeles, R. H., 1969, A derangement in B_{12} metabolism leading to homocystinemia, cystathioninemia, and methylmalonic aciduria, *Biochem. Biophys. Res. Commun.* **35**:121.

Mudd, S. H., Uhlendorf, B. W., Hinds, K. R., and Levy, H. L., 1970a, Deranged B_{12} metabolism: Studies of fibroblasts grown in tissue culture, *Biochem. Med.* **4**:215.

Mudd, S. H., Edwards, W. A., Loeb, P. M., Brown, M. S., and Laster, L., 1970b, Homocystinuria due to cystathionine synthase deficiency: The effect of pyridoxine, *J. Clin. Invest.* **49**:1762.

Mudd, S. H., Uhlendorf, B. W., Freeman, J. M., Finkelstein, J. D., and Shih, V. E., 1972, Homocystinuria associated with decreased methylenetetrahydrofolate reductase activity, *Biochem. Biophys. Res. Commun.* **46**:905.

Murthy, P. N. A., and Mistry, S. P., 1974, *In vitro* synthesis of propionyl-CoA holocarboxylase by a partially purified mitochondrial preparation from biotin-deficient chicken liver, *Can. J. Biochem.* **52**:800.

O'Donnell, J. J., Sandman, P. P., and Martin, S. R., 1978, Gyrate atrophy of the retina: Inborn error of L-ornithine: 2-oxoacid aminotransferase, *Science* **200**:200.

Perry, T. L., Hansen, S., Love, D. L., Crawford, L. E., and Tischler, B., 1968, Treatment of homocystinuria with a low-methionine diet, supplemental cystine, and a methyl donor, *Lancet* **2**:474.

Pletsch, Q. A., and Coffey, J. W., 1971, Intracellular distribution of radioactive vitamin B_{12} in rat liver, *J. Biol. Chem.* **246**:4619.

Poole, J. R., Mudd, S. H., Conerly, E. B., and Edwards, W. A., 1975, Homocystinuria due to cystathionine synthase deficiency. Studies of nitrogen balance and sulfur excretion, *J. Clin. Invest.* **55**:1033.

Rasmussen, H., and Anast, C., 1978, Familial hypophosphatemic (vitamin D-resistent) rickets and vitamin D-dependent rickets, in: *The Metabolic Basic of Inherited Disease* (J. B. Stanbury, J. B. Wyngaarden, and D. S. Fredrickson, eds.), pp. 1537–1562, McGraw-Hill, New York.

Rogers, L. E., Porter, F. S., and Sidbury, J. B., 1969, Thiamine-responsive megaloblastic anemia, *J. Pediatr.* **74**:494.

Rosenberg, L. E., 1976, Vitamin-responsive inherited metabolic disorders, *Adv. Hum. Genet.* **6**:1.

Rosenberg, L. E., 1978, Disorders of propionate, methylmalonate, and cobalamin metabolism, in: *The Metabolic Basis of Inherited Disease* (J. B. Stanbury, J. B. Wyngaarden, and D. S. Fredrickson, eds.), pp. 411–429, McGraw-Hill, New York.

Rosenberg, L. E., Lilljeqvist, A.-C., and Hsia, Y. E., 1968, Methylmalonic aciduria: Metabolic block localization and vitamin B_{12} dependency, *Science* **162**:805.

Rosenberg, L. E., Lilljeqvist, A.-C., Hsia, Y. E., and Rosenbloom, F. M., 1969, Vitamin B_{12} dependent methylmalonicaciduria: Defective B_{12} metabolism in cultured fibroblasts, *Biochem. Biophys. Res. Commun.* **37**:607.

Rosenblatt, D. S., Cooper, B. A., Lue-Shing, S., Wong, P. W. K., Berlow, S., Narisawa, K., and Baumgartner, R., 1979, Folate distribution in cultured human cells. Studies on 5,10-CH_2-H_4PteGlu reductase deficiency, *J. Clin. Invest.* **63**:1019.

Santiago-Borrero, P. J., Santini, R., Jr., Perez-Santiago, E., and Maldonado, N., 1973, Congenital isolated defect of folic acid absorption, *J. Pediatr.* **82**:450.

Schulman, J. D., Lustberg, T. J., Kennedy, J. L., Museles, M., and Seegmiller, J. E., 1970, A new variant of maple syrup urine disease (branched chain ketoaciduria), *Am. J. Med.* **49**:118.

Schulman, J. D., Mudd, S. H., Schneider, J. A., Sheetz, M., Spielberg, S. P., Boxer, L. A., Oliver, J., and Corash, L. M., 1980a, Inborn errors of glutathione and sulfur amino acid metabolism, *Ann. Intern. Med.* **93**:330.

Schulman, J. D., Corash, L., Bartsocas, C., Spielberg, S., Boxer, L., Sheetz, M., Steinherz, R., and Papadatos, C., 1980b, in: *Progress in Clinical and Biological Research,* Vol. 34, *Management of Genetic Disorders* (C. Papadatos and C. Bartsocas, eds.), pp. 381–387, Alan R. Liss, New York.

Scott, C. R., Hakami, N., Teng, C. C., and Sagerson, R. N., 1972, Hereditary transcobalamin II deficiency: The role of transcobalamin II in vitamin B_{12}-mediated reactions, *J. Pediatr.* **81**:1106.

Scriver, C. R., 1964, Pyridoxine-dependent seizures in infancy (a metabolically and genetically determined form of epilepsy), in: *Yearbook of Pediatrics* (S. S. Gellis, ed.), pp. 44–48, Year Book, Chicago.

Scriver, C. R., 1973, Progress in endocrinology and metabolism. Vitamin-responsive inborn errors of metabolism, *Metab. Clin. Exp.* **22**:1319.

Scriver, C. R., and Hutchison, J. H., 1963, The vitamin B_6 deficiency syndrome in human infancy: Biochemical and clinical observations, *Pediatrics* **31**:240.

Scriver, C. R., and Whelan, D. T., 1969, Glutamic acid decarboxylase (GAD) in mammalian tissue outside the central nervous system, and its possible relevance to heredity vitamin B_6 dependency with seizures, *Ann. N.Y. Acad. Sci.* **166**:83.

Scriver, C. R., MacKenzie, S., Clow, C. L., and Delvin, E., 1971, Thiamine-responsive maple syrup urine disease, *Lancet* **1**:310.

Shih, V. E., Salam, M. Z., Mudd, S. H., Uhlendorf, B. W., and Adams, R. D., 1972, A new form of homocystinuria due to $N^{5,10}$-methylenetetrahydrofolate reductase deficiency, *Pediatr. Res.* **6**:135.

Shih, V. E., Berson, E. L., Mandell, R., and Schmidt, S. Y., 1978, Ornithine ketoacid transaminase deficiency in gyrate atrophy of the choroid and retina, *Am. J. Hum. Genet.* **30**:174.

Smith, L. H., Jr., and Williams, H. E., 1967, Treatment of primary hyperoxaluria, *Mod. Treat.* **4**:522.

Snape, B. M., Badawy, A. A.-B., and Evans, M., 1980, Stabilization of rat liver tyrosine aminotransferase *in vivo* by pyridoxine administration, *Biochem. J.* **186**:625.

Spielberg, S. P., Boxer, L. A., Corash, L. M., and Schulman, J. D., 1979, Improved erythrocyte survival with high-dose vitamin E in chronic hemolyzing G6PD and glutathione synthetase deficiency, *Ann. Intern. Med.* **90**:53.

Tada, K., Yokayama, Y., Nakagawa, H., Yoshida, T., and Arakawa, T., 1967, Vitamin B_6-dependent xanthurenic aciduria, *Tokohu J. Exp. Med.* **93**:115.

Takki, K., 1974, Gyrate atrophy of the choroid and retina associated with hyperornithinaemia, *Br. J. Ophthamol.* **58**:3.

Uhlendorf, B. W., Conerly, E. B., and Mudd, S. H., 1973, Homocystinuria: Studies in tissue culture, *Pediatr. Res.* **7**:645.

Walker, G. A., Murphy, S., and Huennekens, F. M., 1969, Enzymatic conversion of vitamin B_{12a} to adenosyl-B_{12}. Evidence for the existence of two separate reducing systems. *Arch. Biochem. Biophys.* **134**:95.

Walton, K. E., Styer, D., and Gruenstein, E. I., 1979, Genetic polymorphism in normal human fibroblasts as analyzed by two-dimensional polyacrylamide gel electrophoresis, *J. Biol. Chem.* **254**:7951.

Weleber, R. G., Kennaway, N. G., and Buist, N. R. M., 1978, Vitamin B_6 in management of gyrate atrophy of choroid and retina, *Lancet* **2**:1213.

Weyler, W., Sweetman, L., Maggio, D. C., and Nyhan, W. L., 1977, Deficiency of propionyl-CoA carboxylase and methylcrotonyl-CoA carboxylase in a patient with methylcrotonylglycinuria, *Clin. Chim. Acta* **76**:321.

Willard, H. F., and Rosenberg, L. E., 1979a, Inborn errors of cobalamin metabolism: Effect of cobalamin supplementation in culture on methylmalonyl CoA mutase activity in normal and mutant human fibroblasts, *Biochem. Genet.* **17**:57.

Willard, H. F., and Rosenberg, L. E., 1979b, Inherited deficiencies of methylmalonyl CoA mutase activity: Biochemical and genetic studies in cultured skin fibroblasts, in: *Models for the Study of Inborn Errors of Metabolism* (F. A. Hommes, ed.), pp. 297–310, Elsevier, New York.

Willard, H. F., Mellman, I. S., and Rosenberg, L. E., 1978, Genetic complementation among inherited deficiencies of methylmalonyl CoA mutase activity: Evidence for a new class of human cobalamin mutant, *Am. J. Hum. Genet.* **30**:1.

Williams, H. E., and Smith, L. H., Jr., 1978, Primary hyperoxaluria, in: *The Metabolic Basis of Inherited Disease* (J. B. Stanbury, J. B. Wyngaarden, and D. S. Fredrickson, eds.), pp. 182–204, McGraw-Hill, New York.

Wolf, B., 1980, Molecular basis for genetic complementation in propionyl CoA carboxylase deficiency, *Exp. Cell Res.* **125**:502.

Wong, P. W. K., Justice, P., Smith, G. F., and Hsia, D. Y. Y., 1972, A case of classical maple syrup urine disease "thiamine non-responsive," *Clin. Genet.* **3**:27.

Wong, P. W. K., Justice, P., Hruby, M., Weiss, E. B., and Diamond, E., 1977, Folic acid nonresponsive homocystinuria due to methylenetetrahydrofolate reductase deficiency, *Pediatrics* **59**:749.

Yoshida, T., Tada, K., and Arakawa, T., 1971, Vitamin B_6-dependency of glutamic acid decarboxylase in the kidney from a patient with vitamin B_6 dependent convulsion, *Tohoku J. Exp. Med.* **104**:195.

Youngdahl-Turner, P., Rosenberg, L. E., and Allen, R. H., 1978, Binding and uptake of transcobalamin II by human fibroblasts, *J. Clin. Invest.* **61**:133.

Youngdahl-Turner, P., Mellman, I. S., Allen, R. H., and Rosenberg, L. E., 1979, Protein mediated vitamin uptake. Adsorptive endocytosis of the transcobalamin II-cobalamin complex by cultured human fibroblasts, *Exp. Cell. Res.* **118**:127.

Chapter 2

Vitamin D Binding Proteins

John G. Haddad, Jr.

1. Introduction

In recent years, considerable growth has occurred in our understanding of the distribution, metabolism, and action of antiricketic sterols. Vitamin D is recognized to be synthesized in the skin, absorbed from the intestine, and transformed to more potent bioactive metabolites. Whether derived from cutaneous transformation of 7-dehydrocholesterol or dietary sources, vitamin D enters the liver where it is metabolized to 25-hydroxyvitamin D (25-OHD) by microsomal (DeLuca and Schnoes, 1976) and/or mitochondrial (Björkhem and Holmberg, 1978) hydroxylases. Subsequently, 25-OHD can be metabolized to an array of dihydroxylated forms of vitamin D. The most potent of these sterols in calcium translocation at the intestine and skeleton is 1,25-dihydroxyvitamin D [1,25-$(OH)_2D$], a product of kidney (Fraser and Kodicek, 1970) cell metabolism. The production of 1,25-$(OH)_2D$ is closely regulated, perhaps most strikingly by the trophic action of parathyroid hormone (Fraser and Kodicek, 1973). Other

Abbreviations used: DBP, plasma binding protein for vitamin D and its metabolites; Gc, group-specific component plasma protein; PAGE, polyacrylamide gel electrophoresis; SDS, sodium dodecylsulfate; 25-OHD, 25-hydroxyvitamin D; 24,25-$(OH)_2D$, 24,25-dihydroxyvitamin D; 1,25-$(OH)_2D$, 1,25-dihydroxyvitamin D; 25,26-$(OH)_2D$, 25,26-dihydroxyvitamin D; DTT, dithiothreitol; EDTA, ethylene diaminetetraacetic acid.

John G. Haddad, Jr. • Endocrine Division, University of Pennsylvania Medical School, Philadelphia, Pennsylvania 19104. The work in the author's laboratory was supported in part by Career Development Award AM11674, Research Grant AM14570, and CRC Grant RR-00036 from the N.I.H.

metabolites have been isolated and identified, but their physiological significance awaits clarification (DeLuca, 1979).

Earlier studies indicated the intense body economy of antiricketic sterols, since protracted periods of vitamin D deprivation and avoidance of ultraviolet light exposure were required to produce deficiency disease. Only small amounts of administered sterol are lost in the excreta (Avioli *et al.*, 1967), and tracer or antiricketic activity has been demonstrated in most of the body tissues (Mawer *et al.*, 1972; Norman and DeLuca, 1963; Bills, 1935). The slow egress of vitamin D from fat (Rosenstreich *et al.*, 1971a) also contributes to the conservation of stores of antiricketic activity.

A remarkable aspect of vitamin D physiology is the participation by many tissues in the synthesis and metabolism of these sterols. The skin provides the substrate for the photochemical synthesis of the vitamin itself. The 25-hydroxylation occurs in the liver, and the 1-hydroxylation step is carried out in the kidney and possibly the placenta as well. The placenta effectively transfers some of these sterols to the fetus. The biological activity of 1,25-$(OH)_2D$ appears to require its association with a high-affinity binding protein in target tissues, including intestine, bone, parathyroid gland, and kidney.

Sterol transport and delivery to cells are clearly important considerations in our understanding of the fates of these substances in various tissues. Much is yet to be learned, but recent studies have provided information worthy of examination. It is the purpose of this chapter to examine the means whereby vitamin D sterols are transported in the blood and the nature of their binding in the tissues.

2. Blood Binding Proteins for Vitamin D and Its Metabolites

2.1. General

Reliable quantitation of vitamin D sterols in plasma reveal the dominant moiety to be 25-OHD at a concentration of 5×10^{-8} M. Under normal conditions in man, vitamin D (4 to 5×10^{-9} M) and 1,25-$(OH)_2D$ (1×10^{-10} M) plasma concentrations are relatively low, as are those of 24,25-$(OH)_2D$ (5 to 9×10^{-9} M) and 25,26-$(OH)_2D$ (2–5×10^{-9} M). Collectively, these substances comprise a plasma vitamin D sterol content of less than 10^{-7} M. During pharmacological vitamin D therapy, 25-OHD levels in plasma increase to levels of 10^{-6} M (Haddad and Stamp, 1974).

Although lipids in nature, the antiricketic sterols are not largely associated with lipoproteins in plasma (De Crousaz *et al.*, 1965; Haddad, 1979; Chen and Lane, 1965). Most of the plasma antiricketic sterol activity sediments at 3–4 S in the ultracentrifuge. However, some of the parent vitamin has been observed to be carried on lipoproteins in chyle and in the plasma shortly after injection (Avioli *et al.*, 1967; Rikkers and DeLuca, 1967; Rikkers *et al.*, 1969) or during pharmacological intake of the vitamin (Silver *et al.*, 1978). Other low-affinity binding sites in plasma have been identified by electrophoretic analyses to be albumin (Haddad and Chyu, 1971a). Since the various antiricketic sterols exhibit different rates of plasma clearance and are distributed differently in the body, the "secondary" binding proteins in plasma might play a role in these processes. Of limited solubility in aqueous solutions, vitamin D is readily oxidized and inactivated under such conditions (Bills, 1954).

Following administration of tracer amounts of vitamin D or its metabolites, serial analyses of blood have provided estimates of their plasma half-lives. The potent metabolite, 1,25-$(OH)_2D$, is cleared most rapidly ($t_{\frac{1}{2}} < 10$ hr) (Gray *et al.*, 1978; Mawer *et al.*, 1976). 25-Hydroxyvitamin D apparently disappears at a much slower rate ($t_{\frac{1}{2}} = 10$ to 20 days) (Bec *et al.*, 1972). Half of the parent vitamin is cleared from plasma in 20–24 hr (Avioli *et al.*, 1967). Concurrent analyses of animal tissues reveal an early uptake of radioactive vitamin D into the liver soon followed by the hepatic egress of radioactive 25-OHD. Both vitamin D and 25-OHD have been observed to accumulate in a wide variety of tissues (Norman and DeLuca, 1963; Mawer *et al.*, 1972), and their slow egress from adipose tissue (Rosenstreich *et al.*, 1971a) suggests their storage there. In contrast, 1,25-$(OH)_2D$ is selectively accumulated, with a large proportion of an administered dose being found in the intestine and bone (Omdahl and DeLuca, 1973). The presence of high-affinity 1,25-$(OH)_2D$ binding proteins in certain tissues appears to influence this sterol's distribution, but little is known about the mechanisms of translocation of the vitamin D sterols into cells. A related substance, cholesterol, is known to enter cells by association with low-density lipoprotein, a plasma protein that undergoes receptor-mediated endocytosis (Goldstein and Brown, 1976; Goldstein *et al.*, 1979). Conceivably, lipoprotein interaction with liver and fat cells might help to explain the distribution of vitamin D to these tissues.

2.2. Phylogenetic Studies

With the availability of high specific activity radioactive sterols, studies of binding of vitamin D sterols in plasma from several species have been carried out in many laboratories (Chen and Lane, 1965; Jacobs and Ray, 1968; Rikkers and DeLuca, 1967; Haddad and Chyu, 1971a; Smith and Goodman, 1971; Rosenstreich *et al.*, 1971b; Hay and Watson, 1976a, b). In almost all mammalian sera studied to date, a single plasma binding protein has been demonstrated to have α-globulin mobility. In some instances, the binding protein has albumin mobility, but careful saturation analyses reveal its affinity and specificity of binding to be characteristic of the α-globulin seen in other mammalian sera (Hay, 1975; Bouillon *et al.*, 1976a). In the chicken, two separate plasma binding proteins of β-globulin mobility have been identified, and they possess different affinities for vitamin D and 25-OHD (Edelstein, 1974; Edelstein *et al.*, 1973).

The purpose of this feature in avian blood has been suggested by the finding of an association between avian vitamin D-binding globulin and an egg yolk protein, phosvitin (Fraser and Emtage, 1976). Vitamin D bound to the binding protein–phosvitin complex apparently gains entry into the egg yolk by virtue of the ability of phosvitin to be concentrated in this tissue. Studies in amphibia are contradictory at this time. The apparently lower concentration of plasma binding protein in this species has led to its being overlooked in some studies (D. R. Fraser, personal communication). Similarly, fish plasma appears to contain much less binding protein than mammals (Hay and Watson, 1976b; Nahm *et al.*, 1979). Variations in electrophoretic mobility and high lipoprotein-to-binding-protein ratios have hidden the binding capacity of specific binding proteins in earlier screening analyses of some sera. It seems probable that all species with a skeleton possess at least one specific plasma binding protein for vitamin D sterols.

2.3. Group-Specific Component

At the time of a report of the distribution of vitamin D bioactivity in human serum (Thomas *et al.*, 1959), a genetically determined, previously unrecognized α-globulin in serum was described (Hirschfield, 1959). This system of electrophoretic variants in human serum was called group-specific component (Gc). Information has accumulated that indicates that Gc^1 and Gc^2 serum proteins are directed by a pair of autosomal codominant alleles (Cleve, 1973). The purification of these proteins was reported (Bowman and Bearn, 1965; Cleve *et al.*, 1963; Heimburger *et al.*, 1964; Simons and Bearn, 1967), and antisera to Gc protein became available for further studies.

Prior to the recognition that human plasma Gc protein and the human plasma binding protein for vitamin D and its metabolites (DBP) were the same protein, the variable electrophoretic expression of Gc in several populations attracted many genetic investigations (Cleve, 1973; Daiger, 1976; Daiger and Cavalli-Sforza, 1977). Three basic patterns of Gc in sera are recognized: Gc^{1-1} serum contains a fast component; Gc^{2-2} serum possesses a protein with slower electrophoretic mobility; Gc^{2-1} serum contains both the fast and slow components. The Gc^1 component is always most frequent in large population studies, but Gc^2 gene frequencies are rarely less than 10%. Over 80,000 sera have been analyzed for the presence of Gc, and an absence of Gc has never been detected. The typing of Gc has been a tool in the determinations of paternity, and many electrophoretic variants have been reported.

A function of Gc protein was not recognized until its incubation with a variety of radioactive ligands. By employing polyacrylamide gel electrophoresis in conjunction with Gc antisera and autoradiography, both vitamin D and 25-OHD were shown to associate with α-globulins previously recognized to be the Gc proteins (Daiger *et al.*, 1975). In spite of polymorphic variation at the Gc locus, all serum variants tested appeared to bind vitamin D sterols (Daiger and Cavalli-Sforza, 1977). Recently, no differential affinity of vitamin D metabolite binding or altered binding capacity was detectable in several types of Gc sera (Kawakami *et al.*, 1979), clearly reducing the likelihood of an adaptive Gc function serving conditions that influence vitamin D bioavailability.

2.4. Purification of Vitamin D Binding Protein

The human plasma DBP has been isolated in several laboratories (Haddad and Walgate, 1976a; Imawari *et al.*, 1976; Bouillon *et al.*, 1976b), and its features resemble those of the Gc proteins which were purified earlier. With one exception (Peterson, 1971), antisera against Gc or purified DBP have been shown to provide identical reactions against purified DBP preparations (Haddad and Walgate, 1976a; Bouillon *et al.*, 1976b; Imawari *et al.*, 1976). Physiochemical and immunologic evidence clearly indicates the identity of these proteins in human plasma, and electrophoretic variants in sera from other species suggests that a similar form of heterogeneity may exist. The DBP in rat serum has been purified in two laboratories (Botham *et al.*, 1976; Bouillon *et al.*, 1978a), and rabbit serum DBP has recently been isolated as well (D. E. M. Lawson, personal communication).

Initial estimates of DBP content in human serum were based on the assumption that the protein was nearly saturated with vitamin D sterols (Smith and Goodman, 1971). This proved to be incorrect, since only 1–3% of binding sites are occupied in normal human serum. With the large excess of apo-DBP to holo-DBP, mammalian serum had proven to be a simple and concentrated source of binding protein for utilization in competitive binding radioassays (Rojanasathit and Haddad, 1977; Edelstein *et al.*, 1974; Haddad *et al.*, 1976b; Bayard *et al.*, 1972). The common procedure has been to identify the DBP by preincubation with radioactive 25-OHD in tracer amounts. The holo- and apo-DBP moieties have different chromatographic and electrophoretic properties, however (Imawari *et al.*, 1976), and this must be considered in the purification steps.

Purification of DBP from plasma requires less than a 200-fold enrichment from total plasma protein, and several physiochemical separation techniques have been applied with success (Haddad, 1980). Table I provides a representative purification scheme. The protein is reasonably stable and can be protected from thermal injuries by preincubation with 25-OHD$_3$ (Bouillon *et al.*, 1976b). Conversely, 25-OHD$_3$ is protected from various insults by its association with DBP (Haddad and Walgate, 1976a). The purified protein stains as one band in SDS-PAGE gels, but under appropriate conditions, the microheterogeneity of DBP can be demonstrated by PAGE in 10-cm separating gels and discontinuous buffer systems (Imawari *et al.*, 1976; Daiger *et al.*, 1975). The reported physiochemical features of DBP or Gc protein in human plasma are indicated in Table II.

Table I. Purification of Human Plasma DBP[a]

Step	Fold purification compared to plasma
1. Cohn IV	2.1[b]
2. DEAE-cellulose	4
3. Sephadex G-200	9
4. DEAE-Sephadex	49
5. DEAE-Sephadex	86
6. Sephadex G-200	109
7. Preparative gel electrophoresis	167

[a]Adapted from Haddad and Walgate (1976a).
[b]Estimate based on saturation analysis of plasma and the Cohn IV material.

2.5. Characteristics of Vitamin D Binding Protein

The features that have interested geneticists studying Gc proteins are centered on the electrophoretic mobility of these proteins. Recently, studies have indicated that the fast components owe their anodal mobility to the presence of sialic acid in the protein (Svasti and Bowman, 1978; Van Baelen *et al.*, 1978; Cleve and Patutschnick, 1979). Furthermore, the occupancy of the binding site by vitamin D sterol alters the electrophoretic mobility of the protein (Van Baelen *et al.*, 1978; Peterson, 1971).

The plasma DBP binding preference for 25-OHD, 24,25-$(OH)_2D$, and 25,26-$(OH)_2D$ over vitamin D and 1,25-$(OH)_2D$ has been demonstrated in several laboratories (Haddad and Walgate, 1976a; Shepard *et al.*, 1979; Preece *et al.*, 1974). Apparently, one sterol-binding site is present in each molecule of DBP (Table II), and no other specific binding protein for these sterols is present in plasma, since most of these sterols are precipitated from plasma by antisera against DBP (Haddad and Walgate, 1976a; Bouillon *et al.*, 1978a). A large variety of non-vitamin-D-type compounds has been tested for competitive binding to DBP, but no competitors have been identified. In mammalian sera, ergocalciferol (vitamin D_2) and cholecalciferol (vitamin D_3) metabolites are bound with equal affinity by DBP (Haddad *et al.*, 1973b, 1976b; Preece *et al.*, 1974). Discrimination against the D_2 series is displayed by avian serum DBP, however (Belsey *et al.*, 1974). Recently, another metabolite, 25-OHD_3-26,23-lactone has been reported to be five times more potent than 25-OHD_3 in the competition for binding to rat serum DBP (Horst, 1979).

Estimates of the binding affinity and capacity of DBP in serum have been made, and considerable variation is apparent in the reported binding affinities for 25-OHD by DBP. In part, this may be because of the nature of the protein material used, incubation conditions, and the method used for separating bound and free sterol. Saturation analyses have also underestimated DBP content com-

Table II. Reported Physicochemical Features of DBP and Gc Proteins

Feature	Peterson (1971)	Haddad and Walgate (1976a,b)	Imawari and co-workers[a]	Bouillon *et al.* (1976b, 1977)	Others[b]
Molecular weight	53,000	58,000	52,000	58,000	50,800
Sedimentation	3.8 S[c]	3.46 S[c]	3.49 S[c]	4.15 S[d]	4.1 S
Electrophoretic mobility	Alpha 1	Inter-alpha	Alpha	Alpha	Alpha
Isoelectric point	—	4.86	4.8	4.89	—
25-OHD-binding force	—	$K_d = 6.4 \times 10^{-8}$ M	—	$K_a = 1.2 \times 10^{10}$ liter/mol	—
Ligand preference	—	$25\text{-OHD}_3 = 24,25\text{-(OH)}_2D_3 > D_3, 1,25\text{-(OH)}_2D_3$	—	—	—
Moles ligand bound/mole DBP	—	0.9	0.97	0.8	—
Serum concentration		$8\text{–}9 \times 10^{-6}$ M	8×10^{-6} M	6×10^{-6} M	1.5×10^{-5} M

[a]Imawari and Goodman (1977); Imawari *et al.* (1976).
[b]Cleve *et al.* (1963); Heimburger *et al.* (1964); Simons and Bearn (1967).
[c]Analytical ultracentrifuge.
[d]Sucrose gradient ultracentrifugation.

pared to direct immunoassay quantitation of the protein in serum (Rojanasathit and Haddad, 1977; Haddad *et al.*, 1976b; Kawakami *et al.*, 1979), and this is likely because of the removal of protein by adsorbent methods in distinguishing bound and free sterol. However, it is possible that a portion of circulating DBP is already occupied by an endogenous ligand or is not capable of sterol binding. Additional studies are required to explore these possibilities.

2.6. Concentrations of Vitamin D Binding Protein

As seen in Table II, quantitations of human serum DBP and Gc are in reasonable accord, indicating that the level is 5 to 9 $\times$ 10^{-6} M. A slightly higher value has been reported for rat serum DBP (Bouillon *et al.*, 1978b). Although uniformly lower levels are observed by saturation analyses (Kawakami *et al.*, 1979; Haddad *et al.*, 1976b), their correlations to the fluctuations observed by immunoassay are intact. The assays employing antisera have utilized radioimmunoassay (Haddad and Walgate, 1976b; Imawari and Goodman, 1977) and radial immunodiffusion (Bouillon *et al.*, 1977, 1978b; Imawari and Goodman, 1977) techniques.

Available data clearly indicate that mammalian DBP represents a high-capacity system with high affinity toward vitamin D metabolites. In contrast to hormone-binding systems in plasma, the binding system for vitamin D and metabolites is remarkably large and mostly unoccupied with ligand. In addition, no correlations have been seen between DBP and vitamin D sterol concentrations (Haddad *et al.*, 1976b). As indicated in Table III, DBP levels in plasma are not altered by vitamin D deficiency or excess. Earlier studies indicated that the liver was the likely organ of Gc or DBP synthesis, and clinical studies have supported this idea (Haddad and Walgate, 1976b; Imawari *et al.*, 1979; Kitchin and Bearn, 1965). Although heavy urinary losses of DBP can occur in the nephrotic syndrome (Barragry *et al.*, 1977; Schmidt-Gayk *et al.*, 1977), this does not seriously detract from the normally huge binding capacity present. Retinol-binding protein (RBP) depends on its hepatic association with retinol for movement from liver to plasma (Smith *et al.*, 1970), and this mechanism clearly does not apply to DBP. Whereas vitamin A toxicity can be associated with supersaturation of plasma RBP, vitamin D toxicity apparently occurs at levels of D sterols below that required for saturation of DBP. However, a redistribution of vitamin D itself toward the lipoprotein classes has been observed during vitamin D intoxication (Silver *et al.*, 1978).

In man, cord blood DBP levels in term infants approximate those seen in adults, whereas premature infants' cord sera contain lower levels (Table III). Recently, serial examinations of premature infants' sera in our laboratory reveal that serum DBP concentrations achieve normal levels at about 40 weeks postconceptual age. Maternal sera and sera from women receiving estrogen–progestogen medication contain increased amounts of DBP, presumably reflecting an estro-

Table III. DBP Concentrations in Human Serum[a]

Group	Number	DBP (μg/ml) Mean ± SEM
Men and women	40	525 ± 24
Men	19	555 ± 32
Women	21	499 ± 36
Normal children	12	524 ± 19
Cord (term)	10	528 ± 78
Cord (premature)	9	359 ± 38[b]
Pregnant women (third trimester)	11	1254 ± 89[b]
Oral contraceptive therapy	10	824 ± 51[b]
Vitamin D deficiency	11	576 ± 33
Vitamin D therapy	12	496 ± 72
Sex-linked hypophosphatemic rickets	14	551 ± 66
Sarcoidosis	6	492 ± 78
Hypoproteinemia	10	234 ± 28[b]
Anticonvulsant therapy	5	485 ± 34
Adrenocorticosteroid therapy	6	660 ± 75

[a] Adapted from Haddad and Walgate (1976b).
[b] $p < 0.001$ compared to adult levels.

gen effect on DBP synthesis (Table III). Interesting variations of these hormonal effects have been demonstrated in the rat plasma DBP system (Bouillon *et al.*, 1978b).

2.7. Metabolism and Cellular Association of Vitamin D Binding Protein

At the present time, almost nothing is known about the rate of synthesis or catabolism of plasma DBP, but recent experiments by the author in the laboratory of D. R. Fraser and D. E. M. Lawson indicate a fairly rapid plasma clearance and widespread tissue accumulation of rabbit plasma DBP. Additional investigations along these lines would provide useful information for the assessment of sterol translocation in the body. Since various mechanisms of ligand delivery to tissues have been recognized for plasma transport proteins (Heller, 1975; Goldstein and Brown, 1976), the impetus for investigating the mechanism(s) relevant to vitamin D sterols is certainly present currently.

A feature of sterol distribution studies and competitive radioassay work was the demonstration of 5–6 S vitamin D metabolite-binding protein in the high-speed supernatant of a wide variety of tissues (Haddad and Birge, 1971, 1975; Lawson *et al.*, 1976). Although the sterol-binding preference by the 5–6 S protein was identical to that of plasma DBP, the binding affinity for 25-OHD was higher (Haddad and Birge, 1975; Lawson *et al.*, 1976; Edelstein, 1974). An aggregate of plasma DBP with a tissue factor was suspected, but plasma DBP was not generated by treatment of the 5-6 S material in high-salt solutions (Haddad and Birge, 1975). The 5-6 S binding protein was reported to yield plasma DBP during heat treatment, and the 5-6 S complex was produced *in vitro* by mixing appropriate dilutions of rat serum and tissue cytosol (Van Baelen *et al.*, 1977). Although it was initially identified by radioactive ligand binding, direct immunologic and radioactivity-labeled DBP analyses were possible by utilizing specific antisera and purified human DBP preparations (Cooke *et al.*, 1979a,b). Another group used antisera to rat serum DBP and noted its cross reaction with the 5-6 S binding protein (Kream *et al.*, 1979).

Presently available data indicate that the 5-6 S complex is a noncovalent, high-affinity ($K_A = 10^7$ liter/mol) bond between plasma DBP and an unidentified tissue protein found in all nucleated cells studied to date. In cultured fibroblasts that had been extensively washed *in vitro*, it was observed that approximately 3% of the cytosol protein could be accounted for by the protein that is capable of binding to DBP. Found in tissues from all species tested thus far, the cellular DBP-binding protein appears to be a well-conserved, major cell protein. Furthermore, the association between this protein and DBP is highly specific, not recognizing competition in a large variety of proteins tested. The DBP-binding protein from cells is heat labile, approximately 40,000 molecular weight, and not retained by concanavalin A columns (Cooke *et al.*, 1979b). More sensitive analyses have revealed that mammalian erythrocytes, previously considered devoid of DBP-binding protein, contain an altered form of the DBP-binding protein (Cooke *et al.*, 1979b).

Initial evaluations of the early information regarding the nature of the 5-6 S binding protein centered on the "artifactual" nature of this material. However, the 5-6 S material was found in rat cartilage (Haddad and Birge, 1975), and a DBP-like factor that inhibited the avian renal 25-OHD 1-hydroxylase had been clearly demonstrated in well-perfused rat kidney (Botham *et al.*, 1974). In fact, these workers demonstrated this inhibitor in well-washed subcellular fractions of rat kidney and later reported it to be rat plasma DBP (Botham *et al.*, 1976; Ghazarian *et al.*, 1978b). Although much of the 5-6 S binding protein could easily be produced by serum DBP contamination of extracted cellular protein, its presence in relatively avascular and easily washed tissues such as cartilage and perfused kidney suggests the possibility of the cellular entry of DBP *in vivo*. Its absence from kidney cells cultured without serum (Van Baelen *et al.*, 1977) did not exclude the possibility of DBP movement on or into cells *in vivo*, and immunoassays of extracts from well-washed leukocytes and cultured fibroblasts indicated a DBP-to-albumin ratio higher than could be explained by serum contamination of these extracts (Cooke, 1979b).

Clearly, much remains to be learned about whether the 5-6 S binding material only represents an "artifact" or provides a clue to a physiological interaction of DBP with cells. The specific and high-affinity binding of DBP by the cellular protein certainly suggests the possibility of a biological role by their interplay (Cooke *et al.*, 1979b), and future investigations of this phenomenon clearly warrant our continued scrutiny.

3. Tissue 1,25-Dihydroxyvitamin D Binding Proteins

3.1. General

Following an injection of labeled 1,25-$(OH)_2D_3$ into the vitamin D-deficient chick, radioactivity can be observed to accumulate in the intestinal nuclear fraction (Chen and DeLuca, 1973; Tsai *et al.*, 1972; Lawson and Wilson, 1974), and at nuclear sites in the bone (Weber *et al.*, 1971) and parathyroid gland (Hughes and Haussler, 1978; Brumbaugh *et al.*, 1975). Although controversy existed about the validity of early observations concerning radioactive vitamin D sterol association with nuclear chromatin (Chen *et al.*, 1970), presently published information strongly indicates that 1,25-$(OH)_2D_3$ action is mediated by its translocation to target cell nuclei, analogous to the mechanism proposed for steroid hormes (Yamamoto and Alberts, 1976; Baxter and Funder, 1979).

The biological expression of 1,25-$(OH)_2D_3$ action involves the transport of calcium and phosphorus across the intestinal epithelium, bone cells, and probably kidney cells as well. At present, RNA polymerase activity (Zerwekh *et al.*, 1974), mRNA synthesis (Spencer *et al.*, 1976), protein synthesis (Wasserman and Feber, 1977; Wilson and Lawson, 1977), enzymatic activity (Haussler *et al.*, 1970; Melancon and DeLuca, 1970), and cyclic AMP levels (Corradino, 1976) have been shown to increase in target tissues following vitamin D sterol administration. Since some of these responses appear to follow the increased ion transport activity (DeLuca, 1979; Haussler and McCain, 1977), the order and importance of these events is not currently clear. Furthermore, the controversy regarding whether or not blockade of transcription and translation abolishes avian or rat intestinal calcium transport response to 1,25-$(OH)_2D$ (DeLuca, 1979) appears to suggest alternate or additional mechanisms whereby 1,25-$(OH)_2D$ exerts its influence.

1,25-Dihydroxyvitamin D circulates in adult human blood at approximately 0.1 nM concentration (Brumbaugh *et al.*, 1974; Eisman *et al.*, 1976). Until recently, the radioactive sterol was not available except by its biosynthesis *in vitro*. Over the past several years, the search for specific 1,25-$(OH)_2D$ binding proteins led to the development of a clinically suitable, albeit difficult, competitive protein binding radioassay for 1,25-$(OH)_2D$ in chromatographically purified extracts of plasma.

3.2. Identification of 1,25-Dihydroxyvitamin D Binding Proteins

One of the complications in the earlier searches for a tissue receptor for 1,25-$(OH)_2D$ was the presence of the serum DBP–tissue protein complex (Haddad, 1979) already discussed. In addition, it became apparent that the tissue 1,25-$(OH)_2D$ binding protein was labile, requiring certain buffer conditions to prevent its degradation during processing of tissue extracts. Studies of 1,25-$(OH)_2[^3H]D$ binding by tissue extracts have yielded various results (Haddad *et al.*, 1973a; Oku *et al.*, 1974; Ghazarian *et al.*, 1978a). In some instances, the 5-6 S complex of serum DBP and tissue protein was the only binding protein observed (Haddad *et al.*, 1973a; Kream *et al.*, 1976; Ulmann *et al.*, 1977; Shimura *et al.*, 1977), whereas other reports indicated that a less dense material with 1,25-$(OH)_2D$ binding ability could be seen in high-speed supernatants and nuclear extracts (Tsai and Norman, 1973; Brumbaugh and Haussler, 1974b; Lawson and Wilson, 1974).

Convincing evidence for the presence of an intestinal cytosol binding protein specific for 1,25-$(OH)_2D_3$ was reported in 1973 and 1974, and the 3-4 S material was apparently confined to "target" tissues, since nuclear localization of administered radioactive sterol was dominant in the intestine and kidney. The time course of binding of 1,25-$(OH)_2D_3$ to cytosol and nuclear components indicated an early event preceding the onset and maximal stimulation of increased intestinal transport activity (Tsai *et al.*, 1972). Considerable effort was made to identify the importance of the cytosol binding protein in the movement of the sterol to nuclear sites, and this was reportedly a tissue-specific and temperature-dependent process (Tsai and Norman, 1973; Brumbaugh and Haussler, 1974a,b). These binding sites were saturable *in vivo* and *in vitro* and exhibited a definite preference for 1,25-$(OH)_2D_3$ over related sterols (Eisman and DeLuca, 1977; Kream *et al.*, 1977a).

A remarkably useful application of the information gathered by analyses of the interaction among intestinal cytosol, 1,25-$(OH)_2D_3$, and nuclear chromatin was the development of a competitive protein-binding radioassay for this sterol (Brumbaugh *et al.*, 1974). The complexity of the procedures and the difficulties arising from sterol availability, lability of binding protein and its contamination with 5-6 S protein, and demanding chromatographic steps in sterol isolation from lipid extracts of plasma collectively contributed to a delay in the widespread application of this assay. Additional studies provided more information about the extraction and stabilization of the cytosol binding protein for 1,25-$(OH)_2D_3$, and simplified, nonnuclear material was shown to be suitable for assay purposes (Eisman *et al.*, 1976).

The field has grown in popularity in recent times, with several laboratories exploring the tissue distribution and characteristics of 1,25-$(OH)_2D$ binding proteins. The availability of high-specific-activity 1,25-$(OH)_2[^3H]D_3$ and more rigorous analyses have led to the recognition that a binding protein for this sterol is present in a variety of tissues including avian intestine, rat intestine (Kream *et al.*, 1977c), human intestine (Wecksler *et al.*, 1979), parathyroid gland (Hughes and Haussler, 1978; Haddad *et al.*, 1976a), bone (Kream *et al.*, 1977b; Manolagas *et al.*, 1979), kidney (Colston and Feldman, 1979; Christakos and Norman, 1979), and pancreas (Christakos and Norman, 1979). The physiological implications for some of these distributions are not known, but suggestions have been made (Chertow *et al.*, 1975).

3.3. Physicochemical Features

At the present time, the purification of this protein has not been achieved. This is easy to understand, since it is a labile protein present in very small concentrations in tissue. All of the available information about its features, therefore, has been derived from experiments dependent on its ability to bind 1,25-$(OH)_2D$ selectively.

Table IV lists some of the features that have been reported for this protein in various tissues of various species. The included data and references are not intended as an exhaustive literature search but are representative and can direct further reading. Whereas most of the identifications have been made by incubating tissues with the sterol *in vitro,* some of the observations followed sterol administration *in vivo* and isolation of subcellular fractions. In general, the specific 1,25-$(OH)_2D$ binding occurs on a 3 to 3.7 S macromolecule that is sensitive to proteolytic enzymes and is thermolabile. The protein's behavior on ion-exchange columns has been reported by one laboratory, and its elution from gel filtration systems has not provided general agreement about its molecular size.

A key factor in its demonstration in some laboratories has been the type of buffer used in the preparation of the tissue extract. The inclusion of 0.1 to 0.3 M KCl has been cited to be beneficial, and the careful rinsing away of surface proteolytic enzymes from mammalian intestinal mucosa has been emphasized. Also, the use of EDTA, dithiothreitol, or thioglycerol has been reported to help in the stabilization of these proteins. It should be emphasized that the utilization of avian intestinal cytosol alone, rather than cytosol–chromatin complexes, in the

Table IV. Ability of Various Sterols to Compete with 1,25-$(OH)_2$-[23,24-^3H]D_3 for Binding to the Avian Intestinal Cytosol Binding Protein[a]

Sterol	Moles required to provide displacement equal to 1 mol of 1,25-$(OH)_2D_3$
1,25-$(OH)_2D_3$	1
1,24(*R*),25-$(OH)_3D_3$	20
5,6-*trans*-25-OHD_3	500
25-OHD_3	1,000
24(*R*),25-$(OH)_2D_3$	8,000
25,26-$(OH)_2D_3$	8,000
Vitamin D_3	1,000,000

[a] Adapted from Eisman and DeLuca (1977).

competitive radioassay of 1,25-$(OH)_2D$ was reportedly made possible by the discovery of suitable protein extraction and stabilization techniques (Eisman *et al.*, 1976).

Agreement exists concerning the high specificity of the binding of 1,25-$(OH)_2D$ by this tissue protein (Haussler and McCain, 1977; Kream *et al.*, 1977a; Eisman and DeLuca, 1977). Its preference for binding 1,25-$(OH)_2D_3$ over 25-OHD_3 is clear (see Table IV); this is the opposite of the preference shown by plasma DBP and of a magnitude which approximates the 500-fold difference in plasma concentration between 25-OHD and 1,25-$(OH)_2D$ (Haddad and Chyu, 1971b; Brumbaugh *et al.*, 1974). Whereas the avian plasma binding of ergocalciferol and cholecalciferol metabolites clearly favors the latter (Belsey *et al.*, 1974), the 1,25-$(OH)_2D$ binding protein in avian intestinal cytosol has variously been reported to recognize 1,25-$(OH)_2D_2$ and 1,25-$(OH)_2D_3$ equally (Eisman and DeLuca, 1977) or to distinguish between these moieties (Hughes *et al.*, 1976).

Several groups have reported their findings regarding the affinity of the association between 1,25-$(OH)_2D_3$ and its tissue binding protein (Table V). In general, there is agreement about the binding system's high affinity and low capacity characteristics. Recently, very high specific activity 1,25-$(OH)_2$ [^{3}H-26,27]D_3 (160 Ci/mmol) was synthesized (Napoli *et al.*, 1979) and utilized in estimates of binding affinity by avian gut crude cytosol (Mellon and DeLuca, 1979). A very high affinity was reported ($K_d = 7.1 \times 10^{-11}$ M by Scatchard plot and 7.5×10^{-12} M by ratio of the dissociation rate constant to the association rate constant), but firm information must await the purification of the 1,25-$(OH)_2D$ binding protein.

3.4. Receptor Candidacy of the 1,25-Dihydroxyvitamin D Binding Proteins

The very nature of the 1,25-$(OH)_2D$ molecule and studies of the time course of its stimulation of calcium transport suggest its action via mechanisms analogous to those described for steroid hormones (Yamamoto and Alberts, 1976). Earlier analyses of the subcellular distribution of tracer sterol indicated a nuclear association (Tsai *et al.*, 1972), and autoradiographic studies lend support to this localization (Zile *et al.*, 1978; Weber *et al.*, 1971). The movement of steroid hormones into the nucleus is regarded to be the job of cytosolic receptors which are high-affinity, low-capacity, high-specificity molecules capable of affecting DNA-directed, RNA-mediated synthesis of proteins that carry out the biological functions characteristic of the response to these hormones. Although other mechanisms of steroid hormone action have been proposed for high concentrations of hormone under special circumstances (Baxter and Funder, 1979), the cytosol receptor–nucleus interaction appears to be the dominant route of influence for steroid hormones that circulate at fairly low concentrations.

The association of the cytosol binding protein for 1,25-$(OH)_2D_3$ and nuclear elements has been reported by several investigators (Tsai and Norman, 1973;

Table V. Study Methods ad Features of 1,25-$(OH)_2D$ Binding Proteins in Tissue Extracts

Species	Tissue	Vitamin D status	Sterol exposure	Extraction buffer	Density and gradient isotonic strength	Size gel filtration	1,25-$(OH)_2D$ binding affinity	Other	Reference
Avian	Intestine	D-deficient	*In vivo, in vitro*	0.05 M Tris-HCl, 0.02 M KCl, 0.005 M $MgCl_2$, pH 7.4 or 0.25 M sucrose	3-3.5 S (0.15 M KCL)	—	K_d 2.2 × 10^{-9} M	Sensitive to heat and pronase	Brumbaugh and Haussler (1974b)
Avian	Intestine	D-deficient or D-sufficient (cytosol); D-deficient (nuclear)	*In vivo, in vitro*	0.01 M Tris-HCl, 0.001 M mercaptoethanol, 0.0015 M EDTA, pH 7.5	3.0 S (cytosol) with, without KCl 3.5 S (nuclear extract)	G25-excluded G150-V_0 and inclusion peaks (nuclear)	K_a 1 × 10^9 M^{-1} (cytosol) K_a 2 × 10^9 M^{-1} (nuclear)	Sensitive to trypsin, heat (nuclear)	Lawson and Wilson (1974)
Avian	Intestine	D-deficient	*In vitro*	0.25 M sucrose, 0.05 M Tris-HCl, 0.02 M KCl, 0.05 M $MgCl_2$, pH 7.5	—	G200 65,000-150,000	K_a 5.3 × 10^{-10} M	Sensitive to pronase and heat	Tsai and Norman (1973)
Avian	Intestine	D-deficient	*In vitro*	0.05 M Tris-HCl, 0.025 M KCl, 0.005 M $MgCl_2$, pH 7.4	3.7 S (0.15 M KCl)	—	—	—	Kream *et al.* (1977b)
Avian	Intestine	D-deficient	*In vitro*	0.25 M sucrose, 0.05 M Tris-HCl, 0.025 M KCl, 0.005 M $MgCl_2$, pH 7.4	3.7 S	Agarose 50,000	—	Holoprotein more stable, ion exchange and adsorption chromatography	McCain *et al.* (1978)
Human	Intestine (jejunum)	D-sufficient	*In vitro*	0.025 M KH_2PO_4, 0.1 M KCl, 0.001 M DTT, pH 7.5	3.5 S (0.3 M KCl)	—	K_d 2 × 10^{-10} M	Sensitive to trypsin and pronase	Wecksler *et al.* (1979)
Rat	Intestine	D-deficient	*In vitro*	0.01 M Tris-HCl, 0.3 M KCl, 0.0015 M EDTA, 0.0005 M DTT, pH 7.4	3.2 S (0.3 M KCl)	—	—	Sensitive to trypsin and heat	Kream *et al.* (1977c)

(continued)

Table V. (*Continued*)

Species	Tissue	Vitamin D status	Sterol exposure	Extraction buffer	Density and gradient isotonic strength	Size gel filtration	$1,25\text{-}(OH)_2D$ binding affinity	Other	Reference
Avian, rat	Bone	Embryonic & fetal calvaria	*In vitro*	As immediately above	3.5 S (0.3 M KCl)	—	—	—	Kream *et al.* (1977b)
Human, avian, bovine	Parathyroid glands	Avian D-deficient	*In vivo, in vitro* (avian); *in vitro* (human, bovine)	0.25 M sucrose, 0.05 M Tris-HCl, 0.3 M KCl, 0.012 M thioglycerol, 0.001 M EDTA pH 7.4	3.1–3.7 S (0.3 M KCl) cytosolic and nuclear	Agarose 37,000	K_d 4.2 × 10^{-9} M	Sensitive to pronase and heat; ion exchange chromotographic analysis	Hughes and Haussler (1978)
Avian	Kidney	D-deficient	*In vitro*	0.01 M Tris-HCl, 0.3 M KCl, 0.001 M EDTA, 0.0005 M DTT, pH 7.4	3.7 S (0.3 M KCl)	—	K_d 1.2 × 10^{-9} M	—	Christakos and Norman (1979)
Mouse	Kidney	D-sufficient (tubule preparation)	*In vitro*	0.01 M Tris-HCl, 0.274 M KCl, 0.005 M DTT, 0.01 M sodium molybdate, 0.001 M EDTA 500 kIU/ml Trasylol, pH 7.4	3.2 S (0.3 M KCl)	—	K_d 2 × 10^{-10} M	—	Colston and Feldman (1979)
Avian	Pancrease	D-deficient	*In vitro*	Same as avian kidney buffer	3.6 S (0.3 M KCl)	—	K_d 4.1 × 10^{-10} M	—	Christakos and Norman (1979)

Brumbaugh and Haussler, 1974a,b; Lawson and Wilson, 1974). It is remarkable that the first successful competitive radioassay for 1,25-$(OH)_2D$ was dependent on the use of a cytosol–chromatin complex reconstituted from subcellular fractions of avian intestinal epithelium (Brumbaugh *et al.*, 1974). Early reports of the *in vivo* binding of 1,25-$(OH)_2D$ to nuclear lipoprotein and chromatin (Tsai *et al.*, 1972; Chen and DeLuca, 1973) were soon followed by an analysis of the *in vitro* movement of the sterol to chromatin in intestinal homogenates (Tsai and Norman, 1973). The latter report indicated that the sterol was bound to a cytoplasmic protein prior to its association with the chromatin fraction, and the *in vivo* and *in vitro* specificity and saturability of this nuclear chromatin accumulation were confirmed by others (Brumbaugh and Haussler, 1974a). A similar process has been reported to occur in homogenates of parathyroid tissue (Hughes and Haussler, 1978) and rat intestinal homogenates (Batchelet *et al.*, 1977). The movement of the sterol to the chromatin fraction was reported to be temperature dependent. Elsewhere, similar studies led to the observation of the association of 1,25-$(OH)_2D_3$ with nuclear membranes and a nuclear acidic protein (Lawson and Wilson, 1974). Extraction of the nuclear proteins has led to identification of a specific binding protein with characteristics similar to those reported for the cytosol binding protein (Table V). The available data are not sufficient to determine whether this nuclear protein is identical to, or a modification of, the cytosol binding protein.

Collectively, present information appears to indicate that the cytosol binding protein for 1,25-$(OH)_2D$ is a receptor for this hormone: (1) high-affinity binding; (2) high-specificity binding; (3) saturability of binding at low concentrations of hormone; (4) tissue specificity of the receptor; (5) biological response dependent on the association of the hormone and receptor. Much remains to be learned about the receptor content in the target tissues under various conditions. Whereas some indication of the stimulation of intestinal messenger RNA biosynthesis has been presented to closely follow 1,25-$(OH)_2D$ administration (Tsai and Norman, 1973; Zerwekh *et al.*, 1974, 1976), the precise roles and disposition of the receptor and the receptor–sterol complex must be defined by further studies and receptor purification.

4. Summary

A high-affinity and high-capacity binding system for vitamin D and its metabolites exists in the plasma of man as well as of all other skeletal species studied. The binding observed is highly specific for the vitamin D structure, exhibiting a preference for 25-OHD = 25,26-$(OH)_2D$ = 24,25-$(OH)_2D$ > 1,25-$(OH)_2D$ and vitamin D. The purified protein is an α-globulin of 58,000 daltons and has high affinity for 25-OHD_3 ($K_d = 5 \times 10^{-8}$ M). The plasma binding protein circulates at a concentration of 6–8 $\times 10^{-6}$ M, remarkably in excess of the normal plasma levels of antiricketic sterols (1–2 × 10^{-7} M). This plasma protein is identical to the previously recognized group-specific component, and its deletion appears to be a lethal mutation. It is apparently synthesized in the liver, and its plasma concentration varies little in a wide variety of disorders of mineral homeostasis. Its metabolic disposition is unknown, but this plasma protein is recognized to bind noncovalently, with high affinity (K_a = 1.8 × 10^7 M^{-1}) and specificity, to an unidentified cellular protein which appears to be a major cell constituent with a molecular weight of approximately 40,000 daltons. The physiological role, if any, for the plasma binding protein–cell protein interaction is not presently known.

A 3–4 S tissue binding protein that has highest binding affinity for 1,25-$(OH)_2D$ has been observed in extracts of intestine, bone, parathyroid gland, kidney, and pancreas. This protein has not been isolated but appears to play a role in the movement of 1,25-$(OH)_2D$ into the nuclei of target tissues. This tissue protein is likely the receptor for 1,25-$(OH)_2D$ and the means whereby DNA-directed, RNA-mediated protein biosyntheses are triggered to permit the biological expression of this hormone.

Since the tissue distribution of antiricketic sterols is different for various sterols, it is probable that their binding proteins govern their conservation, plasma-to-cell movement, and intracellular disposition. Very little is understood about the nature of the cellular entry of these sterols, and the conjecture about the existence of a plasma membrane receptor-mediated process deserves investigation. Additional information about the functions of blood and tissue binding proteins will enable us to better understand the metabolic dispositions of the entire family of antiricketic sterols.

ACKNOWLEDGMENTS. Ms. B. Kaplan's expert secretarial assistance is gratefully acknowledged.

References

Avioli, L. V., Lee, S. W., McDonald, J. E., Lund, J., and DeLuca, H. F., 1967, Metabolism of vitamin D_3-^3H in human subjects: Distribution in blood, bile, feces and urine, *J. Clin. Invest.* **46**:983.

Barragry, M. D., Carter, N. D., Beer, M., France, M. W., Anton, J. A., Boucher, B. J., and Cohen, R. D., 1977, Vitamin D metabolism in nephrotic syndrome, *Lancet* **2**:629.

Batchelet, M., Ulmann, A., Cloix, J. F., and Funck-Brentano, J. L., 1977, Nuclear uptake of cholecalciferol metabolites in rat duodenal mucosa, *J. Steroid Biochem.* **8**:1047.

Baxter, J. D., and Funder, J. W., 1979, Hormone receptors, *N. Engl. J. Med.* **301**:1149.

Bayard, F., Bec, P., and Louvet, J. P., 1972, Measurement of plasma 25-hydroxycholecalciferol in man, *Eur. J. Clin. Invest.* **2**:195.

Bec, P., Bayard, F., and Louvet, J. P., 1972, 25-Hydroxycholecalciferol dynamics in human plasma, *Rev. Eur. Etudes Clin. Biol.* **17**:793.

Belsey, R. E., DeLuca, H. F., and Potts, J. T., Jr., 1974, Selective binding properties of vitamin D transport protein in chick plasma *in vitro, Nature* **247**:208.

Bills, C. E., 1935, Physiology of the sterols, including vitamin D, *Physiol. Rev.* **15**:1.

Bills, C. E., 1954, Chemistry of vitamin D group, in: *The Vitamins* (W. H. Sebrell, Jr. and R. S. Harris, eds.), p. 132, Academic Press, New York.

Björkhem, I., and Holmberg, I., 1978, Assay and properities of a mitochondrial 25-hydroxylase active on vitamin D_3, *J. Biol. Chem.* **253**:842.

Botham, K. M., Tanaka, Y., and DeLuca, H. F., 1974, 25-Hydroxyvitamin D_3-1-hydroxylase. Inhibition in vitro by rat and pig tissues, *Biochemistry* **13**:4961.

Botham, K. M., Ghazarian, J. G., Kream, B. E., and DeLuca, H. F., 1976, Isolation of an inhibitor of 25-hydroxyvitamin D_3-1-hydroxylase from rat serum, *Biochemistry* **15**:10.

Bouillon, R., Van Baelen, H., and DeMoor, P., 1976a, The transport of vitamin D in the serum of primates, *Biochem. J.* **150**:463.

Bouillon, R., Van Baelen, H., Rombauts, W., and DeMoor, P., 1976b, The purification and characterization of the human-serum binding protein for the 25-hydroxycholecalciferol. Identity with group-specific component, *Eur. J. Biochem.* **66**:285.

Bouillon, R., Van Baelen, H., and DeMoor, P., 1977, The measurement of the vitamin D-binding protein in human serum, *J. Clin. Endocrinol. Metab.* **45**:225.

Bouillon, R., Van Baelen, H., Rombuts, W., and DeMoor, P., 1978a, The isolation and characterization of the vitamin D-binding protein from rat serum, *J. Biol. Chem.* **253**:4426.

Bouillon, R., Vandoren, G., Van Baelen, H., and DeMoor, P., 1978b, Immunochemical measurement of the vitamin D-binding protein rat serum, *Endocrinology* **102**:1710.

Bowman, B. H., and Bearn, A. G., 1965, The presence of subunits in the inherited group-specific protein of human serum, *Proc. Natl. Acad. Sci. U.S.A.* **53**:722.

Brumbaugh, P. F., and Haussler, M. R., 1974a, 1α,25-Dihydroxycholecalciferol receptors in intestine. I. Association of 1α,25-dihydroxycholecalciferol with intestinal mucosa chromatin, *J. Biol. Chem.* **249**:1251.

Brumbaugh, P. F., and Haussler, M. R., 1974b, 1α,25-Dihydroxycholecalciferol receptors in intestine. II. Temperature-dependent transfer of the hormone to chromatin via a specific cytosol receptor, *J. Biol. Chem.* **249**:1258.

Brumbaugh, P. F. Haussler, D. H., Bressler, R., and Haussler, M. R., 1974, Radioreceptor assay for 1α-25-dihydroxyvitamin D_3, *Science* **183**:1089.

Brumbaugh, P. F., Hughes, M. R., and Haussler, M. R., 1975, Cytoplasmic and nuclear binding components for 1α,25-dihydroxyvitamin D_3 in chick parathyroid glands, *Proc. Natl. Acad. Sci. U.S.A.* **72**:4871.

Chen, P. S., Jr., and Lane, K., 1965, Serum protein binding of vitamin D_3, *Arch. Biochem. Biophys.* **112**:70.

Chen, T. C., and DeLuca, H. F., 1973, Receptors of 1,25-dihydroxycholecalciferol in rat intestine, *J. Biol. Chem.* **248**:4890.

Chen, T. C., Weber, J. C., and DeLuca, H. F., 1970, On the subcellular location of vitamin D metabolites in intestine, *J. Biol. Chem.* **245**:3776.

Chertow, B. S., Baylink, D. J., Wergedal, J. E., Su, M. H. H., and Norman, A. W., 1975, Decrease in serum immunoreactive parathyroid hormone in rats and in parathyroid hormone secretion in vitro by 1,25-dihydroxycholecalciferol, *J. Clin. Invest.* **56**:668.

Christakos, S., and Norman, A. W., 1979, Studies on the mode of action of calciferol XVIII. Evidence for a specific high-affinity binding protein for 1,25-dihydroxyvitamin D_3 in chick kidney and pancreas, *Biochem. Biophys. Res. Commun.* **89**:56.

Cleve, H., 1973, The variants of the group-specific component: A review of their distribution in human populations, *Israel J. Med. Sci.* **9**:1133.

Cleve, H., and Patutschnick, W., 1979, Neuraminidase treatment reveals sialic acid differences in certain genetic variants of the Gc system (vitamin D-binding protein), *Hum. Genet.* **47**:193.

Cleve, H., Prunier, J. H., and Bearn, A. G., 1963, Isolation and partial characterization of the two principal inherited group-specific components of human serum, *J. Exp. Med.* **118**:711.

Colston, K. W., and Feldman, D., 1979, Demonstration of a 1,25-dihydroxycholecalciferol cytoplasmic receptor-like binder in mouse kidney, *J. Clin. Endocrinol. Metab.* **49**:798.

Cooke, N. E., Walgate, J., and Haddad, J. G., 1979a, Human serum binding protein for vitamin D and its metabolites I. Physiochemical and immunological identification in human tissues, *J. Biol. Chem.* **254**:5958.

Cooke, N. E., Walgate, J., and Haddad, J. G., 1979b, Human serum binding protein for vitamin D and its metabolites II. Specific, high-affinity association with a protein in nucleat-d tissue, *J. Biol. Chem.* **254**:5965.

Corradino, R. A., 1976, Embryonic chick intestine organ culture: Earliest action of 1,25-$(OH)_2D_3$ is the stimulation of cyclic AMP production, *Fed. Proc.* **35**:339.

Daiger, S. P., 1976, *The genetics of Transport Proteins in Human Plasma and Serum*, Ph.D. Thesis, Stanford University, Stanford.

Daiger, S. P., and Cavalli-Sforza, L. L., 1977, Detection of genetic variation with radioactive ligands. II. Genetic variants of vitamin D-labelled group-specific component (Gc) proteins, *Am. J. Hum. Genet.* **29**:593.

Daiger, S. P., Schanfield, M. S., and Cavalli-Sforza, L. L., 1975, Group specific components (Gc) proteins bind vitamin D and 25-hydroxyvitamin D, *Proc. Natl. Acad. Sci. U.S.A.* **72**: 2076.

DeCrousaz, P., Blanc, B., and Antener, I., 1965, Vitamin D activity in normal human serum and serum proteins, *Helv. Odontol. Acta* **9**:151.

DeLuca, H. F., 1979, Recent advances in our understanding of the vitamin D endocrine system, *J. Steroid Biochem.* **11**:35.

DeLuca, H. F., and Schnoes, H. K., 1976, Metabolism and mechanisms of action of vitamin D, *Annu. Rev. Biochem.* **45**:631.

Edelstein, S., 1974, Vitamin D-binding proteins, *Biochem. Soc. Spec. Publ.* **3**:43.

Edelstein, S., Lawson, D. E. M., and Kodicek, E., 1973, The transporting proteins of cholecalciferol and 25-hydroxycholecalciferol in serum of chick and other species, *Biochem. J.* **135**:417.

Edelstein, S., Charman, M., Lawson, D. E. M., and Kodicek, E., 1974, Competitive protein-binding assay for 25-hydroxycholecalciferol, *Clin. Sci. Mol. Med.* **46**:231.

Eisman, J. A., and DeLuca, H. F., 1977, Intestinal 1,25-dihydroxyvitamin D_3 binding protein: Specificity of binding, *Steroids* **30**:245.

Eisman, J. A., Haustra, A. J., Kream, B. E., and DeLuca, H. F., 1976, A sensitive, precise and convenient method for determination of 1,25-dihydroxyvitamin D in human plasma, *Arch. Biochem. Biophys.* **176**:235.

Fraser, D. R., and Emtage, J. S., 1976, Vitamin D in the avian egg. Its molecular identity and mechanisms of incorporation into yolk, *Biochem. J.* **160**:671.

Fraser, D. R., and Kodicek, E., 1970, Unique biosynthesis by kidney of a biologically active vitamin D metabolite, *Nature* **228**:764.

Fraser, D. R., and Kodicek, E., 1973, Regulation of 25-hydroxycholecalciferol 1-hydroxylase activity in kidney by parathyroid hormone, *Nature* [*New Biol.*] **241**:163.

Ghazarian, J. G., Hsu, P.-Y., Gviotti, A. W., and Winkelhake, J. L., 1978a, Purification of calciferol-binding proteins from kidney: Physicochemical and immunological properties, *J. Lipid Res.* **19**:601.

Ghazarian, J. G., Kream, B., Botham, K. M., Mickells, M., and DeLuca, H. F., 1978b, Rat plasma 25-hydroxyvitamin D_3 binding protein: An inhibitor of the 25-hydroxyvitamin D_3-1*l*-hydroxylase, *Arch. Biochem. Biophys.* **189**:212.

Goldstein, J. L., and Brown, M. S., 1976, The LDL pathway in human fibroblasts: A receptor-mediated mechanisms for the regulation of cholesterol metabolism, *Curr. Top. Cell. Regul.* **11**:147.

Goldstein, J. L., Anderson, R. G. W., and Brown, M. S., 1979, Coated pits, coated vesicles, and receptor-mediated endocytosis, *Nature* **279**:679.

Gray, R. W., Caldas, A. E., Wilz, D. R., Lemann, J., Jr., Smith, G. A., and DeLuca, H. F., 1978, Metabolism and excretion of ^{3}H-1,25-$(OH)_2$-vitamin D_3 in healthy adults, *J. Clin. Endocrinol. Metab.* **46**:756.

Haddad, J. G., 1979, Transport of vitamin D metabolites, *Clin. Orthop.* **142**:249.

Haddad, J. G.,1980, Purification, characterization and quantitation of the human serum binding protein for vitamin D and its metabolites, *Methods Enzymol.* **67**:449.

Haddad, J. G., and Birge, S. J., 1971, 25-Hydroxycholecalciferol: Specific binding by ricketic tissue extracts, *Biochem. Biophys. Res. Commun.* **45**:829.

Haddad, J. G., and Birge, S. J., 1975, Widespread, specific binding of 25-hydroxycholecalciferol in rat tissues, *J. Biol. Chem.* **250**:299.

Haddad, J. G., and Chyu, K. J., 1971a, 25-Hydroxycholecalciferol-binding globulin in human plasma, *Biochim. Biophys. Acta* **248**:471.

Haddad, J. G., and Chyu, K. J., 1971b, Competitive protein binding radioassay for 25-hydroxycholecalciferol, *J. Clin. Endocrinol. Metab.* **33**:992.

Haddad, J. G., and Stamp, T. C. B., 1974, Circulating 25-hydroxyvitamin D in man, *Am. J. Med.* **57**:57.

Haddad, J. G., and Walgate, J., 1976a, 25-Hydroxyvitamin D transport in human plasma. Isolation and partial characterization of calcifidiol-binding protein, *J. Biol. Chem.* **251**:4803.

Haddad, J. G., and Walgate, J., 1976b, Radioimmunoassay of the binding protein for vitamin D and its metabolites in human serum. Concentrations in normal subjects and patients with disorders of mineral homeostasis, *J. Clin. Invest.* **58**:1217.

Haddad, J. G., Birge, S. J., and Hahn, T. J., 1973a, Vitamin D metabolites: Specific binding by rat intestinal cytosol, *Biochim. Biophys. Acta* **329**:93.

Haddad, J. G., Chyu, K. J., Hahn, T. J., and Stamp, T. C. B., 1973b, Serum concentrations of 25-hydroxyvitamin D in sex-linked hypophysphatemic vitamin D-resistant rickets, *J. Lab. Clin. Med.* **81**:22.

Haddad, J. G., Walgate, J., Min, C., and Hahn, T. J., 1976a, Vitamin D metabolite-binding proteins in human tissue, *Biochim. Biophys. Acta* **444**:921.

Haddad, J. G., Hillman, L., and Rojanasathit, S., 1976b, Human serum binding capacity and affinity for 25-hydroxyergocalciferol and 25-hydroxycholecalciferol, *J. Clin. Endocrinol. Metab.* **43**:86.

Haussler, M. R., and McCain, T. A., 1977, Basic and clinical concepts related to vitamin D metabolism and action, *N. Engl. J. Med.* **297**:974, 1041.

Haussler, M. R., Nagode, L. A., and Rasmussen, H., 1970, Induction of intestinal brush border alkaline phosphatase by vitamin D and identity with Ca-ATPase, *Nature* **228**:1199.

Hay, A. W. M., 1975, The transport of 25-hydroxycholecalciferol in a New World monkey, *Biochem. J.* **151**:193.

Hay, A. W. M., and Watson, G., 1976a, The plasma transport proteins of 25-hydroxycholecalciferol in mammals, *Comp. Biochem. Physiol.* **53B**:163.

Hay, A. W. M., and Watson, G., 1976b, The plasma transport proteins of 25-hydroxycholecalciferol in fish, amphibians, reptiles and birds, *Comp. Biochem. Physiol.* **53B**:167.

Heimburger, N., Heide, K., Haupt, H., and Schultze, H. E., 1964, Baustein Analysen von human serum Proteinen, *Clin. Chim. Acta* **10**:293.

Heller, J., 1975, Interactions of plasma retinol-binding protein with its receptor, *J. Biol. Chem.* **250**:3613.

Hirschfield, J., 1959, Immunoelectrophoretic demonstration of quantitative differences in human sera and their relation to the hepatoglobins, *Acta Pathol. Microbiol. Scand.* **47**:160.

Horst, R. L., 1979, 25-OHD_3-26,23-Lactone: A metabolite of vitamin D_3 that is 5 times more potent than 25-OHD_3 in the rat plasma competitive protein binding radioassay, *Biochem. Biophys. Res. Commun.* **89**:286.

Hughes, M. R., and Haussler, M. R., 1978, 1,25-Dihydroxyvitamin D_3 receptors in parathyroid glands. Preliminary characterization of cytoplasmic and nuclear binding components, *J. Biol. Chem.* **253**:1065.

Hughes, M. R., Baylink, D. J., Hones, P. G., and Haussler, M. R., 1976, Radiology and receptor assay for 25-hydroxyvitamin D_2/D_3 and 1,25-dihydroxyvitamin D_2/D_3, *J. Clin. Invest.* **58**:61.

Imawari, M., and Goodman, D. S., 1977, Immunological and immunoassay studies of the binding protein for vitamin D and its metabolites in human serum, *J. Clin. Invest.* **59**:432.

Imawari, M., Kida, K., and Goodman, D. S., 1976, The transport of vitamin D and its 25-hydroxy metabolite in human plasma, *J. Clin. Invest.* **58**:514.

Imawari, M., Akanuma, Y., Itakura, H., Muto, Y., Kasaka, K., and Goodman, D. S., 1979, The effects of diseases of the liver on serum 25-hydroxyvitamin D and on the serum binding protein for vitamin D and its metabolites, *J. Lab. Clin. Med.* **93**:171.

Jacobs, R. L., and Ray, R. D., 1968, Studies of vitamin D binding in normal and rachitic serum, *Clin. Orthop.* **56**:275.

Kawakami, M., Imawari, M., and Goodman, D. S., 1979, Quantitative studies of the interaction of cholecalciferol and its metabolites with different genetic variants of the serum binding protein for these sterols, *Biochem. J.* **179**:413.

Kitchin, F. D., and Bearn, A. G., 1965, Quantitative determination of the group specific protein in normal human serum, *Proc. Soc. Exp. Biol. Med.* **118**:304.

Kream, B. E., Reynolds, R. D., Knutson, J. C., Eisman, J. A., and DeLuca, H. F., 1976, ntestinal cytosol binders of 1,25-dihydroxyvitamin D and 25-hydroxyvitamin D, *Arch. Biochem. Biophys.* **176**:779.

Kream, B. E., Jose, M. J. L., and DeLuca, H. F., 1977a, The chick intestinal cytosol binding protein for 1,25-dihydroxyvitamin D_3: A study of analog binding, *Arch. Biochem. Biophys.* **179**:462.

Kream, B. E., Jose, M., Yamada, S., and DeLuca, H. F., 1977b, A specific high-affinity binding macromolecule for 1,25-dihydroxyvitamin D_3 in fetal bone, *Science* **197**:1086.

Kream, B. E., Yamada, S., Schnoes, H. K., and DeLuca, H. F., 1977c, Specific cytosol-binding protein for 1,25-dihydroxyvitamin D_3 in rat intestine, *J. Biol. Chem.* **252**:4501.

Kream, B. E., DeLuca, H. F., Moriarity, D. M., Kendrick, N. C., and Ghazarian, J. G., 1979, Origin of 25-hydroxyvitamin D_3 binding protein from tissue cytosol preparations, *Arch. Biochem. Biophys.* **192**:318.

Lawson, D. E. M., and Wilson, P. W., 1974, Intranuclear localization and receptor proteins for 1,25-dihydroxycholecalciferol in chick intestine, *Biochem. J.* **144**:573.

Lawson, D. E. M., Charman, M., Wilson, P. W., and Edelstein, S., 1976, Some characteristics of new tissue-binding proteins for metabolites of vitamin D other than 1,25-dihydroxyvitamin D, *Biochim. Biophys. Acta* **437**:403.

Manalagos, S. C., Taylor, C. M., and Anderson, D. C., 1979, Highly specific binding of 1,25-dihydroxycholecalciferol in bone cytosol, *J. Endocrinol.* **80**:35.

Mawer, E. B., Backhouse, J., Holman, C. A., Lumb, G. A., and Stanbury, S. W., 1972, The distribution and storage of vitamin D and its metabolites in human tissues, *Clin. Sci.* **43**:413.

Mawer, E. B., Backhouse, J., Davies, M., Hill, J. L., and Taylor, C. M., 1976, Metabolic fate of administered 1,25-dihydroxycholecalciferol in controls and patients with hypoparathyroidism, *Lancet* **1**:1203.

McCain, T. A., Haussler, M. R., Okrent, D., and Huges, M. R., 1978, Partial purification of the chick intestinal receptor for 1,25-dihydroxyvitamin D by ion exchange and blue dextran-sepharose chromatography, *FEBS Let.* **86**:65.

Melancon, M. J., Jr., and DeLuca, H. F., 1970, Vitamin D stimulation of calcium-dependent adenosive triphosphatase in chick intestinal brush borders, *Biochemistry* **9**:1658.

Mellon, W. S., and DeLuca, H. F., 1979, An equilibrium and kinetic study of 1,25-dihydroxyvitamin D_3 binding to chicken intestinal cytosol employing high specific activity 1,25-dihydroxy{^{3}H-26,27}vitamin D_3, *Arch. Biochem. Biophys.* **197**:90.

Nahm, T. H., Lee, S. W., Fausto, A., Sonn, Y., and Avioli, L. V., 1979, 25-OHD, a circulating vitamin D metabolite in fish, *Biochem. Biophys. Res. Commun.* **89**:396.

Napoli, J. L., Fivizzani, M. A., Hamstra, A. J., Schnoes, H. K., and DeLuca, H. F., 1979, Synthesis of 25-hydroxy{26,27-^{3}H}vitamin D_3 with high specific activity, *Anal. Biochem.* **96**:481.

Norman, A. W., and DeLuca, H. F., 1963, The preparation of ^{3}H-vitamins D_2 and D_3 and their localization in the rat, *Biochemistry* **2**:1160.

Oku, T., Ooizumi, K., and Hosoya, N., 1974, Binding proteins for 1,25-dihydroxycholecalciferol and 25-hydroxycholecalciferol, *J. Nutr. Sci. Vitaminol.* **20**:9.

Omdahl, J. L., and DeLuca, H. F., 1973, Regulation of Vitamin D metabolism and function, *Physiol. Rev.* **53**:327.

Peterson, P. A., 1971, Isolation and partial characterization of a human vitamin D-binding plasma protein, *J. Biol. Chem.* **246**:7748.

Preece, M. A., O'Riordan, J. L. H., Lawson, D. E. M., and Kodicek, E., 1974, A competitive protein binding assay for 25-hydroxycholecalciferol and 25-hydroxyergocalciferol in serum, *Clin. Chem. Acta* **54**:235.

Rikkers, H., and DeLuca, H. F., 1967, An *in vivo* study of the carrier proteins of ^{3}H-vitamins D_3 and D_4 in rat serum, *Am. J. Physiol.* **213**:380.

Rikkers, H., Kletziens, R., and DeLuca, H. F., 1969, Vitamin D binding globulin in the rat: Specificity for the vitamins D, *Proc. Soc. Exp. Biol. Med.* **130**:1321.

Rojanasathit, S., and Haddad, J. G., 1977, Ontogeny and effect of vitamin D deprivation on rat serum 25-hydroxyvitamin D binding protein, *Endocrinology* **100**:642.

Rosenstreich, S. J., Rich, C., and Volwiler, W., 1971a, Deposition in and release of vitamin D_3 from body fat: Evidence for a storage site in the rat, *J. Clin. Invest.* **50**:679.

Rosenstreich, S. J., Volwiler, W., and Rich, C., 1971b, Metabolism and plasma protein transport of vitamin D_3 in the baboon, *Am. J. Clin. Nutr.* **24**:897.

Schmidt-Gayk, H., Grawunder, C., Tschope, W., Schmitt, W., Ritz, E., Dietsch, V., Andrassy, K., and Bouillon, R., 1977, 25-Hydroxyvitamin D in nephrotic syndrome, *Lancet* **2**:105.

Shepard, R. M., Horst, R. L., Hamstra, A. J., and DeLuca, H. F., 1979, Determination of vitamin D and its metabolites in plasma from normal and anephric man, *Biochem. J.* **182**:55.

Shimura, F., Moriuchi, S., and Hosoya, N., 1977, Some characteristics of cytosol-binding protein for 1*l*,25-dihydroxycholecalciferol and 25-hydroxycholecalciferol in rat intestinal mucosa, *J. Nutri. Sci. Vitaminol.* **23**:187.

Silver, J., Shvil, Y., and Fainaru, M., 1978, Vitamin D transport in an infant with vitamin D toxicity, *B. Med. J.* **2**:93.

Simons, K., and Bearn, A. G., 1967, The use of preparative polyacrylamide column electrophoresis in isolation of electrophoretically distinguishable components of the serum group-specific protein, *Biochem. Biophys. Acta* **133**:499.

Smith, F. R., Raz, A., and Goodman, D. S., 1970, Radioimmunoassay of human plasma retinol-binding protein, *J. Clin. Invest.* **49**:1754.

Smith, J. E., and Goodman, D. S., 1971, The turnover and transport of vitamin D and of a polar metabolite with the properties of 25-hydroxycholecalciferol in human plasma, *J. Clin. Invest.* **50**:2159.

Spencer, R., Charman, M., Emtage, J. S., and Lawson, D. E. M., 1976, Production and properties of vitamin D-induced mRNA for chick calcium-binding protein, *Eur. J. Biochem.* **71**:399.

Svasti, J., and Bowman, B. H., 1978, Human group-specific component, *J. Biol. Chem.* **253**:4188.

Thomas, W. C., Morgan, H. F., Connar, T. B., Haddock, L., Bills, C. E., and Howard, J. E., 1959, Studies of antiricketic activity in sera from patients with disorders of calcium metabolism and preliminary observations in the mode of transport by vitamin D in human serum, *J. Clin. Invest.* **38**:1078.

Tsai, H. C., and Norman, A. W., 1973, Studies on calciferol metabolism VIII. Evidence for a cytoplasmic receptor for 1,25-dihydroxyvitamin D_3 in the intestinal mucosa, *J. Biol. Chem.* **248**:5967.

Tsai, H. C., Wong, R. C., and Norman, A. W., 1972, Studies on calciferol metabolism. IV. Subcellular localization of 1,25-dihydroxyvitamin D_3 in intestinal mucosa and correlation with increased calcium transport, *J. Biol. Chem.* **247**:5511.

Ulmann, A., Brami, M., Pezant, E., Garabedian, M., and Funck-Bretano, J. L., 1977, Binding of cholecalciferol metabolites to rat duodenal mucosa cytosol, *Acta Endocrinol.* (*Kbh.*) **84**:439.

Van Baelen, H., Bouillon, R., and DeMoor, P., 1977, Binding of 25-hydroxycholecalciferol in tissues, *J. Biol. Chem.* **252**:2515.

Van Baelen, H., Bouillon, R., and DeMoor, P., 1978, The heterogeneity of human Gc-globulin, *J. Biol. Chem.* **253**:6344.

Wasserman, R. H., and Feber, J. J., 1977, Vitamin D-dependent calcium binding proteins, in: *Calcium-Binding Proteins and Calcium Function* (R. H. Wasserman, R. A. Corradino, E. Carafoli, R. H. Kretsinger, D. H. MacLennan, and S. L. Siegel, eds.), p. 292, Elsevier-North-Holland, Amsterdam.

Weber, J. C., Pons, V., and Kodicek, E., 1971, The localization of 1,25-dihydroxycholecalciferol in bone cell nuclei of rachitic chicks, *Biochem. J.* **125**:147.

Wecksler, W. R., Mason, R. S., and Norman, A. W., 1979, Specific cytosol receptors for 1,25-dihydroxyvitamin D_3 in human intestine, *J. Clin. Endocrinol. Metab.* **48**:715.

Wilson, P. W., and Lawson, D. E. M., 1977, 1,25-Dihydroxyvitamin D stimulation of specific membrane proteins in chick intestine, *Biochim. Biophys. Acta* **497**:805.

Yamamoto, K. R., and Alberts, B. M., 1976, Steroid receptors: Elements for modulation of eukaryotic transcription, *Annu. Rev. Biochem.* **45**:932.

Zerwekh, J. E., Haussler, M. R., and Lindell, T. J., 1974, Rapid enhancement of chick intestinal DNA-dependent RNA polymerase II activity by 1*l*,25-dihydroxyvitamin D_3 *in vivo*, *Proc. Natl. Acad. Sci. U.S.A.* **71**:2337.

Zerwekh, J. E., Lindell, T. J., and Haussler, M. R., 1976, Increased intestinal chromatin template activity influence of 1,25-dihydroxyvitamin D_3 and hormone–receptor complexes, *J. Biol. Chem.* **251**:2388.

Zile, M., Bunge, E. C., Barsness, L., Yamada, S., Schnoes, H. F., and DeLuca, H. F., 1978, Localization of 1,25-dihydroxyvitamin D in intestinal nuclei *in vivo*, *Arch. Biochem. Biophys.* **186**:15.

Chapter 3

Vitamin D Compounds in Human and Bovine Milk

Bruce W. Hollis, Bernard A. Roos, and Phillip W. Lambert

1. Introduction

During the past decade considerable progress has been made toward an understanding of the transport, measurement, distribution, and molecular action of the antirachitic sterols. As a result of these major advances in our knowledge of vitamin D, many excellent reviews have been written dealing with each of these specific areas of study (Lawson and Emtage, 1974; Favus, 1978; DeLuca and Schnoes, 1976; J. G. Haddad, Jr., this volume).

However, one subject that has largely been ignored in the recent literature is the origin of the antirachitic properties of an important biological fluid: milk. Early studies, constituting the majority of published work, dealt primarily with the determination by the rat line biological test of the antirachitic activity present

Abbreviations used: DBP, vitamin D binding protein; 25-OHD, 25-hydroxyvitamin D; 24,25-$(OH)_2D$, 24,25-dihydroxyvitamin D; 25-OHD_3-26,23-lactone, 25-hydroxyvitamin D_3-26,23-lactone; 25,26-$(OH)_2D$, 25,26-dihydroxyvitamin D; 1,25-$(OH)_2D$, 1,25-dihydroxyvitamin D; GC, gas chromatography; HPLC, high-performance liquid chromatography; CaBP, calcium binding protein; RDA, recommended daily allowance.

Bruce W. Hollis, Bernard A. Roos, and Phillip W. Lambert • Department of Medicine, Case Western Reserve University; and Department of Medicine, Division of Endocrinology and Mineral Metabolism, Veterans Administration Medical Center, Cleveland, Ohio 44106. This work was supported by grants from the Veterans Administration and a National Institutes of Health Fellowship (AM 06403) to B. W. Hollis.

in human and bovine milk (Harris and Bunker, 1939; Polskin *et al.*, 1945). Recent vitamin D research has been directed mainly toward blood, and milk has been neglected.

The antirachitic activity of native milk with respect to actual levels of vitamin D metabolites and the mechanism(s) by which these metabolites are secreted in milk are clearly important considerations for a better understanding the metabolism of vitamin D and the role of these compounds in neonatal nutrition. It is the purpose of this chapter to bring to light new data dealing with levels of vitamin D and its metabolites in native milk as well as the means by which these antirachitic sterols gain access to this biological fluid. An additional aim is to reconcile the results of earlier biological assays on milk with recent findings on the concentrations of vitamin D and its metabolites in milk.

2. Binding Proteins for Vitamin D and Its Metabolites

2.1. Blood

The vitamin D binding protein (DBP) in plasma has been extensively reviewed in the preceding chapter (J. G. Haddad, Jr., this volume). Briefly, the DBP is largely responsible for the transport of the antirachitic sterols throughout the body. This plasma protein expresses a binding preference for 25-hydroxyvitamin D (25-OHD), 24,25-dihydroxyvitamin D [24,25-$(OH)_2$D], and 25,26-dihydroxyvitamin D [25,26-$(OH)_2$D] when compared to the parent vitamin and 1,25-dihydroxyvitamin D [1,25-$(OH)_2$D] (Belsey *et al.*, 1974; Lawson *et al.*, 1976; Hollis *et al.*, 1977a,b). It is interesting to note that a newly discovered vitamin D_3 metabolite, 25-hydroxyvitamin D_3-26,23-lactone (25-OHD_3-26,23-lactone), has the greatest demonstrated affinity for the plasma DBP (Horst, 1979; Hollis *et al.*, 1980a), although the significance of this finding is unknown.

During the last decade a number of in-depth studies on the DBP have resulted in a more complete characterization of its physical properties (Haddad and Walgate, 1976; Imawari *et al.*, 1976; Bouillon *et al.*, 1978). In man, the DBP appears to be an α-globulin possessing a molecular weight of approximately 60,000 with a sedimentation coefficient of 4.0 S. Further, the DBP possesses an isoelectric point of from 4.7 to 4.9 and is capable of binding only one molecule of vitamin D per molecule of DBP.

Determinations of DBP in human plasma by radioimmunoassay (Haddad and Walgate, 1976) or radial immunodiffusion (Bouillon *et al.*, 1977) have shown the molecule to circulate at concentrations of 525 μg/ml in normal individuals and up to 1254 μg/ml in pregnant women. These levels of plasma DBP, even in normal individuals, make it apparent that the plasma has a great capacity to transport vitamin D and its metabolites, especially in light of the fact that the total amount of antirachitic sterols circulating in a normal individual is about 35 ng/ml (Shepard *et al.*, 1979; P. W. Lambert and B. W. Hollis, unpublished data). It has been calculated that under normal circumstances as much as 98% of the plasma DBP circulates with its binding sites unoccupied by any vitamin D metabolite (Haddad *et al.*, 1976).

2.2. Milk

For many years milk has been known to contain antirachitic activity, although this activity was far less than that observed in plasma (Polskin *et al.*, 1945). Several observations have indicated that milk could contain the DBP. Human and rat milk have been shown to contain corticosteroid, folate, and vitamin B_{12} binding proteins identical to those occurring in blood (Payne *et al.*, 1976; Waxman, 1975; Burger and Allen, 1974). It has also been demonstrated that from 3 to 10% of the protein in milk is derived from the blood (Larson and Jorgensen, 1974).

A recent study has shown that human milk actually contains two DBPs (Van Baelen *et al.*, 1977), one of which appears to be identical to the plasma DBP; the second DBP corresponds to a 6.0 S DBP that had previously been isolated from a number of different tissues (Haddad and Birge, 1975). Characterization of this tissue DBP has revealed physical properties similar to those of the plasma DBP with the exception of its sedimentation coefficient and its absence from blood (Kream *et al.*, 1979; Cooke *et al.*, 1979). Hollis and Draper (1979) also demonstrated the presence of two distinct DBPs in human milk whey (Fig. 1). In con-

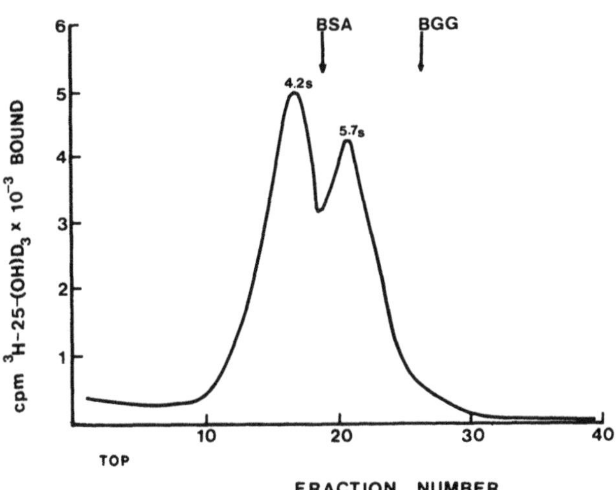

Fig. 1. Demonstration of two separate DBPs in human milk whey by sucrose gradient ultracentrifugation (Hollis and Draper, 1979).

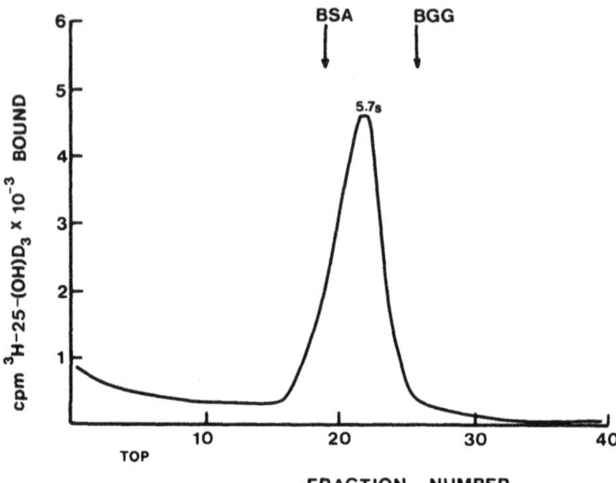

Fig. 2. Demonstration of a single DBP in bovine milk whey by sucrose gradient ultracentrifugation (Hollis and Draper, 1979).

trast, bovine milk whey was shown to contain only the 6.0 S DBP (Fig. 2), even though bovine plasma contains only the 4.0 S DBP (B. W. Hollis and H. H. Draper, unpublished results).

Additional data, including isoelectric focusing (Table I) and Ouchterlony immunodiffusion (Hollis *et al.*, 1981b), indicated that the 4.0 S globulin in human milk was derived from plasma. The source of the 6.0 S tissue DBP in human and bovine milk is unclear. For years the 6.0 S tissue DBP was assumed to be a cell receptor for 25-OHD in tissues throughout the body (Haddad and Birge, 1975). Van Baelen *et al.* (1977) recently reported that the 6.0 S tissue DBP results from the association of the 4.0 S plasma DBP with a cytosolic factor derived from the homogenization of various tissues. It was postulated that the 6.0 S DBP in milk is formed when the 4.0 S plasma DBP comes into contact with cytosol generated from cellular disruption during lactation (Hollis and Draper, 1979). Subsequent studies have demonstrated that this cytosolic factor is actin (Van Baelen *et al.*, 1980). The significance of this highly specific association between the 4.0 S plasma DBP and actin remains unknown.

Table I. Isoelectric Points of DBP of Plasma and Whey[a]

Species	Whey	Plasma
Human	4.7	4.7
Bovine	4.5	4.5
Rhesus monkey	4.8	4.8
Porcine	4.7	4.7

[a]Hollis and Draper (1979).

The actual levels of DBP occurring in human milk have recently been determined (Hollis *et al.*, 1981b) using a specific radioimmunoassay (Haddad and Walgate, 1976). This study demonstrated that the level of DBP in milk is dependent on the stage of lactation, being highest during early lactation and declining as lactation continues. Three weeks following the initiation of lactation, the DBP content of milk was 1–2% of the plasma DBP level reported in normal women (Haddad and Walgate, 1976).

2.3. Mammary Gland

Because of the massive movement of calcium from blood to milk during the lactational process, the mammary gland would be a likely target organ for the hormonal form of vitamin D, 1,25-$(OH)_2D$. A recent study has demonstrated that both lactating and nonlactating bovine mammary tissues contain a cytosolic receptor highly specific for 1,25-$(OH)_2D$ (Fig. 3) (Reinhardt and Conrad, 1980). This cytosolic receptor protein possesses a sedimentation coefficient of 3.7 S which is characteristic of cytosolic receptors for 1,25-$(OH)_2D$ in other target tissues (Brumbaugh and Haussler, 1975; Kream *et al.*, 1977; Christakos and Norman, 1979). T. A. Reinhardt and H. R. Conrad (personal communication) have shown that the 3.7 S cytosolic receptor protein also occurs in bovine milk whey (Fig. 4). This cytosolic receptor protein probably gains access to milk as a result of cellular disruption during lactation. The function of this 1,25-$(OH)_2D$ receptor protein in mammary tissue remains unknown, as vitamin D-deficient lactating rats have recently been found to transfer calcium from blood to milk in a normal fashion (Toverud and Boass, 1979).

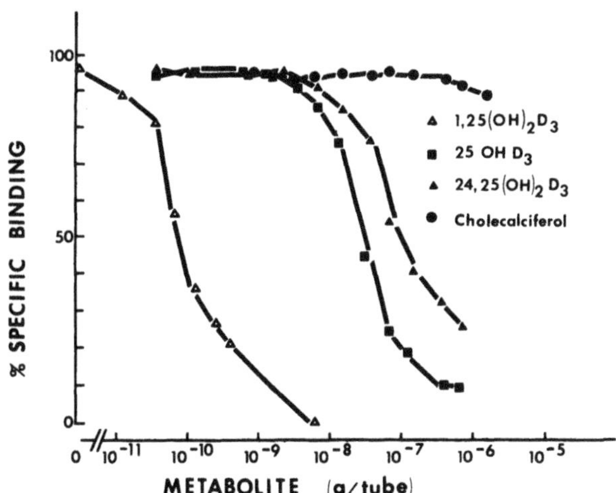

Fig. 3. Competitive binding assays comparing the specificity of the 1,25-$(OH)_2D$ cytosol receptor from bovine mammary gland cytosol with various vitamin D metabolites (Reinhardt and Conrad, 1980).

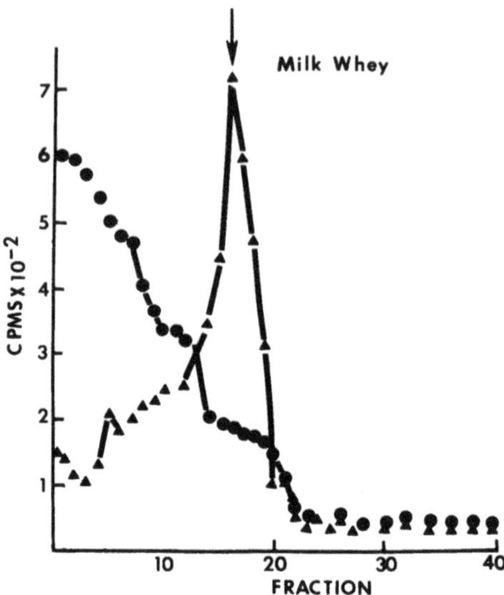

Fig. 4. Demonstration of the 3.7 S cytosol receptor for 1,25-$(OH)_2D$ in bovine milk whey using sucrose gradient ultracentrifugation (T. A. Reinhardt and H. R. Conrad, personal communication).

3. The Question of the Existence of Vitamin D Sulfate in Milk

Utilizing a colorimetric assay that reportedly detected only vitamin D sulfate, Sahashi *et al.* (1967a) first reported the existence of vitamin D sulfate in human and bovine milk whey. The levels of antirachitic activity found by this assay were 950 and 204 IU/liter, respectively. These estimates of biological activity were based on the premise that the antirachitic activity of vitamin D sulfate is equivalent to that of vitamin D (Sahashi *et al.*, 1967b). Subsequent studies employing similar colorimetric assay techniques yielded similar estimates of vitamin D sulfate in human and bovine milk whey (LeBoulch *et al.*, 1974; Lakdawala and Widdowson, 1977).

Because neither human nor bovine milk (Harris and Bunker, 1939; Polskin *et al.*, 1945; Gast *et al.*, 1977) has been found to contain the amounts of antirachitic activity reported to be present as vitamin D sulfate, it was desirable to develop new techniques to reassess the level of vitamin D sulfate in this fluid.

Recent technological advances in high-performance liquid chromatography (HPLC) have permitted the qualitative and quantitative estimation of vitamin D and its metabolites in plasma after appropriate purification of extracts (Lambert *et al.*, 1977, 1980; Hughes *et al.*, 1976; Shepard *et al.*, 1979). With these advances, a reverse-phase HPLC system was developed in our laboratory that is capable of detecting $\geqslant 1$ μg/liter of vitamin D sulfate in milk whey by integrated peak area for UV absorption (Hollis *et al.*, 1979, 1981a). The detection limit of this assay is significantly below the 10–20 μg/liter of this compound reported to occur in milk whey (Sahashi *et al.*, 1967a; LeBoulch *et al.*, 1974; Lakdawala and Widdowson, 1977). Figure 5A illustrates the fact that endogenous vitamin D sulfate could not be detected in human milk whey with this technique although exogenous vitamin D sulfate added to the same sample at a concentration of 1 μg/liter was readily detected (Fig. 5B). Supportive evidence for the lack of a significant amount of vitamin D sulfate in milk includes a recent study utilizing bioassay techniques (Leerbeck and Sondergaard, 1980) and the finding that vitamin D sulfate, based on bone ash or X-ray skeletal analysis, possesses less than 5% of the antirachitic activity of the parent vitamin (D. E. M. Lawson, personal communication). Our laboratory has been unable to produce HPLC-purified [^{3}H]vitamin D_3 sulfate using [^{3}H]-D_3 as a substrate in a liver homogenate system (unpublished results) previously reported to produce vitamin D sulfate (Higaki *et al.*, 1965). In addition, human mammary cell cultures that actively secrete α-lactalbumin failed to show production of vitamin D_3 sulfate from [^{3}H]-D_3 (unpublished data).

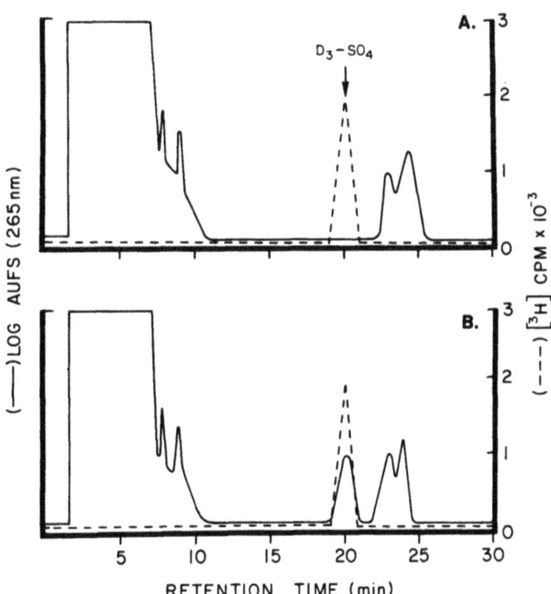

Fig. 5. Elution profile of human milk whey extract on isocratic reverse-phase HPLC. Panel A represents the elution profile of 10 ml of human milk whey with no detectable endogenous vitamin D sulfate present. Panel B represents the elution profile from the same whey sample with 10 ng of exogenously added vitamin D sulfate present (Hollis *et al.*, 1981a).

4. Methods of Analysis and Levels of Antirachitic Activity Determined in Native Milk

4.1. Bioassay Techniques

Previous studies evaluating the antirachitic activity of human or bovine milk have employed the rat line bioassay method (Harris and Bunker, 1939; Polskin *et al.*, 1945; Gast *et al.*, 1977; Leerbeck and Sondergaard, 1980). The findings from these studies are in general agreement that the bioassayable activity of native milk is quite low relative to the RDA of 400 IU/liter (Table II). In general, this bioassay technique is acceptable for crude evaluation of antirachitic activity in milk. However, factors such as lactose in milk can influence calcium absorption and, therefore, results obtained using the bioassay technique (Schaafsma and Visser, 1980). As a result, some of the antirachitic properties of milk could be attributed to factors other than vitamin D and its metabolites.

4.2. Chemical Assay Techniques

Chemical techniques such as the antimony trichloride method (Sahashi *et al.*, 1967a) used for the assessment of purported vitamin D sulfate in milk whey, have been criticized as being nonspecific as well as insensitive (Kodicek and Lawson, 1967).

4.3. Chromatographic Analysis

The use of gas chromatography (GC) or HPLC for the direct quantitation of vitamin D and its metabolites in bovine and/or human milk has been reported in

Table II. Composite Results of Bioassayable Antirachitic Activity in Native Milk[a]

Milk fraction assayed	Animal species	Antirachitic activity detected (IU/liter)	Study
Lipid	Human	4 (pooled sample)	Harris and Bunker (1939)
Whole milk	Human	0–40 ($N = 21$)	Polskin *et al.*, (1945)
Whole milk	Bovine	19 ($N = 5$)	Gast *et al.*, (1977)
Whole milk	Bovine	38 ($N = 1$)	Leerbeck and Sondergaard (1980)
Whole milk	Human	20 ($N = 1$)	Leerbeck and Sondergaard (1980)

[a] Rat line test used in all studies.

two previous studies (Adachi and Kobayashi, 1979; LeBoulch *et al.*, 1974). However, it must be pointed out that these systems are relatively insensitive to the small amount of antirachitic activity shown by bioassay to be present in native milk.

LeBoulch *et al.* (1974), using GC, estimated that native bovine milk contained 1.5 μg/liter (60 IU/liter) of vitamin D_3. In a similar study, Adachi and Kobayashi (1979), using HPLC, reported that native cow milk contained an average of 1.6 μg/liter (64 IU/liter) of the parent vitamin. In both of these studies, 200 ml or more of whole milk was required for a single analysis.

In the only other study involving direct chromatographic analysis of milk, Hollis *et al.* (1979, 1981a) failed to detect vitamin D sulfate in either human or bovine milk whey.

4.4. Ligand-Binding Analysis

Following the development of ligand-binding assays for vitamin D and its metabolites nearly a decade ago (Haddad and Chyu, 1971; Belsey *et al.*, 1971), these assays received wide application in both clinical and basic research. Although widely utilized, the application of the assays is almost totally confined to the analysis of plasma. Initial attempts to apply these ligand-binding assays to milk samples led to the detection of a "25-OHD-like" substance that possessed no apparent biological activity (Osborn and Norman, 1977). Subsequent studies demonstrated that vitamin D metabolite levels in milk were quite low and that the reportedly high levels of the "25-OHD-like" substance in milk were attributable to a compound located in the lipid portion of human milk that interferes with the ligand binding assay (Hollis and Draper, 1978; Hollis *et al.*, 1980b). This substance can be removed from the milk by appropriate extraction and chromatographic steps and appears not to be a vitamin D-related compound (B. W. Hollis and P. W. Lambert, unpublished results).

More recent work with native human and bovine milk utilizing ligand-binding assay methodology coupled with appropriate purification procedures (Hollis *et al.*, 1981b) has provided considerable insight into the specific vitamin D metabolites responsible for antirachitic activity. Table III illustrates the levels of vitamin D and its metabolites in native human milk. As previously demonstrated (Hollis *et al.*, 1980b), 25-OHD is the major antirachitic factor in native human milk. It is also important to note that the levels of the various metabolites are essentially equal in milk whey and whole milk, provided the milk sample is fractionated immediately after collection. R. L. Horst (personal communication) has recently obtained similar results following the analysis of native bovine milk (Table IV). Saponification of bovine milk prior to analysis still resulted in low levels of vitamin D and its metabolites. This observation suggests that the concentrations of esterified forms of vitamin D in milk are insignificant.

On the basis of the levels of vitamin D and its metabolites in milk (Tables III

Table III. Distribution of Vitamin D and Its Metabolites in Normal Human Milk and Milk Whey[a]

Metabolite	Whole Milk	Milk Whey[b]
Vitamin D	39 ± 9	41 ± 10
25-OHD	311 ± 31	310 ± 34
24,25-$(OH)_2$D	52 ± 8	52 ± 7
25,26-$(OH)_2$D	32 ± 9	29 ± 10
1,25-$(OH)_2$D	5.1 ± 0.3	5.4 ± 0.5

[a]Hollis *et al.* (1981b). Data expressed as pg/ml ± SD; $N = 5$.
[b]Obtained by centrifuging freshly obtained whole milk and discarding top lipid layer and precipitated casein.

and IV) and the relative biological potencies of the metabolites (Norman and Henry, 1974; Miravet *et al.*, 1976), it is possible to estimate the total antirachitic activity present. In bovine and human milk, this activity corresponds to 15 IU/liter and 26 IU/liter, respectively. These values are in close agreement with the antirachitic activity determined by bioassay (Table II). These specific vitamin D analyses failed to confirm previous reports that the parent vitamin is present in high concentrations and is the major source of antirachitic activity in native human and bovine milk (LeBoulch *et al.*, 1974; Adachi and Kobayashi, 1979).

5. The Interrelationship between Plasma and Milk Levels of Vitamin D and Its Metabolites

To understand the interrelationship between human plasma and milk levels of vitamin D and its metabolites, it is important first to know the normal circulating levels of vitamin D and its metabolites: vitamin D, 3 ng/ml (B. W. Hollis and P. W. Lambert, unpublished observation; Shepard *et al.*, 1979); 25-OHD, 25

Table IV. Quantitation of Vitamin D and Its Metabolites in Native Bovine Whole Milk[a]

Metabolite[b]	Level (pg/ml, $n = 10$)
Vitamin D	25–100
25-OHD	100–300
24,25-$(OH)_2$D	N.D.[c]
25,26-$(OH)_2$D	N.D.
1,25-$(OH)_2$D	1–3

[a]R. L. Horst (personal communication).
[b]Each sample was saponified prior to analysis.
[c]Not detectable.

ng/ml (Haddad and Chyu, 1971; Lambert *et al.*, 1977; Hollis *et al.*, 1977a); 24,25-$(OH)_2D$, 1.5 ng/ml (Lambert *et al.*, 1980); 25,26-$(OH)_2D$, 0.8 ng/ml (Shepard *et al.*, 1979; Horst *et al.*, 1979; P. W. Lambert and B. W. Hollis, unpublished observation); and 1,25-$(OH)_2D$, 35 pg/ml (Eisman *et al.*, 1976; Hughes *et al.*, 1976; Lambert *et al.*, 1978). Although normal levels of plasma 1,25-$(OH)_2D$ in human adults are about 35 pg/ml, lactating women have levels of about 100 pg/ml (Kumar *et al.*, 1979). When these values are compared with those of normal human milk (Table III), it can be calculated that, with the exception of 1,25-$(OH)_2D$, vitamin D and its metabolites are present in milk at 1.5 to 3% of the levels found in plasma. This ratio is similar to that between the DBP in milk and plasma (Hollis *et al.*, 1981b).

In contrast, the level of 1,25-$(OH)_2D$ in milk is approximately 6% of that found in the plasma of women during lactation. The origin of the DBP in milk is likely the plasma, and therefore, it is not unreasonable to suggest that these sterols cross the mammary complex attached to the DBP during apocrine and halocrine secretion processes. The reason that there is a greater milk–plasma ratio of 1,25-$(OH)_2D$ than of the other metabolites could be that mammary tissue contains the cytosol receptor for this metabolite (Reinhardt and Conrad, 1980). As a result, 1,25-$(OH)_2D$ may be secreted into milk by dual pathways, one involving attachment to the DBP and the other to the cytosolic receptor in milk (Fig. 4). If the above speculations concerning the processes of vitamin D metabolite secretion are correct, one would expect the antirachitic sterols to be contained in the soluble protein portion of milk, i.e., the whey. Recent studies in our laboratory on milk fractionated immediately after collection have shown this to be true (Table III).

When whole milk is allowed to stand following collection, there is a gradual transfer of vitamin D and its metabolites from the aqueous to the fat phase. This transfer has been determined from both quantitative analysis as well as the distribution of $[^3H]$vitamin D_3 metabolites incubated in whole milk (Hollis *et al.*, 1981b). This phenomenon appears to reflect an ability of milk fat to "strip" the vitamin from the DBP. Interestingly, $[^3H]$-1,25-$(OH)_2D_3$ bound to its cytosol receptor in mammary tissue is much more resistant to this "stripping" action during incubation with bovine whole milk than are the metabolites attached to the DBP (Hollis *et al.*, 1981b). The transfer of vitamin D metabolites from the DBP to the lipid portion of milk may explain why this portion of milk ultimately contains a greater amount of antirachitic activity than does the soluble protein portion (Leerbeck and Sondergaard, 1980).

From the above observations, one could postulate that the antirachitic activity of milk is not related to total fat concentration but rather to the DBP in milk derived from the plasma. Under these circumstances, colostrum, which contains a much greater concentration of plasma protein than does regular milk (Jelliffe and Jelliffe, 1978), should possess greater antirachitic activity than milk. An early study showed this to be true (Hibbs and Pounden, 1955). The levels of vitamin D and its metabolites bound to DBP are in theoretical agreement with what would be expected from the fact that plasma proteins constitute 3–10% of the total protein in milk and colostrum (Larson and Jorgensen, 1974).

6. Factors in Milk that Affect the Intestinal Transport of Calcium in the Neonate

The mechanisms controlling calcium absorption in the neonate remain unclear: 1,25-$(OH)_2D$ is currently believed to be the physiologically active form of vitamin D associated with the active transport of intestinal calcium by means of its induction of the synthesis of the calcium binding protein (CaBP) (Lawson and Emtage, 1974) or by other possible mechanisms (Morrissey *et al.*, 1978).

Milk has been considered a poor source of vitamin D relative to the RDA for infants of 400 IU. Lakdawala and Widdowson (1977) state that breast-fed infants rarely develop rickets. This contention, however, is not universally accepted (Toverud and Boass, 1979). The antirachitic activity of human milk has been attributed mainly to vitamin D sulfate (Lakdawala and Widdowson, 1977), although recent studies seriously question the validity of this contention (Hollis *et al.*, 1979, 1981a). It seems probable that milk from a lactating woman in normal vitamin D status contains sufficient antirachitic activity in the form of other vitamin D metabolites to satisfy the requirement of the infant. On the basis of the levels of vitamin D and its metabolites found in normal human milk (Table III) it appears that the amount of antirachitic activity required for the prevention of rickets in neonates is much lower than the RDA of 400 IU. A formula-fed infant receives 400 IU of antirachitic activity in the form of added parent vitamin. In contrast, the breast-fed infant receives only about 25 IU, primarily as 25-OHD. This last point is important because there is evidence suggesting that neonates, especially preterm infants, have impaired hepatic conversion of vitamin D to 25-OHD (Hoff *et al.*, 1979; Kano *et al.*, 1980). Problems for the nursing infant could arise if the mother's plasma levels of antirachitic sterols reached a critically low level, thus limiting their secretion into the milk and restricting their availability to the infant. The threshold level of vitamin D and its metabolites in the mother's plasma necessary for adequate antirachitic activity in the milk for the nursing infant is currently unknown.

Other studies have also raised questions regarding the importance of vitamin D and its role in calcium absorption in the neonate. First, Ueng *et al.* (1979) showed that the intestine of newborn rats lacks the capacity to produce CaBP when 1,25-$(OH)_2D_3$ is administered by an intraperitoneal route. Second, oral administration of 1,25-$(OH)_2D_3$ to newborn rats results in rapid esterification and probable inactivation of this metabolite (Noff and Edelstein, 1978). Third, rat pups suckled on milk from vitamin D-depleted mothers exhibit normal plasma calcium and phosphorus levels (Halloran and DeLuca, 1979). Finally, although nursing newborn rats possess low plasma levels of 1,25-$(OH)_2D$ and high levels of 24,25-$(OH)_2D$, they maintain normal concentrations of plasma calcium and phosphorus (Halloran *et al.*, 1979). When these same animals are weaned, plasma 1,25-$(OH)_2D$ increases dramatically while levels of 24,25-$(OH)_2D$ decline. These authors suggested that the factor mainly responsible for the intestinal absorption of calcium by suckling rats is not 1,25-$(OH)_2D$ but lactose.

Lactose is known to increase the absorption of calcium from the small intestine (Armbrecht and Wasserman, 1976). These workers suggested that lactose increases calcium uptake through an interaction directly with the brush border membrane of the intestinal cells to increase their permeability to this cation. This disaccharide is known to induce an increased flux of intestinal calcium independent of CaBP (Pansu *et al.*, 1979). Lactose has also been shown to improve body calcium retention and bone mass (Schaafsma and Visser, 1980).

Human milk contains almost twice as much lactose as cow's milk (Porter, 1978). This difference may indicate that the human infant can absorb calcium from the intestine by a mechanism that is independent of vitamin D. However, it must be remembered that the aforementioned studies were carried out in rats and that extrapolation to humans is speculative at best. Although vitamin D may not play a significant role in calcium absorption in the early neonatal period, it undoubtedly plays a major role in calcium metabolism as the nursing newborn matures throughout the first year and beyond. Long-term studies are needed to resolve questions about the quantity and form of the antirachitic sterols available to the nursing human neonate and how these factors are affected by the vitamin D status of the nursing mother. These studies could ultimately lead to reevaluation of the RDA and provide an insight into the form of vitamin D best suited for dietary supplementation of the newborn infant.

7. Conspectus

The presence of antirachitic activity in milk has been a subject of research for many years. On the basis of present knowledge, it appears that the antirachitic sterols present in milk are the same as those in plasma. However, they are present in milk in much lower concentrations. These sterols are secreted into the milk bound to their plasma and/or cytosol binding proteins. Lactose may be the most significant factor in the absorption of dietary calcium in the nursing neonate, although the role of vitamin D and its metabolites cannot be discounted.

ACKNOWLEDGMENTS. The authors gratefully acknowledge the secretarial help of G. Galloway and J. M. Taylor. We are especially grateful to Dr. J. G. Haddad, Jr. for performing immunological DBP determinations in milk and also to Dr. T. A. Reinhardt for providing the mammary cytosol receptor for 1,25-$(OH)_2D$.

References

Adachi, A., and Kobayashi, T., 1979, Identification of vitamin D_3 and 7-dehydrocholesterol in cow's milk by gas chromatography-mass spectrometry and their quantitation by high-performance liquid chromatography, *J. Nutr. Sci. Vitaminol.* **25**:67.

Armbrecht, H. J., and Wasserman, R. H., 1976, Enhancement of Ca^{++} uptake by lactose in the rat small intestine, *J. Nutr.* **106**:1265.

Belsey, R., DeLuca, H. F., and Potts, J. T., 1971, Competitive binding assay for vitamin D and 25-OH-vitamin D, *J. Lab. Clin. Med.* **33**:554.

Belsey, R., Clark, M. B., Bernat, M., Glowacki, J., Holick, M. F., DeLuca, H. F., and Potts, J. T., 1974, The physiologic significance of plasma transport of vitamin D and metabolites, *Am. J. Med.* **57**:50.

Bouillon, R., Van Baelen, H., and De Moor, P., 1977, The measurement of the vitamin D-binding protein in human serum, *J. Clin. Endocrinol. Metab.* **45**:225.

Bouillon, R., Van Baelen, H., Rombauts, W., and De Moor, P., 1978, The isolation and characterization of the vitamin-D-binding protein from rat serum, *J. Biol. Chem.* **253**:4426.

Brumbaugh, P. F., and Haussler, M. R., 1975, Nuclear and cytoplasmic binding components for vitamin D metabolites, *Life Sci.* **16**:353.

Burger, R. L., and Allen, R. H., 1974, Characterization of vitamin B_{12}-binding proteins isolated from human milk and saliva by affinity chromatography, *J. Biol. Chem.* **249**:7220.

Christakos, S., and Norman, A. W., 1979, Studies on the mode of action of calciferol XVIII. Evidence for a specific high affinity binding protein for 1,25-dihydroxyvitamin D_3 in chick kidney and pancreas, *Biochem. Biophys. Res. Commun.* **89**:56.

Cooke, N. E., Walgate, J., and Haddad, J. G., 1979, Human serum binding protein for vitamin D and its metabolites. I. Physiochemical and immunological identification in human tissues, *J. Biol. Chem.* **254**:5958.

DeLuca, H. F., and Schnoes, H. K., 1976, Metabolism and mechanism of action of vitamin D, *Annu. Rev. Biochem.* **45**:631.

Eisman, J. A., Hamstra, A. J., Kream, B. E., and DeLuca, H. F., 1976, A sensitive, precise, and convenient method for determination of 1,25-dihydroxyvitamin D in human plasma, *Arch. Biochem. Biophys.* **176**:235.

Favus, M. J., 1978, Vitamin D physiology and some clinical aspects of the vitamin D endocrine system, *Med. Clin. North Am.* **62**:1291.

Gast, D. R., Marquardt, J. P., Jorgensen, N. A., and DeLuca, H. F., 1977, Efficacy and safety of 1α-hydroxyvitamin D_3 for prevention of parturient paresis, *J. Dairy Sci.* **60**:1910.

Haddad, J. G., and Birge, S. J., 1975, Widespread, specific binding of 25-hydroxycholecalciferol in rat tissues, *J. Biol. Chem.* **250**:299.

Haddad, J. G., and Chyu, K. J., 1971, Competitive protein-binding radioassay for 25-hydroxycholecalciferol, *J. Clin. Endocrinol. Metab.* **33**:992.

Haddad, J. G., and Walgate, J., 1976, Radioimmunoassay of the binding protein for vitamin D and its metabolites in human serum. Concentrations in normal subjects and patients with disorders of mineral homeostasis, *J. Clin. Invest.* **58**:1217.

Haddad, J. G., Hillman, L., and Rojanasathit, S., 1976, Human serum binding capacity and affinity for 25-hydroxyergocalciferol and 25-hydroxycholecalciferol, *J. Clin. Endocrinol. Metab.* **43**:86.

Halloran, B. P., and DeLuca, H. F., 1979, Vitamin D deficiency and reproduction in rats, *Science* **204**:73.

Halloran, B. P., Barthell, E. N., and DeLuca, H. F., 1979, Vitamin D metabolism during pregnancy and lactation in the rat, *Proc. Natl. Acad. Sci. U.S.A.* **76**:5549.

Harris, R. S., and Bunker, J. W. M., 1939, Vitamin D potency of human breast milk, *Am. J. Public Health* **29**:744.

Hibbs, J. W., and Pounden, W. D., 1955, Studies on milk fever in dairy cows. IV. Prevention by short-time, prepartum feeding of massive doses of vitamin D, *J. Dairy Sci.* **38**:65.

Higaki, M., Takahashi, M., Suzuki, T., and Sahashi, Y., 1965, Metabolic activities of vitamin D in animals. III. Biogenesis of vitamin D sulfate in animal tissues, *J. Nutr. Sci. Vitaminol.* **11**:261.

Hoff, N., Haddad, J., Teitelbaum, S., McAlister, W., and Hillman, L. S., 1979, Serum concentrations of 25-hydroxyvitamin D in rickets of extremely premature infants, *J. Pediatr.* **94**:460.

Hollis, B. W., and Draper, H. H., 1978, Vitamin D metabolites in milk, *Fed. Proc.* **37**:409.

Hollis, B. W., and Draper, H. H., 1979, A comparative study of vitamin D binding globulins in milk, *Comp. Biochem. Physiol.* **64B**:41.

Hollis, B. W., Burton, J. H., and Draper, H. H., 1977a, A binding assay for 25-hydroxycalciferols and 24*R*,25-dihydroxycalciferols using bovine plasma globulin, *Steroids* **30**:285.

Hollis, B. W., Hibbs, J. W., and Conrad, M. R., 1977b, Vitamin D binding factors in bovine blood, *J. Dairy Sci.* **60**:1605.

Hollis, B. W., Lambert, P. W., and Draper, H. H., 1979, On the occurrence of vitamin D sulfate in milk, *J. Nutr.* **109**:XXIV.

Hollis, B. W., Roos, B. A., and Lambert, P. W., 1980a, 25,26-Dihydroxycholecalciferol: A precursor in the renal synthesis of 25-hydroxycholecalciferol-26,23-lactone, *Biochem. Biophys. Res. Commun.* **95**:520.

Hollis, B. W., Roos, B. A., and Lambert, P. W., 1980b, Isolation and quantitation of vitamin D metabolites in human and bovine milk, *Clin. Res.* **28**:395a.

Hollis, B. W., Roos, B. A., Draper, H. H., and Lambert, P. W., 1981a, On the occurrence of vitamin D sulfate in human milk whey, *J. Nutri.* **111**:384.

Hollis, B. W., Roos, B. A., Draper, H. H., and Lambert, P. W., 1981b, Vitamin D metabolites in human and bovine milk, *J. Nutr.* **111**:1240.

Horst, R. L., 1979, 25-OHD_3-26,23-Lactone: A metabolite of vitamin D_3 that is 5 times more potent than 25-OHD_3 in the rat plasma competitive protein binding radioassay, *Biochem. Biophys. Res. Commun.* **89**:286.

Horst, R. L., Shepard, R. M., Jorgensen, N. A., and DeLuca, H. F., 1979, The determination of 24,25-dihydroxyvitamin D and 25,26-dihydroxyvitamin D in plasma from normal and nephrectomized man, *J. Lab. Clin. Med.* **93**:277.

Hughes, M. R., Baylink, D. J., Jones, P. G., and Haussler, M. R., 1976, Radioligand receptor assay for 25-hydroxyvitamin D_2/D_3 and 1,25-dihydroxyvitamin D_2/D_3. Application to hypervitaminosis D, *J. Clin. Invest.* **58**:61.

Imawari, M., Kida, K., and Goodman, D. S., 1976, The transport of vitamin D and its 25-hydroxy metabolite in human plasma. Isolation and partial characterization of vitamin D and 25-hydroxyvitamin D binding protein, *J. Clin. Invest.* **58**:514.

Jelliffe, D. B., and Jelliffe, E. F., 1978, *Human Milk In the Modern World*, Oxford University Press, New York.

Kano, K., Yoshida, H., Yata, J., and Suda, T., 1980, Age and seasonal variations in the serum levels of 25-hydroxyvitamin D and 24,25-dihydroxyvitamin D in normal humans, *Endocrinol. J.* **27**:215.

Kodicek, E., and Lawson, D. E. M., 1967, Vitamin D, in: *The Vitamins* (P. Gyorgy and W. N. Pearson, eds.), pp. 211–244, Academic Press, New York.

Kream, B. E., Jose, M., Yamada, S., and DeLuca, H. F., 1977, A specific high-affinity binding macromolecule for 1,25-dihydroxyvitamin D_3 in fetal bone, *Science* **197**:1086.

Kream, B. E., DeLuca, H. F., Moriarity, D. M., Kendrick, N. C., and Ghazarian, J. G., 1979, Origin of 25-hydroxyvitamin D_3 binding protein from tissue cytosol preparations, *Arch. Biochem. Biophys.* **192**:318.

Kumar, R., Cohen, W. R., Silva, P., and Epstein, F. H., 1979, Elevated 1,25-dihydroxyvitamin D plasma levels in normal human pregnancy and lactation, *J. Clin. Invest.* **63**:342.

Lakdawala, D. R., and Widdowson, E. M., 1977, Vitamin D in human milk, *Lancet* **1**:67.

Lambert, P. W., Syverson, B. J., Arnaud, C. D., and Spelsberg, T. C., 1977, Isolation and quantitation of endogenous vitamin D and its physiologically important metabolites in human plasma by high pressure liquid chromatography, *J. Steroid Biochem.* **8**:929.

Lambert, P. W., Toft, D. O., Hodgson, S. F., Lindmark, E. A., Witrak, B. J., and Roos, B. A., 1978, An improved method for the measurement of 1,25-$(OH)_2$-D_3 in human plasma, *Endocr. Res. Commun.* **5**:293.

Lambert, P. W., Hollis, B. W., Bell, N. H., and Epstein, S., 1980, Demonstration of a lack of change in serum 1,25-dihydroxyvitamin D in response to parathyroid extract in pseudohypoparathyroidism, *J. Clin. Invest.* **66**:782.

Larson, B. L., and Jorgensen, G. N., 1974, Biosynthesis of the milk proteins, in: *Lactation: Biosynthesis and Secretion of Milk* (B. L. Larson and V. R. Smith, eds.), pp. 115–146, Academic Press, New York.

Lawson, D. E. M., and Emtage, J. S., 1974, Molecular action of vitamin D in the chick intestine, *Vitam. Horm.* **32**:277.

Lawson, D. E. M., Charman, M., Wilson, P. W., and Edelstein, S., 1976, Some characteristics of new tissue-binding proteins for metabolites of vitamin D other than1,25-dihydroxyvitamin D, *Biochem. Biophys. Acta* **437**:403.

LeBoulch, N., Gulat-Marnay, C., and Raoul, Y., 1974, Derives de la vitamine D_3 des laits de femme et de vache: Ester sulfate de cholecalciferol et hydroxy-25 cholecalciferol, *Int. J. Vitam. Nutr. Res.* **44**:167.

Leerbeck, E., and Sondergaard, H., 1980, The total content of vitamin D in human milk and cow's milk, *Br. J. Nutr.* **44**:7.

Miravet, L., Redel, J., Carre, M., Queille, M. L., and Bordier, P., 1976, The biological activity of synthetic 25,26-dihydroxycholecalciferol and 24,25-dihydroxycholecalciferol in vitamin D-deficient rats, *Calcif. Tissue Res.* **21**:145.

Morrissey, R. L., Zolock, D. T., Bikle, D. D., Empson, R. N., and Bucci, T. J., 1978, Intestinal response to 1,25-dihydroxycholecalciferol. I. RNA polymerase, alksline phosphatase, calcium and phosphorus uptake *in vitro,* and *in vivo* calcium transport and accumulation, *Biochim. Biophys. Acta* **538**:23.

Noff, D., and Edelstein, S., 1978, Vitamin D and its hydroxylated metabolites in the rat. Placental and lacteal transport, subsequent metabolic pathways and tissue distribution, *Horm. Res.* **9**:292.

Norman, A. W., and Henry, H. H., 1974, 1,25-Dihydroxycholecalciferol: A hormonally active form of vitamin D_3, *Recent Prog. Horm. Res.* **30**:431.

Osborn, T. W., and Norman, A. W., 1977, A "25-hydroxycholecalciferal-like compound" in milk as determined by specific competitive binding radioassays, in: *Vitamin D: Biochemical, Chemical and Clinical Aspects Related to Calcium Metabolism* (A. W. Norman, ed.), pp. 523–525, deGruyter, Berlin.

Pansu, D., Bellaton, C., and Bronner, F., 1979, Effect of lactose on duodenal calcium-binding protein and calcium absorption, *J. Nutr.* **109**:508.

Payne, D. W., Peng, L. H., and Pearlman, W. H., 1976, Corticosteroid-binding proteins in human colostrum and milk and rat milk, *J. Biol. Chem.* **251**:5272.

Polskin, L. J., Kramer, B., and Sobel, A. E., 1945, Secretion of vitamin D in milks of women fed fish liver oil, *J. Nutr.* **30**:451.

Porter, J. W. G., 1978, Milk as a source of lactose, vitamin and minerals, *Proc. Nutr. Soc.* **37**:225.

Reinhardt, T. A., and Conrad, H. R., 1980, Specific binding protein for 1,25-dihydroxyvitamin D_3 in bovine mammary gland, *Arch. Biochem. Biophys.* **203**:108.

Sahashi, Y., Suzuki, T., Higaki, M., and Asano, T., 1967a, Metabolism of vitamin D in animals. V. Isolation of vitamin D sulfate from mammlian milk, *J. Vitaminol.* **13**:33.

Sahashi, Y., Suzuki, T., Higaki, M., Takahashi, M., Asano, T., Hasegawa, T., and Miyazawa, E., 1967b, Metabolic activities of Vitamin D in animals. VI. Physiological activities of vitamin D sulfate, *J. Vitaminol.* **13**:37.

Schaafsma, G., and Visser, R., 1980, Nutritional interrelationships between calcium, phosphorus and lactose in rats, *J. Nutr.* **110**:1101.

Shepard, R. M., Horst, R. L., Hamstra, A. J., and DeLuca, H. F., 1979, Determination of vitamin D and its metabolites in plasma from normal and anephric man, *Biochem. J.* **182**:55.

Toverud, S. U., and Boass, A., 1979, Hormonal control of calcium metabolism in lactation, *Vitam. Horm.* **37**:303.

Ueng, T.-H., Golub, E. E., and Bronner, F., 1979, The effect of age and 1,25-dihydroxyvitamin D_3 treatment on the intestinal calcium-binding protein of suckling rats, *Arch. Biochem. Biophys.* **196**:624.

Van Baelen, H., Bouillon, R., and DeMoor, P., 1977, Binding of 25-hydroxycholecalciferol in tissues, *J. Biol. Chem.* **252**:2515.

Van Baelen, H., Bouillon, R., and DeMoor, P., 1980, Vitamin D-binding protein (GC-globulin) binds actin, *J. Biol. Chem.* **255**:2270.

Waxman, S., 1975, Folate binding proteins, *Br. J. Haematol.* **29**:23.

Chapter 4

Dietary Protein, Metabolic Acidosis, and Calcium Balance

John T. Brosnan and Margaret E. Brosnan

1. Sources of Acid and Base

A very large number of metabolic reactions that produce or remove hydrogen ions occur in the body. However, in the great majority of these, the production and removal of protons are merely intermediary steps in processes that are essentially neutral. A ready example comes from the production and utilization of ATP in cells. ATP hydrolysis in cells produces a proton.

$$ATP^{4-} + H_2O \rightarrow ADP^{3-} + Pi^{2-} + H^+ \quad (1)$$

The production of protons by this route may be quite large since a 68-kg man utilizes more than 40 kg of ATP per "restful" day (Erecinska and Wilson, 1978). The same man carrying a 25-kg load up a vertical ladder at a speed of 11.9m/min utilizes approximately 0.6 kg of ATP/min. This enormous rate of ATP utilization produces very large quantities of hydrogen ions (about 80 g of hydrogen ions in a "restful" day), yet there is no net production of acid because the utilization of each molecule of ATP is balanced by the synthesis of one molecule of ATP, a process that utilizes a proton.

$$H^+ + Pi^{2-} + ADP^{3-} \rightarrow ATP^{4-} + H_2O \quad (2)$$

John T. Brosnan and Margaret E. Brosnan • Department of Biochemistry, Memorial University of Newfoundland, St. John's, Newfoundland A1B 3X9, Canada. Work from the authors' laboratories was supported by the Medical Research Council of Canada.

Similar arguments apply to the vast majority of metabolic reactions that produce or utilize protons (Krebs *et al.*, 1975). There are, however, a limited number of processes that produce net acid. Credit goes to Claude Bernard for realization that whether an animal excretes acid or base in its urine depends not on the species of animal but on the diet consumed by the animal. Bernard realized that herbivores generally excrete an alkaline urine and carnivores an acid urine. He was surprised to find that some rabbits brought to the laboratory excreted an acid urine. This was associated with their being fasted, and he reasoned that these animals had been transformed, by fasting, into carnivores, living on their own tissues. When he gave the rabbits grass to eat, their urine became alkaline within hours. Similar results were obtained with another herbivore, the horse. Bernard then fed meat to his rabbits and found that as long as this diet was maintained, the rabbits excreted an acid urine (Bernard, 1865).

Principal among the processes that produce net acid are the oxidation of proteins, the production of organic acids, and the formation of bone. Among the processes that produce net base are the oxidation of the inorganic salts of organic anions and the resorption of bone.

1.1. Protein Oxidation

That foodstuffs rich in protein such as lean meats, fish, and eggs have an excess of acid-forming elements has been known for over 60 years (Sherman and Gettler, 1912). The oxidation of sulfur-containing amino acids was shown to be a major source of endogenous acid production by Hunt (1956) who studied the effects of varying the dietary intake of sulfur on the urinary output of acid. A quantitative picture of the relationship between the oxidation of sulfur-containing amino acids and net acid production was provided by Lemann and Relman (1959). In these studies, the effect of adding 13.9 g of methionine to the diet of healthy young males was studied. These authors found that "after final restoration of acid-base balance, accumulated increments in sulfate excretion were virtually equalled by the cumulative increments in the excretion of 'net acid.'" Thus, the production of sulfate and hydrogen ions from the oxidation of methionine was as predicted from equations (1) and (2) in Table I. The endogenous production of sulfuric acid is rather minor in infants because of the preferential utilization of methionine and cysteine for protein synthesis (Kildeberg *et al.*, 1969).

The production of acid from phosphoproteins is a more complex process. Most, if not all, phosphate in proteins is present in monoester form. Whether or not net acid is liberated in the hydrolysis of the phosphoester bond depends on the nature of the cations neutralizing the phosphate anion (Relman *et al.*, 1961; Lennon *et al.*, 1962, 1966). Hydrolysis of the monoester at pH 7.4 yields acid if the phosphates are neutralized by hydrogen ions as shown in equation 3 of Table I. However, hydrolysis of the monoester at pH 7.4 actually removes a small

Table I. Some Reactions Generating Acid or Base *in Vivo*

Reactant			Products	
$2C_5O_2NH_{11}S$ methionine	$+15O_2$	$\longrightarrow$	$(NH_2)_2CO + 9CO_2 + 7H_2O + 4H^+ + 2SO_4^{2-}$ urea	(1)
$2C_3O_2NH_7S$ cysteine	$+9O_2$	$\longrightarrow$	$(NH_2)_2CO + 5CO_2 + 3H_2O + 4H^+ + 2SO_4^{2-}$ urea	(2)
RH_2PO_4 monophosphate ester	$+H_2O$	$\xrightarrow[\text{pH 7.4}]{}$	$ROH + 0.8HPO_4^{2-} + 0.2H_2PO_4^- + 1.84H^+$	(3)
RK_2PO_4 potassium salt of monophosphate ester	$+H_2O + 0.2H^+$	$\xrightarrow[\text{pH 7.4}]{}$	$ROH + 0.8HPO_4^{2-} + 0.2H_2PO_4^- + 2K^+$	(4)
$R'R''KPO_4$ potassium salt of diphosphate ester	$+2H_2O$	$\xrightarrow[\text{pH 7.4}]{}$	$R'OH + R''OH + 0.8HPO_4^{2-} + 0.2H_2PO_4^- + K^+ + 0.8H^+$	(5)
$C_{16}H_{32}O_2$[a] palmitic acid	$+7O_2$	$\longrightarrow$	$4C_4H_5O_3^- + 4H^+ + 4H_2O$ acetoacetate	(6)
$C_{16}H_{32}O_2$[a] palmitic acid	$+5O_2$	$\longrightarrow$	$4C_4H_8O_3^- + 4H^+$ β-hydroxybutyrate	(7)
$C_6H_{12}O_6$ glucose		$\longrightarrow$	$2C_3H_5O_3^- + 2H^+$ lactate	(8)
$C_6O_2N_4H_{15}Cl$ arginine hydrochloride	$+5\frac{1}{2}O_2$	$\longrightarrow$	$2(NH_2)_2CO + 4CO_2 + 3H_2O + H^+ + Cl^-$ urea	(9)
$C_3O_3H_5K$ potassium lactate	$+3O_2$	$\longrightarrow$	$3CO_2 + 2H_2O + K^+ + OH^-$	(10)
$10Ca^{2+} + 4.8HPO_4^{2-} + 1.2H_2PO_4^- + 2H_2O$		$\longrightarrow$	$[Ca_3(PO_4)_2]_3 \cdot Ca(OH)_2 + 9.2H^+$ hydroxyapatite	(11)
$2NH_4Cl + CO_2$ ammonium chloride		$\longrightarrow$	$(NH_2)_2CO + H_2O + 2H^+ + 2Cl^-$ urea	(12)

[a] Palmitic acid is written as the undissociated acid.

quantity of acid if the phosphates are neutralized by an inorganic cation, e.g., K^+, as shown in equation 4 of Table I. Hydrolysis of a diphosphate ester yields acid even if it is neutralized by an inorganic cation (Table I, equation 5). Should a phosphate monoester be neutralized by a combustible cation (e.g., arginine) (Relman *et al.*, 1961), its metabolism will yield the same amount of acid as if the isoelectric protein were fed, although in this case the protons are not produced in the hydrolysis reaction but, rather, in the subsequent oxidation of the combustible cation.

That phosphoproteins can provide acid *in vivo* was convincingly shown by Lennon *et al.* (1962) in experiments with young males in which a purified soy phosphoprotein prepared by isoelectric precipitation and virtually free of minerals except phosphorus was employed as the sole protein source. A liquid diet containing this protein was fed to the subjects. The effect of an extra load of this protein was then compared to the effect of an equivalent amount of nitrogen in the form of beefsteak. The experiments showed a comparable increase in net acid excretion in both groups of subjects. The increased acid excretion in the subjects fed beef as the protein load was predictable from the increase in the excretion of sulfate and organic acids. This was not the case for the group of subjects that ingested the phosphoprotein load. These subjects excreted far more acid than could be accounted for from the sum of excreted sulfate plus organic acids. This discrepancy disappeared when the acid produced from hydrolysis of the phosphoprotein's phosphate ester was included. Thus, ''the extra phosphoprotein provided a third source of endogenous acid production in addition to the oxidation of sulfur to sulfate and the production of organic acids'' (Lennon *et al.*, 1962).

Thus, the catabolism of protein produces acid as a result of the oxidation of methionine and cysteine and may produce acid from the hydrolysis of phosphate monoesters in proteins.

1.2. Organic Acids and the Salts of Organic Acids

That the excretion of organic acids is an important component of net acid production was demonstrated in 1961 by Relman and colleagues. In subjects fed a liquid-formula diet, organic acid excretion accounted for about half of the endogenous fixed acid produced. The precise acids were not identified and, indeed, are probably variable from time to time. However, it is certain that uric acid, produced from purine catabolism, is a significant component in addition to variable amounts of lactic, citric, and other acids that escape reabsorption. It is well appreciated that in certain disease states the endogenous production of organic acids becomes the major source of the acid load. In certain gouty patients, very large amounts of uric acid are excreted. In diabetic ketoacidosis, large quantities of acetoacetic and β-hydroxybutyric acids are produced, accumulate in plasma, and are lost in the urine (Table I, reactions 6 and 7). Similarly, in lactic acidosis, lactic acid excretion constitutes a major part of the acid load (Table I, reaction 8).

The salts of organic cations and anions may also contribute to the production of net acid and base, respectively. For example, the oxidation of arginine hydrochloride produces acid (Table I, reaction 9), whereas the oxidation of potassium lactate produces base (Table I, reaction 10). It is for this reason that inorganic cations are considered to be potential alkali-forming elements and inorganic anions potential acid-forming elements. Most western diets contain an excess of inorganic cations over inorganic anions, and hence, the net effect on acid–base production from these compounds will be the production of alkali.

That such considerations are important was shown by the studies of Lennon *et al.* (1966). These workers showed that the two sources of endogenous acid production identified by Relman *et al.* (1961), the oxidation of dietary sulfur and the excretion of organic acids produced from neutral precursors, were not sufficient to account for acid excretion when subjects ate whole-food diets. In normal North American diets, the daily intake of dietary inorganic cations exceeds that of dietary inorganic anions by about 1 mEq/kg body weight (Lemann and Lennon, 1972). This represents an equal quantity of "potential base." However, the feces contain an excess of inorganic cations over anions of about 0.5 mEq/kg body weight per day, representing a loss of "potential base." Lennon *et al.* (1966) fed their subjects either a liquid-formula diet or one of two whole-food diets designed to supply either an excess of inorganic cations or an excess of inorganic anions. Urinary net acid excretion by the subjects fed whole-food diets could not be accounted for merely by summing sulfate and organic acid excretion. The subjects who received the diet with an excess of inorganic cations over anions excreted less acid than would be predicted from the sum of urinary sulfate plus organic acids. The reverse was true in the case of subjects ingesting the diet with an excess of inorganic anions over cations. Good agreement between the actual renal excretion of acid and the predicted excretion of acid was obtained only when: (1) the "potential base" and "potential acid" of the diets was taken into account and (2) the loss of "potential base" in the stool was accounted for. In all situations, the stool contained an excess of inorganic cations over inorganic anions, even when the reverse was true of the dietary intake.

The intestine appears to play a major role in the acid–base homeostasis of the growing infant by determining the amount of "potential base" absorbed. In fact, the dietary intake of unidentified anions ("potential base") and the fecal excretion of unidentified anions are the two largest single variables of the net acid balance of infants (Kildeberg *et al.*, 1969). Between the 12th and 19th day of life of a healthy 1.4-kg premature infant, the net acid excretion averaged 3.49 mEq/day. During the same period, the average dietary intake of "potential base" averaged 8.76 mEq/day, and the fecal loss of "potential base" averaged 8.30 mEq/day. Clearly then, the intestine was an important determinant of net acid balance, since if all of the "potential base" had been absorbed and oxidized and if no "potential base" were excreted in the feces, urinary excretion of base rather than of acid would be required. Since the "potential base" in these infant studies was known to be derived largely from citrate in the milk formula, and since citrate is readily absorbable, Kildeberg *et al*. (1969) argued that fecal base excretion was actively regulated.

1.3. Bone Formation and Resorption

An additional source of hydrogen ion production was identified by Kildeberg *et al.* (1969) from their studies in growing infants. The deposition of hydroxyapatite during the process of skeletal mineralization is associated with the production of hydrogen ions according to equation 11 in Table I. The ratio (0.92) of protons released to calcium deposited is slightly higher than the ratio obtained by direct titration of dissolved bone mineral. Kildeberg *et al.* consider the difference to result from differences in the composition of the bone crystal layer and the intercrystalline inorganic material. The quantitative importance of acid produced as a result of calcification was shown in Kildeberg's (1969) studies of rapidly growing healthy premature infants. Hydrogen ion production attributable to calcification, calculated from the calcium balance, was found to amount to about one-third of the net acid excreted. More recently (Wamberg *et al.*, 1976), a similar phenomenon has been observed in rapidly growing weanling rats.

Kildeberg's study also showed that the pattern of acid excretion in infants was quite different from that seen in adults. In particular, the endogenous production of sulfuric acid was responsible for very little (about 6%) of net acid excretion, whereas in adults it is a very major component. There is very little oxidation of methionine and cysteine, presumably because they are required in large amounts by the rapidly growing infant for protein synthesis.

Since acid is released during the deposition of bone mineral, it is obvious that base will be released during bone resorption. Indeed, as discussed below, such dissolution of bone mineral can serve to neutralize acid loads. The stoichiometry of such base release is not simply the reverse of equation 11 (Table I), since apatite and nonapatite components may be involved in varying proportions.

1.4. Acid from Ammonium Chloride

Ammonium chloride ingestion was introduced by J. B. S. Haldane (1921) as a means of introducing an acid load to the body. Net acid is released as a result of the hepatic conversion of NH_3 to urea as described by equation 12 (Table I). Ammonium chloride has proven very useful for experimental production of acidosis, but it should be noted that there are differences between an ammonium chloride-induced acidosis and other acidoses. First, ammonium chloride-induced acidosis is hyperchloremic, whereas most sponteneous acidoses are hypochloremic. Second, hydrochloric acid-induced acidosis in dogs causes a more marked acidemia than acidosis caused by sulfuric or nitric acids (DeSousa *et al.*, 1974). This may occur because chloride is much more readily reabsorbed than sulfate or nitrate.

2. Calcium Balance

Calcium represents some 2% of the total body weight in the adult. Of this, approximately 99% is present in the bones and teeth, chiefly as a poorly crystallized apatite (Posner, 1973). There is also a noncrystalline or amorphous calcium phosphate present. It appears that calcium phosphate is first deposited in the more labile amorphous phase and then a portion of it is later converted to the apatite of bone. Termine and Posner (1966) have observed that young bone is richer in amorphous than apatitic calcium phosphate, whereas mature bone contains more crystalline than amorphous phase. There is also a small amount of calcium carbonate present, probably on the crystal mineral surface of the apatite phase (Posner, 1973). The remaining 1% is present in intra- and extracellular fluid compartments. Intracellular calcium concentration varies from 10^{-7} M in cytosol to 10^{-3} M in mitochondria, although much of the mitochondrial calcium is probably present as insoluble phosphate complexes (Borle, 1973). Extracellular fluid contains a total calcium concentration of 2.5×10^{-3} M, of which approximately 50% is present as ionized calcium. A further 40 to 45% is protein bound, chiefly to albumin, and the remaining 5 to 10% is complexed with anions such as phosphate, bicarbonate, and citrate. Ionized calcium is in equilibrium with calcium complexed with protein or other anions, and thus, any factor that shifts this equilibrium can readily increase or decrease ionized calcium. Calcium in apatite crystals in bone and teeth does not readily equilibrate with ionized calcium, and thus, it does not readily enter the pool of ionizable calcium unless both the inorganic salts and organic matrix of bone are first resorbed (Ham, 1974). It appears to be ionized calcium which is transported across membranes into the body (e.g., intestine, kidney) and within cells, and it is also this fraction which is important in the many physiological functions attributed to calcium (e.g., activation of enzymes, muscle contraction).

It is obvious that any change in the concentration of calcium or of its counteranion, be it protein or inorganic base, will shift this equilibrium, thus changing the concentration of ionized calcium. The hydrogen ion concentration of extra- or intracellular fluid can also cause a shift in the equilibrium, with an increased pH resulting in an increase in complexed calcium and a fall in pH increasing ionized calcium. A decrease of 0.1 pH unit increases ionized calcium by about 0.04 mEq/liter. It should also be noted that insoluble calcium salts found in the labile phase of bone and also possibly in intracellular organelles can be solubilized by increased hydrogen ion concentration according to the following equations.

$$Ca_3(PO_4)_2 + 2H^+ \rightarrow 3Ca^{2+} + 2HPO_4^{2-} \quad (3)$$

$$CaCO_3 + H^+ \rightarrow Ca^{2+} + HCO_3^- \quad (4)$$

In both cases, the insoluble calcium salt buffers an increased hydrogen ion concentration, is rendered soluble in so doing, and generates anions (HPO_4^{2-} and HCO_3^-) that may buffer additional protons.

2.1. Whole-Body Calcium Balance

There is a variable daily loss of calcium from the body in the urine, feces, and sweat, and in milk during lactation. These losses must be balanced by dietary intake and carefully regulated by the body. Calcium enters the gastrointestinal tract in the food and also in the secretions and sloughed cells that enter the tract. Calcium is absorbed by a vitamin-D-dependent process. Many recent reviews are available on this subject (Haussler and McCain, 1977; DeLuca, 1979), and thus, it will not be discussed here. However, it is important to note that other components of the diet can markedly influence the absorption of calcium. A low pH, as mentioned earlier, leads to increases in ionized calcium and in soluble complexes of calcium and thus to a higher rate of absorption. The presence of lactose increases calcium absorption and thus greatly increases the absorbability of calcium in breast milk. The presence of some amino acids (e.g., lysine and arginine; Wasserman *et al.*, 1956) also increases calcium absorption. On the other hand, substances that form insoluble calcium salts (e.g., fatty acids and phylate or oxalate present in some plant products) result in less calcium absorption. The increasing use of unusual phosphates as food additives (e.g., polyphosphates in nondairy creamers) could also pose a potential problem for calcium absorption (Draper and Bell, 1979).

Ionized calcium and soluble calcium salts (about 55 to 60% of the total calcium in plasma) are filtered in the kidney. The amount of calcium entering the tubule depends on both glomerular filtration rate and on the concentration of calcium present in these filterable forms. About 9 to 10 g of calcium are filtered each day by the kidneys of a normal adult. Of this, 98 to 99% is reabsorbed by the tubules. Normal men on a diet containing 800 mg of calcium excrete less than 300 mg per day in the urine (Epstein, 1968). About two-thirds of filtered calcium is normally reabsorbed in the proximal tubule, about 20 to 30% in the loop of Henle, and the remainder in the distal tubule and collecting ducts (Sutton and Dirks, 1977). There is normally a close relationship between the reabsorption of sodium and calcium.

2.2. Hormonal Control of Calcium Concentration in Plasma

The concentration of calcium in plasma depends in part on the same factors as whole-body calcium balance, since entry of calcium to, and exit from, the body is through this pool. However, it may also be markedly affected by interchange of calcium between plasma and bone. These processes are known to be subject to hormonal regulation.

Calcitonin has been shown to decrease plasma calcium concentration by increasing bone matrix synthesis and calcium phosphate deposition (reviewed by Vaughan, 1970) and by inhibiting bone resorption (O'Riordan and Aurbach, 1968). The plasma concentration of this hormone responds to elevated plasma calcium concentration and to dietary calcium (Cooper *et al.*, 1978).

1α,25-Dihydroxycholecalciferol elevates plasma calcium and phosphate concentrations by increasing intestinal absorption of calcium, by increasing bone resorption, and possibly by increasing distal tubular reabsorption of calcium (DeLuca, 1979).

Parathyroid hormone increases plasma calcium concentration principally by increasing bone resorption and by increasing the production of 1α,25-dihydroxycholecalciferol by the kidney (Garabedian *et al.*, 1972). It also causes phosphaturia and increases calcium reabsorption by kidney, although the increased filtered load of calcium can often result in calciuria (Nordin *et al.*, 1967). Parathyroid hormone concentration in plasma is usually inversely proportional to the concentration of ionized calcium, although it has also been reported that parathyroid hormone secretion is subject to feedback inhibition by 1α,25-dihydroxycholecalciferol (Brumbaugh *et al.*, 1975). Thus, a fall in ionized calcium concentration in plasma would be expected to result in low calcitonin and elevated parathyroid hormone concentrations in plasma and an increased production of 1α,25-dihydroxycholecalciferol. These alterations in hormone concentrations would cause increased absorption of calcium and phosphate from intestine and increased resorption of bone together with normal or increased excretion of calcium in urine and phosphaturia. The net result would be an increase in plasma calcium concentration, whereas plasma phosphate would remain normal.

A decrease in phosphate concentration in plasma (with no change in calcium) has been reported to cause an increased production of 1α,25-dihydroxycholecalciferol by kidney (Tanaka and DeLuca, 1973) and, therefore, to result in a decrease in plasma parathyroid hormone concentration (Brumbaugh *et al.*, 1975). Hughes and co-workers (1975) postulated that dual control of 1α,25-dihydroxycholecalciferol synthesis by both calcium and phosphate would permit homeostatic regulation of these ions. The increase in 1α,25-dihydroxycholecalciferol induced by hypophosphatemia would be expected to increase calcium and phosphate movement into plasma from the gastrointestinal tract but would probably have relatively little effect on bone resorption in the face of a depressed parathyroid hormone concentration (DeLuca, 1979). Decreased parathyroid hormone concentration would also tend to favor loss of excess dietary calcium in urine while preventing phosphaturia. Thus, plasma phosphate concentration would be increased, and calcium would remain normal.

3. The Buffering of Acid

A great amount is now known about the means by which animals deal with strong acids. Much of our information comes from the manner in which animals respond to an acid load, either chronic or acute. It appears that the mechanisms by which the kidney eliminates an acute acid load are essentially similar to the manner in which it deals with the much smaller amounts of acid produced by normal metabolism. However, the buffering of an acute acid load may also call into play mechanisms not generally involved in the day-to-day acid–base homeostasis of normal animals.

The plasma hydrogen ion concentration is regulated between very fine limits. The normal pH of 7.4 corresponds to a hydrogen ion concentration (strictly, activity) of 39.8 nmol/liter. A doubling of this concentration corresponds to a pH of 7.1, and a halving corresponds to a pH of 7.7. Changes of this magnitude are seldom seen because of the homeostatic mechanisms regulating $[H^+]$. There are two principal aspects to the regulation of plasma pH—the renal adjustment of $[HCO_3^-]$ and the pulmonary adjustment of P_{CO2}. In addition, the ability of the erythrocyte to buffer the carbonic acid that arises as a consequence of CO_2 produced by tissues is an important factor as is the ability of many tissues to buffer loads of fixed acid.

3.1. Buffering of an Acute Acid Load

A great deal of information is now available regarding the buffering of an acute acid load in the body. Van Slyke and Cullen (1917) calculated that only one-sixth of infused acid is neutralized by blood buffers and they postulated that the remainder is neutralized by bicarbonate in interstitial fluid and intracellularly, presumably by organic phosphates and proteins. Swan and Pitts (1955) examined acid buffering by infusing HCl into nephrectomized dogs. About 40% of the acid load was buffered by the conversion of bicarbonate to CO_2. The remainder was buffered by the exchange of hydrogen ions for tissue cations. In support of this postulate, sodium equivalent to over one-third of the infused acid load and potassium equivalent to about one-sixth of the infused acid load appeared in the extracellular fluid. The sodium and potassium were thought to arise principally from soft tissues in which proteins and phosphates were believed to be responsible for the intracellular buffering of the hydrogen ions. However, some of the cations may have arisen from bone surfaces. Bergstrom and Wallace (1954) have shown that bone contains much more sodium and potassium than can be accounted for by extracellular and intracellular fluid. These cations may be deposited on the crystal surface and may engage in ion exchange with hydrogen ions. Bergstrom and Wallace (1954) also showed that a large portion of these cations were readily mobilized in rats within 2 days of acidosis. Studies by Schwartz *et al.* (1954) on the buffering of an acute load led to conclusions similar to those of

Swan and Pitts (1955), although in their subjects (sodium-depleted humans made acidotic by ingestion of ammonium chloride), potassium release into the extracellular fluid played a greater role in buffering the hydrogen ions than did sodium release. Further studies by Schwartz *et al.* (1957) showed that the distribution of hydrogen ions among the various body buffers is not affected by the magnitude of the acid load provided that sufficient time is given for their equilibration between intracellular and extracellular buffers.

Recent experiments by Fraley and Adler (1979) have implicated parathyroid hormone in the disposal of acute acid loads. These workers showed that intracellular buffering was greatly attenuated in thyroparathyroidectomized rats and dogs. In fact, acute infusions of hydrochloric acid that were well tolerated in intact animals were lethal in thyroparathyroidectomized animals. However, the ability of such animals to buffer an acute acid load was restored if they were injected with synthetic (1–34) parathyroid hormone some hours before the infusion. Since the animals were nephrectomized, the kidney could not be the target organ of PTH in this instance. Measurements of pH change in skeletal muscle, liver, and heart showed only very small changes after acid infusion. Fraley and Adler (1979) therefore suggested that bone may be the organ principally reponsible for the buffering of an acute acid load.

The manner in which PTH acts in this regard in dogs is particularly puzzling in view of the fact that in acute acidosis no significant mobilization of calcium or phosphate occurs (Burnell, 1971; Sutton *et al.*, 1979). In rats, however, a mobilization of calcium from bone does take place in acute acidosis, but it does not require the presence of the parathyroid glands (Beck and Webster, 1976). Neither is there definitive evidence for an increase in PTH levels during acidosis. Parathyroid hormone has been reported to decrease during acute metabolic acidosis in sheep (Kaplan *et al.*, 1971), to increase slightly during acute metabolic acidosis in man (Coe *et al.*, 1975), to increase substantially during chronic metabolic acidosis in man when hypercalciuria was occurring but not to increase if the hypercalciuria was prevented by a low sodium intake (Coe *et al.*, 1975), and not to change during stable chronic metabolic acidosis in man (Weber *et al.*, 1976). However, it may not be necessary to postulate an increased serum PTH level in acute acidosis to account for its role in buffering the acute load, since some form of synergy between circulating levels of PTH and local acidity may be important (Wachman and Bernstein, 1970). It is already known that a given dose of PTH is more effective in mobilizing bone calcium in acidotic rats than in normal rats (Beck and Webster, 1976).

3.2. Buffering of a Chronic Acid Load

An important role for bone in the buffering of acids was deduced in experiments from Relman's laboratory (Goodman *et al.*, 1965; Lemann *et al.*, 1965). One of the remarkable features of acidosis of renal origin is that once developed it may remain quite stable for long periods of time. Low but stable plasma bicarbonate levels may be evident for months or even years. Goodman *et al.* (1965) showed that such patients were in a state of positive acid balance, i.e., that they excreted significantly less acid than was produced by metabolic processes. This phenomenon results from the acidosis rather than from other effects of the renal disease, since an acid balance of zero was evident when the acidosis of these patients was corrected by the ingestion of $NaHCO_3$. Furthermore, a similar phenomenon could be demonstrated in healthy subjects made acidotic by ammonium chloride ingestion (Lemann *et al.*, 1965). These latter experiments showed that a lowered but stable blood pH and blood bicarbonate level were maintained in the face of a continuously positive acid balance of about 24 mEq of hydrogen ions per day.

Goodman *et al.* (1965) argued that bone must be the organ of acid buffering, since retention of acid might proceed for months or years in patients with chronic renal acidosis, and thus, the tissue supplying the base would need to have a large supply of alkali that could be released slowly. Many years earlier, Albright and Reifenstein (1948) had recognized that bone demineralization could release alkali. Goodman *et al.* (1965) calculated that the adult skeleton contains approximately 35,000 mEq of potentially available base. In a subsequent investigation using whole-food diets, Lemann *et al.* (1966) carried out metabolic balance studies during and after ammonium chloride acidosis. A summary of their results is presented in Table II. Again, an accumulation of acid was observed during the acidosis which was accompanied by losses of cations. Initially, sodium and potassium losses were substantial, but calcium balances were maintained. Subsequently, a change in the pattern of cation loss took place such that in the final acidotic period (13–18 days) calcium was the only cation lost. Phosphate losses occurred throughout the acidotic regimen. During recovery from metabolic acidosis, sodium and potassium losses were rapidly replaced, and plasma bicarbonate rapidly rose. Since calcium continued to be lost during the first recovery period, Lemann *et al.* (1966) suggested that continuing liberation of base from bone may have contributed to the restoration of intracellular and extracellular buffers. Nevertheless, notwithstanding the restoration of these buffer stores, 193 mEq of hydrogen ions had been retained at the end of the second recovery period. This accompanied a loss of 185 mEq of calcium and 79 mmole of phosphate.

Many other studies over the last half century have demonstrated the phenomenon of calcium loss during chronic metabolic acidosis. A number of studies have shown that correction of the acidosis attenuates the calcium loss. For example, the mild acidosis that occurs in prolonged starvation in obese women is accompanied by an average urinary loss of 190 mg calcium per day, and this was decreased to 90 mg per day when the acidosis was corrected by $NaHCO_3$ therapy (Reidenberg *et al.*, 1966). Litzow *et al.* (1967) found that correction of acidosis by $NaHCO_3$ therapy in patients with chronic uremic renal disease reduced both urinary and fecal calcium excretion so that the daily calcium balance became

Table II. Mean Changes in Some Parameters of Acid and Electrolyte Balance in Man during and after Ammonium Chloride Acidosis

Period[a]	NH_4Cl fed (mEq)	Blood pH	Blood $[HCO_3^-]$	Balances				
				Acid (mEq)	Na (mEq)	K (mEq)	Ca (mEq)	P (mmole)
I	733	7.39	26.5	+296	−100	−138	+13	−21
II	1369	7.31	19.9	+167	+35	−166	−54	−44
III	1402	7.30	18.8	+57	−3	−9	−94	−26
IV	0	7.31	19.3	−444	+162	+252	−19	+9
V	0	7.44	31.1	+117	−63	+17	+2	+3
Cumulative balance at end of recovery				+193[b]	+31	−44	−185[b]	−79[b]

[a]Each period was of 6 days duration. The balances are for each entire period. The blood pH and HCO_3^- are for the morning of the first day of each 6-day period. Data from Lemann *et al.* (1966).
[b]Difference significantly different from zero.

indistinguishable from zero. Cochran and Wilkinson (1975) also showed that correction of blood pH in patients with renal osteomalacia was associated with a significant increase in the rates of bone mineralization and an improvement in calcium balance.

The most striking alteration in bone during acidosis is a fall in carbonate. There are also losses of calcium and of sodium. Irving and Chute (1933) found a 6–13% loss of bone carbonate in rats and guinea pigs fed HCl for 1–4 weeks. Lesser losses of calcium were found. Pellegrino and Blitz (1965) also observed losses of calcium and of carbonate in bones from patients in uremia. The magnitude of the changes were proportional to the duration of the disease. These investigators felt that there is a separate labile $CaCO_3$ phase in bone that is involved in the buffering of an acid load. More recently, Kaye *et al.* (1970) have confirmed that a fall in bone carbonate takes place in chronic renal failure. However, they found this to be caused by a carbonate–phosphate interchange within the apatite crystal rather than by loss of a separate calcium carbonate phase. In support of this, they showed that synthetic apatites formed in bicarbonate-deficient media were identical to apatite crystals from uremic bone.

Lemann and Lennon (1972) have suggested that such carbonate–phosphate exchanges could provide for the continuing buffering of hydrogen ions by bone even in the absence of a negative calcium balance.

$$15Ca_{10}(PO_4)_6CO_3 + 8Na_2HPO_4 + 2NaH_2PO_4 + 3CO_2 + 3H_2O \rightarrow 50Ca_3(PO_4)_2 + 18NaHCO_3 \quad (5)$$

The bone changes that occur in acidosis of shorter duration are less obvious because of the smaller magnitude of the change. Nevertheless, in dogs made acidotic with ammonium chloride for 5–10 days, Burnell (1971) found substantial falls in bone carbonate (9.5%) and in bone sodium (6.3%). It was not determined whether the loss of carbonate was through exchange with phosphate.

4. Renal Handling of Hydrogen Ion and of Calcium during Metabolic Acidosis

4.1. Elimination of Hydrogen Ions

The general mechanisms by which the kidney excretes strong acid are now well understood (Pitts, 1974). The buffering of a strong acid in blood involves a loss of bicarbonate by virtue of its conversion to carbonic acid and subsequent volatilization as CO_2.

$$H^+A^- + Na^+HCO_3^- \rightarrow Na^+A^- + H_2CO_3 \quad (6)$$

The renal excretion of acid is essentially a process that reverses the above reaction so that $NaHCO_3$ is regenerated and leaves the kidney in the venous blood while the strong acid, HA, is excreted in the urine. Only trivial amounts of acid may be excreted as free protons since the kidney is unable to produce urine that is more than about 3 pH units more acid than blood (this imposes a lower limit on urine pH of about 4.0). Thus, the excretion of one liter of urine per day at pH 5.0 (a low urinary pH) effects the excretion of about 0.01 mEq of protons as free hydrogen ions, whereas the daily net acid production of man on a typical North American diet is about 30–100 mEq per day (Lennon *et al.*, 1966).

To accomplish the excretion of these larger quantities of acid within the limits imposed by the pH of the urine and the volume of fluid excreted, it is necessary to buffer the hydrogen ions. The kidney accomplishes this through the production of titratable acid and of ammonia. In the production of titratable acid, filtered dibasic phosphate accepts a proton and is converted to monobasic phosphate.

$$HPO_4^{2-} + H^+ \rightarrow H_2PO_4^- \qquad (7)$$

However, the quantity of acid excreted by this means is generally limited by the quantity of phosphate that is being excreted daily in the course of the renal adjustment of normal phosphate homeostasis.

Ammonia production, however, permits the kidney to produce the extra buffer required. Ammonia produced in the tubular cells diffuses into the tubular fluid where it accepts a proton and is excreted as ammonium.

$$NH_3 + H^+ \rightarrow NH_4^+ \qquad (8)$$

Ammonium excretion is quantitatively more important than titratable acid production in normal man (Pitts, 1974) and, furthermore, has the capacity to increase greatly in metabolic acidosis. Urinary ammonia is derived from the metabolism of glutamine within the tubular cells. A number of metabolic adaptations are known to take place in kidneys of acidotic animals. Thus, in acidotic rats there are increased renal activities of glutaminase (Rector *et al.*, 1955), glutamate dehydrogenase (Seyama *et al.*, 1977), and phosphoenolpyruvate carboxykinase (Alleyne and Scullard, 1969). In addition, kidneys from acidotic rats display an increased rate of gluconeogenesis *in vitro* (Goodman *et al.*, 1966), and isolated mitochondria from kidneys of acidotic rats produce ammonia from glutamine at a more rapid rate than do mitochondria from kidneys of normal rats (Brosnan and Hall, 1977). This latter effect has been attributed to an increased rate of glutamine entry into mitochondria. The changes in glutaminase and in glutamate dehydrogenase activities are clearly related to the increased metabolism of glutamine.

$$\text{glutamine} + H_2O \xrightarrow{\text{glutaminase}} \text{glutamate}^- + NH_4^+ \tag{9}$$

$$\text{glutamate}^- + NAD^+ \xrightarrow[\text{dehydrogenase}]{\text{glutamate}} \alpha\text{-ketoglutarate}^{2-} + NH_4^+ + (NADH + H^+) \tag{10}$$

The changes in the activity of phosphoenolpyruvate carboxykinase and in renal gluconeogenesis are also related to ammonia production, although less obviously so. They may be explained as follows. The reactions catalyzed by glutaminase and glutamate dehydrogenase do not produce ammonia but, rather, ammonium. Ammonium cannot serve as a urinary buffer since it cannot accept hydrogen ions. Thus, the renal metabolism of glutamine must proceed beyond α-ketoglutarate to a neutral end product. Only then will the nitrogenous end product be ammonia. The two most likely neutral end products are CO_2 (i.e., total oxidation of the glutamine carbon) and glucose. The equations describing these processes are as follows.

$$2\ (C_5H_{10}O_3N_2) + 2H_2O + 3O_2 \rightarrow C_6H_{12}O_6 + 4CO_2 + 4NH_3 \tag{11}$$

$$C_5H_{10}O_3N_2 + 4\tfrac{1}{2}O_2 \rightarrow 5CO_2 + 2H_2O + 2NH_3 \tag{12}$$

Since renal gluconeogenesis *in vitro* increases during metabolic acidosis in rats, it is probable that glutamine carbon is metabolized by this pathway *in vivo*. However, the increase in phosphoenolpyruvate carboxykinase would be equally compatible with total oxidation of glutamine to CO_2 (Brosnan *et al.*, 1978).

Essentially all of the metabolic adaptations found in kidneys from acidotic rats are also found in kidneys from rats fed a high-protein diet (Table III). This is clearly related to the need of these animals to excrete more acid than control animals. An important point is that these renal adaptations are not directly related to a detectable acidemia. The blood bicarbonate and pH in the animals fed the high-protein diet for 7 days were indistinguishable from those in the control animals. Thus, the kidney perceives the acid load and continues to excrete it in the absence of detectable alteration in either blood pH or bicarbonate (Brosnan *et al.*, 1978).

The metabolic adaptations decribed above in the acidotic rat are adaptations occurring in the cells of the proximal tubule, primarily of the proximal convoluted tubule. This segment of the nephron produces the bulk of the kidneys' ammonia (Glabman *et al.*, 1963). Other segments of the nephron are also important, and the final acidification of urine occurs in the distal tubule and collecting duct (Malnic *et al.*, 1972; Warnock and Rector, 1979).

4.2. Renal Handling of Calcium during Acidosis

The renal mechanisms involved in the hypercalciuria of metabolic acidosis have been intensively studied. There is some disagreement as to whether respiratory acidosis results in hypercalciuria (Stacy and Wilson, 1970). However, there

Table III. Metabolic Adaptations in Kidneys from NH_4Cl-Acidotic Rats and from Rats Fed a High-Protein Diet

Experimental group	Control	Acidotic	13% Casein	55% Casein
Urinary ammonia excretion (μmol/day per 100 g)	236	1348	294	897
Glutaminase (μmol/hr per 100 g)	1070	3020	1180	2080
Phosphoenolpyruvate carboxykinase (μmol/hr per 100 g)	230	1107	409	996
Mitochondrial ammonia production from 2 mM glutamine (μmol/hr per mg protein)	0.88	3.04	1.50	2.02
Glucose production from 2 mM glutamine by kidney slices (μmol/hr per g kidney)	6.26	15.32	6.39	9.06

[a]Data are taken from Brosnan *et al.* (1978) and were obtained on the seventh day of the various dietary regimens. The control animals ate a stock diet and drank water, whereas the acidotic group ate the same diet but drank 1.5% NH_4Cl. The 13% casein group ate a purified diet containing 13% casein as the sole protein source, whereas the 55% casein group ate an isocaloric diet containing 55% casein as sole protein source. All of the data from acidotic rats were significantly different from the control rats, and all of the data from the 55% casein rats were significantly different from the 13% casein group.

does appear to be good evidence for increased calcium excretion during acute respiratory acidosis in sheep (Stacy and Wilson, 1970) and in dogs (Williamson and Freeman, 1957). In chronic respiratory acidosis, on the other hand, reduced quantities of calcium appear in the urine. In their studies of human volunteers exposed to 1.5% CO_2 for 42 days, Schaefer *et al.* (1963) showed that urinary calcium excretion was about 55% of control even during the first 3 weeks of their protocol when a significant acidemia was present. Two important conclusions may be drawn from these studies. First, the effects of respiratory acidosis on the renal handling of calcium can be qualitatively different from the effects of metabolic acidosis. Second, a fall in blood pH is not necessarily accompanied by an increased urinary excretion of calcium.

That the hypercalciuria of metabolic acidosis has a renal component is demonstrated by the fact that it occurs in the face of a decrease in the filtered load of calcium in man (Lemann *et al.*, 1967) and dog (Sutton *et al.*, 1979). Thus, there is decreased tubular reabsorption of calcium. Furthermore, hypercalciuria of metabolic acidosis has been observed in hypoparathyroid patients (Lemann *et al.*, 1967) and in thyroparathyroidectomized rats (Reidenberg *et al.*, 1968). Sutton *et al.* (1979) have thoroughly examined the effect of metabolic acidosis on calcium transport in the dog kidney. They found no hypercalciuria in dogs made acutely acidotic by infusion of HCl even though blood pH decreased to 7.13, again indicating that there is no necessary relationship between acidemia and hypercalciuria. Acute acidosis did result in altered calcium handling within the kidney since there was a decreased proximal fluid reabsorption with an augmented delivery of calcium and sodium to the distal tubule; however, the distal tubular reabsorption of these ions was increased so that no hypercalciuria occurred. Hypercalciuria does occur during acute metabolic acidosis in the rat (Beck and Webster, 1976). In chronic NH_4Cl-induced metabolic acidosis in the dog, there was actually an increase in the proximal reabsorption of both calcium and sodium (Sutton *et al.*, 1979). However, there was a depressed reabsorption of calcium beyond the proximal tubule, resulting in a hypercalciuria. Acute correction of the chronic acidosis by infusion of sodium bicarbonate increased the reabsorption of calcium by the distal tubule in both intact and hypoparathyroidectomized dogs. Thus, there is a component of calcium reabsorption situated in the distal nephron that is inhibited by chronic acidosis and is enhanced by metabolic alkalosis independently of parathyroid hormone. The mechanism by which these alterations in calcium reabsorption are brought about is not known.

Borle (1978) and Studer and Borle (1979) have shown that acidosis markedly influences calcium transport and distribution in isolated renal cells *in vitro*. The principal effects of acidosis are a decrease in the total cell calcium and a decreased influx of calcium into the cell. However, it is difficult to explain the effects of metabolic acidosis *in vivo* in terms of these effects of acidosis *in vitro*, since respiratory and metabolic acidosis had similar effects and since the cellular origin of the renal cells is uncertain. Lemann *et al.* (1979) argue that a critical factor may be the delivery of bicarbonate to the distal tubule. Urinary calcium excretion can increase with increments in fixed acid production too small to produce detectable changes in blood acid–base parameters. Lemann *et al.* (1979) suggest that very slight changes in blood bicarbonate may be magnified in the distal tubule and produce an effect on calcium reabsorption at this level. In support of this thesis they note that patients with proximal-tubule acidosis show neither hypercalciuria nor bone disease whereas patients with distal-tubule acidosis often have hypercalciuria. The former group of patients enjoy normal or enhanced delivery of bicarbonate to the distal tubule, whereas the latter group may have a reduced delivery of bicarbonate to the distal tubule. It is unlikely, however, that delivery of bicarbonate to the distal tubule can be the only regulatory factor involved, since acute metabolic acidosis does not provoke hypercalciuria in dogs (Sutton *et al.*, 1979) or in man (Coe *et al.*, 1975) and since, during ammonium chloride-induced acidosis in man, restriction of sodium intake abolished the hypercalciuria even though the degree of acidemia became more pronounced (Coe *et al.*, 1975).

5. Protein Intake and Calcium Balance

Osteoporosis, defined as an absolute loss of bone substance, is markedly prevalent in the Western Hemisphere, especially in postmenopausal women. The high protein content of the Western diet has been implicated as a factor in its etiology. It has been suggested by Wachman and Bernstein (1968) that the calciuric effects of high-protein diets are secondary to the acid produced during their metabolism. These acids were held to be constantly buffered by bone, and "the increased incidence of osteoporosis with age may represent, in part, the result of a life-long utilization of the buffering capacity of the basic salts of bone for the constant assault against pH homeostasis" (Wachman and Bernstein, 1968). Inherent in this proposal is the assumption that the kidneys' role in acid-base homeostasis is imperfect, that is, that the kidneys are unable to fully excrete the extra fixed acid produced as a result of the ingestion of a high-protein diet. Wachman and Bernstein's hypothesis has stimulated much work both in man and in experimental animals. The most favored experimental animal has been the rat, although there are a number of important differences in calcium metabolism between man and rat. Man may excrete via the urine almost half of a normal dietary calcium intake, whereas the rat excretes a much smaller fraction. In addition, rat long bone differs from human long bone in a number of significant ways such the absence of a Haversian system and in the continued epiphyseal growth throughout life.

In 1972, Ellis *et al.* reported that the density of bones of the hand was greater in vegetarians than in omnivores and attributed the difference to the alkaline ash of the vegetarian diet. However, a subsequent study by the same group failed to substantiate the original finding (Ellis *et al.*, 1974). Bone mineral content has also been directly studied in North Alaskan Eskimos (Mazess and Mather, 1974), and it has been shown that in Eskimos over 40 years of age, there is a deficit of 10 to 15% of bone mineral relative to United State whites of similar age. Mazess and Mather (1974) considered the acidic effect of a meat diet to be important in causing this bone loss. Draper and Bell (1979) have pointed out that the Canadian Eskimo diet is rich in phosphorus and low in calcium, that blood phosphorous is elevated and calcium decreased compared to the Canadian population as a whole, and that the bone loss may result from increased PTH activity consequent upon these alterations.

5.1. Acidosis and Osteoporosis

The proposition that prolonged ingestion of acid by experimental animals can produce osteoporosis is also disputed. Jaffe *et al.* (1932) reported that osteoporosis could be caused in adult dogs by either feeding a low-calcium diet or by the prolonged ingestion of ammonium chloride. When ammonium chloride and a low-calcium diet were fed simultaneously, the osteoporosis was more pronounced. The situation in rats is controversial, however. Barzel (1969) and Barzel and Jowsey (1969) reported ammonium chloride feeding to cause osteoporosis in rats. Barzel (1969) substituted 2% NH_4Cl for drinking water for 6 months and then examined the composition of the right femora of these rats.

There were significant decreases in bone density and in bone calcium content in the acidotic rats. A very low-calcium diet (20 mg/100 g diet) caused similar change, and these were even more pronounced when both ammonium chloride and a low-calcium diet were fed. Histological evidence of increased bone resorption was also present in the acidotic rats. When rats were fed a smaller amount of ammonium chloride (1.5% NH_4Cl in lieu of drinking water), similar though smaller decreases in bone calcium and bone ash were evident (Barzel and Jowsey, 1969).

Draper and Bell (1979) have critized some of these conclusions. In particular, they point out that in the study of Barzel and Jowsey (1969), the NH_4Cl-fed rats weighed less than their controls (because of depressed food intake) and that when bone calcium content is expressed as a percent of final body weight, the significant difference in bone calcium content may disappear. However, Barzel (1975) showed distinct decreases in bone density, fat-free weight, ash content, and calcium content in the femora of female rats that drank 1.5% NH_4Cl for 300 days. These changes were not attributable to differences in weight. Delling and Dornath (1973) have also concluded from microscopic examination that chronic ammonium-chloride-induced acidosis mimics the effect of parathyroid hormone on bone resorption, although PTH does not mediate the effect since similar changes were evident in parathyroidectomized acidotic rats. In a recent study, however, Newell and Beauchene (1975) found no effect of acid feeding (2 g NH_4Cl/100 g diet for 9 months) on bone composition in rats. This lack of effect of acidosis was observed in old and in young rats and in rats fed on two levels of dietary calcium. However, it should be noted that feeding NH_4Cl as 2% of diet results in an intake of acid only about one-third to one-half of that ingested when given as a 2% solution in drinking water. Upton and L'Estrange (1977) have also failed to observe any change in femur composition in rats fed substantial quantities of HCl (the highest intake of 900 mmol per kg of diet dry matter amounts to approximately 1.5 to 2.0 mmol/day per 100 g body weight which is slightly less than the acid intake of rats fed 1.5% NH_4Cl in their drinking water). However, the experiments of Upton and L'Estrange (1977) were carried out for only 9 to 12 weeks, a period that may be insufficient to produce discernable changes in bone composition. In summary, it seems that acid feeding does produce osteoporosis in rats provided that very high quantities are fed for quite prolonged periods.

5.2. High-Protein Diets and Calcium Balance

The calciuric effect of dietary protein has been investigated by a number of groups. Bell *et al.* (1975) studied the effects of 10, 20, and 40% protein on calcium excretion by rats "deep-labeled" with ^{45}Ca by administration of the isotope 1 month before the experiments. The excretion of ^{45}Ca in the urine and feces, therefore served as a measure of bone resorption. Their experiments over a period of 100 days did show an increased urinary excretion of ^{45}Ca in rats fed the 40% protein diet, but this was accompanied by a decreased fecal excretion of ^{45}Ca so that the total ^{45}Ca excreted, and hence the rate of bone resorption, was unaffected by the increase in dietary protein. Rather, the effect was of a shift in the pattern of excretion from feces to urine.

A study by Allen and Hall (1978) also showed no sustained effect of high-protein diets on calcium balance in rats. These workers compared the effects of an 18% casein diet to those of a 36% casein diet, both on calcium balance and on the characteristics of the rapidly exchanging and slowly exchanging calcium pools. These studies confirmed that feeding the high-protein diet did cause a calciuria, but not a sustained one. Increased urinary calcium excretion was still evident after 2 weeks on the high-protein diet but not after 4 weeks. There were no changes in apparent calcium absorption. Similarly the sizes of the two kinetically identifiable calcium pools were unchanged. Overall calcium retention was not affected throughout the experiment, even when increased quantities of calcium were excreted in the urine because, in these experiments, urinary excretion was found to be an extremely minor (1–2%) component of the total calcium excreted.

In a recent study, Whiting and Draper (1980) compared the calciuric effects of different types of protein. The degree of hypercalciuria seemed to be related to the methionine + cysteine content of the proteins since the order of urinary calcium excretion (lactalbumin > egg white > casein > gelatin) was the same as the order of sulfur-containing amino acid content. This study also demonstrated an attenuation with time of the calciuria. There was a rapid fall in urinary calcium excretion over the first few weeks such that after 4 weeks there was no significant hypercalciuria in the rats fed extra protein in the form of gelatin or casein. These results, therefore, account for the discrepancy between the mild, transient calciuria reported by Allen and Hall (1978) and the marked and more persistent calciuria reported by Bell *et al.* (1975).

Whiting and Draper also directly examined the role of sulfur-containing amino acids by adding methionine and cystine to the control (18% casein) diet. The additional methionine + cystine caused an increased urinary calcium excretion. In a subsequent experiment they added sulfate to the 18% casein diet and found a marked calciuric effect. This suggested that the calciuric effect of sulfur-containing amino acids is caused by sulfate per se. Walser and Browder (1959) had earlier showed that sulfate ions can form a calcium complex that is not readily reabsorbed. However, the possibility of an acid–base effect cannot be ruled out in the experiments of Whiting and Draper (1980) since sulfate was added to their diets as $CaSO_4$ and $MgSO_4$ in lieu of appropriate amounts of $CaCO_3$ and $MgCO_3$. Thus, neutral salts replaced alkaline salts so that the diet became more acidic.

A number of studies have dealt with the effects of high-protein diets on calcium balance in humans. Sherman (1920) showed that the addition of meat to the diet caused an increased urinary excretion of calcium. McCance *et al.* (1942) showed a calciuretic effect by supplementing the diet with peptone, gluten, gelatin, or egg white. The calciuretic effect of high-protein diets in man has occasionally been attributed to an enhanced absorption of calcium. However, although several older studies showed a beneficial effect of increased protein intake on calcium absorption (Adolph and Chen, 1932; Kunerth and Pittman, 1939; McCance *et al.*, 1942), it has been pointed out by Margen *et al.* (1974) and Chu *et al.* (1975) that these early studies involved modest protein supplementation of diets that supplied a rather low basal protein intake and that no further improvement in calcium absorption occurs at higher protein intakes. In a series of studies from Linkswiler's laboratory (Johnson *et al.*, 1970; Walker and Linkswiler, 1972; Anand and Linkswiler, 1974) summarized by Linkswiler *et al.* (1974), the interactions among protein intake, calcium intake, and calcium balance in young adult males were investigated. In these studies, diets containing 500, 800, or 1400 mg of calcium per day were studied at three levels of dietary protein: 47 g, 95 g, and 142 g per day. Magnesium intake was 400 mg per day (in one study, 490 mg per day), and the phosphorus intakes were equal to or slightly higher than the calcium intake. The increased protein intakes were supplied by casein, lactalbumin, gluten, and gelatin. The subjects consumed each diet for a 15-day period. The subjects were always in calcium balance at the low (47 g/day) protein intake. On the 95 g/day protein diet, however, subjects were in negative calcium balance when consuming 500 mg calcium per day but were in balance when consuming 800 mg per day or 1400 mg per day. On the highest protein intake (142 g/day), negative calcium balance was evident at all calcium intakes.

The reason for these negative calcium balances was a greatly increased urinary excretion of calcium at the higher protein intakes. Indeed, increased protein intake stimulated the apparent absorption of calcium, the maximal effect already being reached at the intake of 95 g. However, the calciuria more than offset the increased absorption, and negative balances ensued. Schwartz *et al.* (1973), in a study in which young boys consumed diets containing either 43 or 93 g protein/day at a calcium intake of about 30 mg/kg body weight per day, found no reduction in overall calcium balance in the group fed the high-protein diet, although increased urinary calcium excretion was evident. A new steady state in urinary calcium was not attained in the 30-day experimental periods, and Schwartz *et al.* pointed out that considerably longer periods are required before calcium equilibrium becomes reestablished after a dietary change.

A series of experiments by Margen and associates are in general agreement with those of Linkswiler's group. Healthy young males were fed a formula-type diet with variable protein content for 15-day periods (Margen *et al.*, 1974). Increased calciuria occurring within 24 hr and complete within 72 hr was evident whenever protein intake was increased. Calciuria was maintained for the full 15 days of the study; i.e., there was no tendency to adapt.

Margen *et al.* (1974) also evaluated the role of individual amino acids by feeding diets in which protein was replaced by different mixtures of crystalline amino acids. A similar degree of calciuria was observed in diets containing quite different quantities of sulfur-containing amino acids. However, it is not possible to conclude from these experiments that diet acidity is not an important factor in the calciuria of protein feeding since it is not possible to calculate the overall acid or base residue of the various amino acid mixtures employed. A subsequent study from the same group (Chu *et al.*, 1975) examined the effects of variable protein intake on the parameters of calcium balance in young males consuming low quantities of calcium (100 mg/day) over 15-day periods. Although calcium intake was very low, every subject excreted more calcium in the urine when protein intake was increased. This finding argues against an important role for increased calcium absorption in determining the calciuria.

In a longer term study, Allen *et al.* (1979b) examined the effects of protein intake on calcium balance in subjects ingesting 1400 mg calcium/day. Young male subjects were fed formula diets containing either 12 or 36 g N per day for 48 days, whereupon the diets were changed such that those consuming the 12-g N diet were fed the 36-g N diet for 48 days and vice versa. Again, a pronounced calciuria was evident in the subjects fed the high-protein diet, and no tendency to adapt was evident; i.e., there was no tendency for urine calcium to decline even after 48 days of feeding the high-protein diet. All of the subjects on the high-protein diet displayed a negative calcium balance, averaging 137 mg/day. Allen *et al.* (1979b) pointed out that this degree of calcium loss would amount to a loss of approximately 50 g per year which amounts to about 4% of the total skeletal calcium per year. Curiously, no increase in hydroxyproline excretion was observed in the subjects fed the high-protein diets. Thus, increased bone resorption has not been proven in these experiments, and it must be considered that the calcium loss may have another origin such as, perhaps, the pools of amorphous calcium carbonate and calcium phosphate.

The calciuria observed in subjects ingesting high-protein diets appears to be caused by a decreased tubular reabsorption of calcium similar to that observed during metabolic acidosis. Allen *et al.* (1979a) during short-term experiments and Kim and Linkswiler (1979) in more prolonged experiments demonstrated significantly decreased fractional tubular reabsorption of calcium by subjects fed high-protein diets. An additional contributory factor during the prolonged experiments was the increased delivery of calcium to the tubules consequent to an increased glomerular filtration rate. The level of circulating parathyroid hormone (Allen *et al.*, 1979b; Kim and Linkswiler, 1979; Adams *et al.*, 1979) and of 1,25-dihydroxycholecalciferol (Adams *et al.*, 1979) do not change during the consumption of a high-protein diet.

In contrast to many of the above studies, a recent investigation by Spencer *et al.* (1978) found no effect of an elevated protein intake on calcium balance. This study differed from most previous studies in three important respects. First, the elevated protein intake was achieved by the ingestion of meat rather than of purified proteins; second, the subjects were considerably older (mean age of 57) than in the other studies; third, the study was continued for longer time periods (for 72 days in some cases). These differences tend to make the study of Spencer *et al.* 1978) more relevant than other studies to the hypothesis that the Western diet, high in meat, plays a role in the development of osteoporosis. The effect of adequate protein intake (1 g/kg body weight per day) was compared to that of a high protein intake (2 g/kg body weight per day) at several levels of dietary calcium (0.2 to 2 g per day). No increase in urinary calcium was found when subjects ingested the high-meat diet together with a low (0.2 g) or normal (0.8 g) calcium intake. Overall calcium balance was also unaffected by the increased meat intake. Increased urinary excretion of calcium was evident in occasional subjects on higher calcium intakes, but it always returned to control or lower levels with time. Spencer *et al.* (1978) attribute the differences between their results and those of previous workers to the higher phosphorus content of the high-meat intake. Whether or not this is the explanation remains to be determined.

6. Conclusions

Prolonged acidosis, either from the continuous ingestion of large quantities of acid or from chronic renal disease, results in substantial skeletal demineralization. During experimental metabolic acidosis, the renal loss is partly caused by a decreased tubular reabsorption of calcium. However, neither this renal phenomenon nor the increased bone resorption is readily explicable on the basis of changes in circulating levels of calcium-regulating hormones.

Ingestion of a high-protein diet by man or rats results in a hypercalciuria in most studies. However, in the rat, urinary calcium excretion is a minor component of total excretion, and negative calcium balances have not been demonstrated. In man, the results are less clear-cut because of differences in the experimental subjects, differences in the amount and type of protein fed, and difficulties in carrying out such studies for a sufficiently long period. In some human studies, no effect on overall calcium balance could be found on ingestion of a high-protein diet, and such hypercalciuria as occurred was transient. In other studies, negative calcium balances were found, but since there was no increase in urinary excretion of hydroxyproline, it is not likely that increased bone resorption was taking place. The calcium was probably derived by mobilization of labile stores. Thus, extrapolation from these losses to the amount of bone loss that could occur over a number of years may not be justified.

This is not to deny that acid produced from the ingestion of a high-protein diet may play a role in the development of osteoporosis. However, there is only a need for continuous internal buffering of fixed acid when the quantity ingested is large relative to the kidneys' ability to excrete the acid load. In general, the quantity of acid produced by the metabolism of the high-protein diets is not as large as that which can be excreted after ingestion of NH_4Cl by humans (e.g., Fig. 5 of Lemann *et al.*, 1979), so it is unlikely that renal capacity to excrete acid is normally an important factor. However, it is known that renal ability to excrete an acute acid load decreases substantially during aging (Adler *et al.*, 1968). This decrease is parallel to the decrease in GFR and thus is attributed to decreased tubular mass rather than to a specific tubular defect. It is conceivable that such a phenomenon, if sufficiently pronounced, might result in acid retention in some older people and thus play a role in the development of osteoporosis.

ACKNOWLEDGMENTS

The authors thank Drs. J. Dirks, H. Draper, and J. Lemann for providing preprints of unpublished papers and Dr. G. F. Herzberg for reading the manuscript.

References

Adams, N. D., Gray, R. W., and Lemann, J., Jr., 1979, The calciuria of increased fixed acid production in humans: Evidence against a role for parathyroid hormone and 1,25$(OH)_2$-vitamin D, *Calcif. Tissue Int.* **28**:233.

Adler, S., Lindeman, R. D., Yiengst, M. J., Beard, E., and Schock, N. W., 1968, Effect of acute acid loading on urinary acid excretion by the aging human kidney, *J. Lab. Clin. Med.* **72**:278.

Adolph, W. H., and Chen, S. C., 1932, The utilization of calcium in soya bean diets, *J. Nutr.* **5**:379.

Albright, F., and Reifenstein, E. C., 1948, *The Parathyroid Glands and Metabolic Bone Disease*, Williams & Wilkins, Baltimore.

Allen, L. H., and Hall, T. E., 1978. Calcium metabolism, intestinal calcium-binding protein, and bone growth of rats fed high protein diets, *J. Nutr.* **108**:967.

Allen, L. H., Bartlett, R. S., and Block, G. D., 1979a, Reduction of renal calcium in man by consumption of dietary protein, *J. Nutr.* **109**:1345.

Allen, L. H., Oddoye, E. A., and Margen, S., 1979b, Protein-induced hypercalciuria: A longer term study, *Am. J. Clin. Nutr.* **32**:741.

Alleyne, G. A. O., and Scullard, G. H., 1969, Renal metabolic response to acid-base changes. I. Enzymatic control of ammoniagenesis in the rat, *J. Clin. Invest.* **48**:364.

Anand, C. R., and Linkswiler, H. M., 1974, Effect of protein intake on calcium balance of young men given 500 mg calcium daily, *J. Nutr.* **104**:695.

Barzel, U. S., 1969, The effect of excessive acid feeding on bone, *Calcif. Tissue Res.* **4**:94.

Barzel, U. S., 1975, Studies in osteoporosis: The long-term effect of oophorectomy and of ammonium chloride ingestion on the bone of mature rats, *Endocrinology* **96**:1304.

Barzel, U. S., and Jowsey, J., 1969, The effects of chronic acid and alkali administration on bone turnover in adult rats, *Clin Sci.* **36**:517.

Beck, N., and Webster, S. H., 1976, Effects of acute metabolic acidosis on parathyroid hormone action and calcium mobilization, *Am. J. Physiol.* **230**:127.

Bell, R. R., Engelman, D. T., Sie, T. L., and Draper, H. H., 1975, Effect of a high protein intake on calcium metabolism in the rat, *J. Nutr.* **105**:475.

Bergstrom W. H., and Wallace, W. M., 1954, Bone as a sodium and potassium reservoir, *J. Clin. Invest.* **33**:867.

Bernard, C., 1865, *An Introduction to the Study of Experimental Medicine,* translated by H. C. Green, p. 152–153, Dover Publications, New York, 1957.

Borle, A. B., 1973, Calcium metabolism at the cellular level, *Fed. Proc.* **32**:1944

Borle, A. B., 1978, Renal handling of calcium, *Fed. Proc.* **37**:2112.

Brosnan, J. T., and Hall, B., 1977, The transport and metabolism of glutamine by kidney cortex mitochondria from normal and acidotic rats, *Biochem. J.* **164**:331.

Brosnan, J. T., McPhee, P., Hall, B., and Parry, D. M., 1978, Renal glutamine metabolism in rats fed high-protein diets, *Am. J. Physiol.* **235**:E261.

Brumbaugh, P. F., Hughes, M. R., and Haussler, M. R., 1975, Cytoplasmic and nuclear binding components for 1α,25-dihydroxy vitamin D_3 in chick parathyroid glands, *Proc. Natl. Acad. Sci. U.S.A.* **72**:4871.

Burnell, J. M., 1971, Changes in bone sodium and carbonate in metabolic acidosis and alkalosis in the dog, *J. Clin. Invest.* **50**:327.

Chu, J. -Y., Margen, S., and Costa, F. M., 1975, Studies in calcium metabolism. II. Effects of low calcium and variable protein intake on human calcium metabolism, *Am. J. Clin. Nutr.* **28**: 1028.

Cochran, M., and Wilkinson, R., 1975, Effect of correction of metabolic acidosis on bone mineralisation rates in patients with renal osteomalacia, *Nephron* **15**:98.

Coe, F. L., Firpo, J. J., Jr., Hollandsworth, D. L., Segil, L., Canterbury, J. M., and Reiss, E., 1975, Effect of acute and chronic metabolic acidosis on serum immunoreactive parathyroid hormone in man, *Kidney Int.* **8**:262.

Cooper, C. W., Bolma, R. M. III, Linehan, W. M., and Wells, S. A., Jr., 1978, Interrelationships between calcium, calcemic hormones and gastrointestinal hormones, *Recent Prog. Horm. Res.* **34**:259.

Delling, G., and Dornath, K., 1973, Morphometrische, elektronenmikroskopische und physikelisch-chemisch Untersuchungen uber die experimentell Osteoporose bei chronischer Acidose. *Virchows Arch. Pathol Anat.* **358**:321.

DeLuca, H. F., 1979, The vitamin D system in the regulation of calcium and phosphorus metabolism, *Nutr. Rev.* **37**:161.

DeSousa, R. C., Harrington, J. T., Ricanati, E. S., Shelkrot, J. W., and Schwartz, W. B., 1974, Renal regulation of acid-base equilibrium during chronic administration of mineral acid, *J. Clin. Invest.* **53**:465.

Draper, H. H., and Bell, R. R., 1979, Nutrition and osteoporosis, in: *Advances in Nutritional Research,* Vol. 2 (H. H. Draper, ed.), pp. 79–106, Plenum Press, New York.

Ellis, F. R., Holesh, S., and Ellis, J. W., 1972, Incidence of osteoporosis in vegetarians and omnivores, *Am. J. Clin. Nutr.* **25**:555.

Ellis, F. R., Holesh, S., and Sanders, T. A. B., 1974, Osteoporosis in British vegetarians and omnivores, *Am. J. Clin. Nutr.* **27**:769.

Epstein, F., 1968, Calcium and the kidney, *Am. J. Med.* **45**:700.

Erecinska, M., and Wilson, D. F., 1978, Homeostatic regulation of cellular energy metabolism, *Trends Biochem. Sci.* **3**:219.

Fraley, D. S., and Adler, S., 1979, An extrarenal role for parathyroid hormone in the disposal of acute acid loads in rats and dogs, *J. Clin. Invest.* **63**:985.

Garabedian, M., Holick, M. F., DeLuca, H. F., and Boyle, I. T., 1972, Control of 25-hydroxycholecalciferol metabolism by parathyroid glands, *Proc. Natl. Acad. Sci. U.S.A.* **69**:1673.

Glabman, S., Klose, R. M., and Giebisch, G., 1963, Micropuncture study of ammonia excretion in the rat, *Am. J. Physiol.* **205**:127.

Goodman, A. D., Lemann, J., Lennon, E. J., and Relman, A. S., 1965, Production, excretion and net balance of fixed acid in patients with renal acidosis, *J. Clin. Invest.* **44**:495.

Goodman, A. D., Fuisz, R. F., and Cahill, G. F., Jr., 1966, Renal gluconeogenesis in acidosis, alkalosis, and potassium deficiency: Its possible role in regulation of renal ammonia production, *J. Clin. Invest.* **45**:612.

Haldane, J. B. S., 1921, Experiments on the regulation of the blood's alkalinity. II. *J. Physiol. (Lond.)* **55**:265.

Ham, A. W., 1974, *Histology*, J. B. Lippincott, Toronto.

Haussler, M. R., and McCain, T. A., 1977, Basic and clinical concepts related to vitamin D metabolism and action, *N. Engl. J. Med.* **297**:974.

Hughes, M. R., Brumbaugh, P. F., Hausser, M. R., Wergedal, J. E., and Baylink, D. J., 1975, Regulation of serum 1α,25-dihydroxy vitamin D_3 by calcium and phosphate in the rat, *Science* **190**:578.

Hunt, J. N., 1956, The influence of dietary sulfur on the urinary output of acid in man, *Clin. Sci.* **5**:119.

Irving, L., and Chute, A. L., 1933, The participation of the carbonates of bone in the neutralization of ingested acid, *J. Cell. Comp. Physiol.* **2**:157.

Jaffe, H. L., Bodansky, A., and Chandler, J. P., 1932, Ammonium chloride acidification, as modified by calcium intake: The relation between generalized osteoporosis and ostitis fibrosa, *J. Exp. Med.* **56**:823.

Johnson, N. E., Alcantera, E. N., and Linkswiler, H., 1970, Effect of level of protein intake on urinary and fecal calcium and calcium retention of young adult males, *J. Nutr.* **100**:1425.

Kaplan, E. L., Hill, B. J., Locke, S., Toth, D. N., and Peskin, G. W., 1971, Metabolic acidosis and parathyroid hormone secretion in sheep, *J. Lab. Clin. Med.* **78**:819.

Kaye, M., Frueh, A. J., and Silverman, M., 1970, A study of vertebral bone powder from patients with chronic renal failure, *J. Clin. Invest.* **49**:442.

Kildeberg, P., Engel, K., and Winters, R. W., 1969, Balance of net acid in growing infants. Endogenous and transintestinal aspects. *Acta Pediatr. Scand.* **58**:321.

Kim, Y., and Linkswiler, H. M., 1979, Effect of protein intake on calcium metabolism and on parathyroid and renal function in the adult human male, *J. Nutr.* **109**:1399.

Krebs, H. A., Woods, H. F., and Alberti, K. G. M. M., 1975, Hyperlactatemia and lactic acidosis, *Essays Med. Biochem.* **1**:81.

Kunerth, B. L., and Pittman, M. S., 1939, A long-time study of nitrogen, calcium and phosphorus metabolism on a low-protein diet, *J. Nutr.* **17**:161.

Lemann, J., Jr., and Lennon, E. J., 1972, Role of diet, gastrointestinal tract and bone in acid-base homeostasis, *Kidney Int.* **1**:275.

Lemann, J., Jr., and Relman, A. S., 1959, The relationship of sulfur metabolism to acid-base balance and of electrolyte excretion: The effects of DL-methionine in normal man, *J. Clin. Invest.* **38**:2215.

Lemann, J., Lennon, E. J., Goodman, A. D., Litzow, J. R., and Relman, A. S., 1965, The net balance of acid in subjects given large loads of acid or alkali, *J. Clin. Invest.* **44**:507.

Lemann, J., Jr., Litzow, J. R., and Lennon, E. J., 1966, The effects of chronic acid loads in normal man: Further evidence for the participation of bone mineral in the defense against chronic metabolic acidosis, *J. Clin. Invest.* **45**:1608.

Lemann, J., Jr., Litzow, J. R., and Lennon, E. J., 1967, Studies of the mechanism by which chronic metabolic acidosis augments urinary calcium excretion in man, *J. Clin. Invest.* **46**:1318.

Lemann, J., Jr., Adams, N. D., and Gray, R. W., 1979, Urinary calcium excretion in human beings, *N. Engl. J. Med.* **301**:535.

Lennon, E. J., Lemann, J., Jr., and Relman, A. S., 1962, The effects of phosphoproteins on acid balance in normal subjects, *J. Clin. Invest.* **41**:637.

Lennon, E. J., Lemann, J., Jr., and Litzow, J. R., 1966, The effects of diet and stool composition on the net external acid balance of normal subjects, *J. Clin. Invest.* **45**:1601.

Linkswiler, H. M., Joyce, C. L., and Anand, C. R., 1974, Calcium retention of young adult males as affected by level of protein and of calcium intake, *Trans, N.Y. Acad. Sci. (Ser. II)* **36**:333.

Litzow, J. R., Lemann, J., and Lennon, E. J., 1967, The effect of treatment of acidosis on calcium balance in patients with chronic azotaemic renal disease, *J. Clin. Invest.* **46**:280.

Malnic G., deMello Aires, M., and Giebisch G., 1972, Micropuncture study of renal tubular hydrogen ion transport in the rat, *Am. J. Physiol.* **222**:147.

Margen, S., Chu, J. -Y., Kaufman, N. A., and Calloway, D. H., 1974, Studies in calcium metabolism. I. The calciuretic effect of dietary protein, *Am. J. Clin. Nutr.* **27**:584.

Mazess, R. B., and Mather, W., 1974, Bone mineral content of North Alaskan Eskimos, *Am. J. Clin. Nutr.* **27**:916.

McCance, R. A., Widdowson, E. M., and Lehmann, H., 1942, The effect of protein intake on the absorption of calcium and magnesium, *Biochem J.* **36**:686.

Newell, G. K., and Beauchene, R. E., 1975, Effects of dietary calcium level, acid stress, and age on renal, serum, and bone responses of rats, *J. Nutr.* **105**:1039.

Nordin, B. E. C, Hodgekinson, A., and Peacock, M., 1967, The measurement and meaning of urinary calcium, *Clin. Orthop.* **52**:293.

O'Riordan, J. L. H., and Aurbach, G. D., 1968, Mode of action of thyrocalcitonin, *Endocrinology* **82**:377.

Pellegrino, E. D., and Biltz, R. M., 1965, The composition of human bone in uremia, *Medicine (Baltimore)* **44**:397.

Pitts, R. F., 1974, *Physiology of the Kidney and Body Fluids,* Third Edition, p. 198–241, Year Book Medical Publishers, Chicago.

Posner, A. S., 1973, Bone mineral on the molecular level, *Fed. Proc.* **32**:1933.

Rector, F. C., Jr., Seldin, D. W., and Copenhaver, J. H., 1955, The mechanism of ammonia excretion during chronic ammonium chloride acidosis, *J. Clin. Invest.* **34**:20.

Reidenberg, M. M., Haag, B. L., Channick, B. J., Shuman, C. R., and Wilson, T. G. G., 1966, The response of bone to metabolic acidosis in man, *Metabolism* **15**:236.

Reidenberg, M. M., Sevy, R. W., and Cucinotta, A. J., 1968, Hypercalciuria during acidosis in hypoparathyroidism, *Proc. Soc. Exp. Biol. Med.* **127**:1.

Relman, A. S., Lennon, E. J., and Lemann, J. Jr., 1961, Endogenous production of fixed acid and the measurements of the net balance of acid in normal subjects, *J. Clin. Invest,* **40**:1621.

Schaefer, K. E., Nichols, G., Jr., and Cerey, C. R., 1963, Calcium phosphorous metabolism in man during acclimitization to carbon dioxide, *J. Appl. Physiol.* **18**:1079.

Schwartz, R., Woodcock, N. A., Blakely, J. D., and MacKeller, I., 1973, Metabolic response of adolescent boys to two levels of dietary magnesium and protein. II. Effect of magnesium and protein level on calcium balance, *Am. J. Clin. Nutr.* **26**:519.

Schwartz, W. B., Jenson, R. L., and Relman, A. S., 1954, The disposition of acid administered to sodium-depleted subjects: The renal response and the role of the whole body buffers, *J. Clin. Invest.* **33**:587.

Schwartz, W. B., Orning, K. J., and Porter, R., 1957, The internal distribution of hydrogen ions with varying degrees of metabolic acidosis, *J. Clin. Invest.* **36**:373.

Seyama, S., Iijima, S., and Katunuma, N., 1977, Biochemical and histocytochemical studies on response of ammonia-producing enzymes for NH_4Cl-induced acidosis, *J. Histochem. Cytochem.* **25**:448.

Sherman, H. C., 1920, Calcium requirement in man, *J. Biol. Chem.* **44**:21.

Sherman, H. C., and Gettler, A. O., 1912, The balance of acid-forming and base-forming elements in foods, and its relation to ammonia metabolism, *J. Biol. Chem.* **11**:323.

Spencer, H., Kramer, L. Osis, D., and Norris, C., 1978, Effect of a high protein (meat) intake on calcium metabolism in man, *Am. J. Clin. Nutr.* **31**:2167.

Stacy, B. D., and Wilson, B. W., 1970, Acidosis and hypercalciuria: Renal mechanisms affecting calcium, magnesium and sodium excretion in the sheep, *J. Physiol (Lond.)* **210**:549.

Studer, R. K., and Borle, A. B., 1979, Effect of pH on the calcium metabolism of isolated rat kidney cells, *J. Memb. Biol.* **48**:325.

Sutton, R. A. L., and Dirks, J. H., 1977, Renal handling of calcium: Overview, *Adv. Exp. Med. Biol.* **81**:15.

Sutton, R. A. L., Wong, N. L. M., and Dirks, J. H., 1979, Effects of metabolic acidosis and alkalosis on sodium and calcium transport in the dog kidney, *Kidney Int.* **15**:520.

Swan, R. C., and Pitts, R. F., 1955, Neutralization of infused acid by nephrectomized dogs, *J. Clin. Invest.* **34**:205.

Tanaka, Y., and DeLuca, H. F., 1973, The control of 25-hydroxy vitamin D metabolism by inorganic phosphorus, *Arch, Biochem. Biophys.* **154**:566.

Termine, J. D., and Posner, A. S., 1966, Infrared analyses of rat bone: Age dependency of amorphous and crystalline mineral fractions, *Science* **153**:1523.

Upton, P. K., and L'Estrange, J. L., 1977, Effects of chronic hydrochloric and lactic acid administrations on food intake, blood acid–base balance and bone composition of the rat, *Q. J. Exp. Physiol.* **62**:223.

Van Slyke, D. D., and Cullen, G. E., 1917, Studies of acidosis, I. The bicarbonate concentration of the blood plasma: Its significance, and its determination as a measure of acidosis, *J. Biol. Chem.* **30**:289.

Vaughan, J. M., 1970, *The Physiology of Bone,* p. 184, Clarendon Press, Oxford.

Wachman, A., and Bernstein, D. S., 1968, Diet and osteoporosis, *Lancet* **1**:958.

Wachman, A., and Bernstein D. S., 1970, Parathyroid hormone and metabolic acidosis. Its role in pH homeostasis, *Clin. Orthop.* **69**:252.

Walker, R. M., and Linkswiler, H. M., 1972, Calcium retention in the adult human male as affected by protein intake, *J. Nutr.* **102**:1297.

Walser, M., and Browder, A. A., 1959, Ion association II. The effect of sulfate infusion on calcium excretion, *J. Clin. Invest.* **38**:1404.

Wamberg, S., Kildeberg, P., and Engel, K., 1976, Balance of net base in the rat II. Reference values in relation to growth rate, *Biol. Neonate,* **28**:171.

Warnock, D. G., and Rector, F. C., Jr., 1979, Proton secretion by the kidney, *Annu. Rev. Physiol.* **41**:197.

Wasserman, R. H., Comar, C. L., and Nold, M. M., 1956, The influence of amino acids and other organic compounds on the gastrointestinal absorption of calcium-45 and strontium-89 in the rat, *J. Nutr.* **59**:371.

Weber, H. P., Gray, R. W., Dominguez, J. H., and Lemann, J. Jr., 1976, The lack of effect of chronic metabolic acidosis on 25-OH-vitamin D metabolism and serum parathyroid hormone in humans, *J. Clin. Endocrinol. Metab.* **43**:1047.

Whiting, S. J., and Draper, H. H., 1980, The role of sulfate in the calciuria of high-protein diets in adult rats, *J. Nutr.* **110**:212.

Williamson, B. J., and Freeman, S., 1957, Effects of acute changes in acid–base balance on renal calcium excretion in dogs. *Am. J. Physiol.* **191**:384.

Chapter 5

The Nutritional Significance, Metabolism, and Function of *myo*-Inositol and Phosphatidylinositol in Health and Disease

Bruce J. Holub

1. Introduction

Inositol (*myo*-inositol) and its derivatives are widely distributed in nature and occur in animals, higher plants, fungi, and some bacteria where they provide important metabolic functions. Interest in inositol as a nutrient was stimulated initially by the work of Woolley (1941) who reported that the alopecia that developed in albino mice raised on a semipurified diet was cured by adding inositol to the ration. Shortly thereafter, however, considerable doubt arose over the status of inositol as a dietary essential. For example, Martin (1941) was unable to detect any role for dietary inositol as a mouse antialopecia factor.

It is quite possible that the importance of inositol from a nutritional viewpoint has been significantly underestimated. For example, the National Research Council does not list inositol as a dietary requirement for various labora-

Bruce J. Holub • Department of Nutrition, College of Biological Science, University of Guelph, Guelph, Ontario NIG 2W1, Canada. The author's research described in this review was supported by grants from the Natural Sciences and Engineering Research Council and the Medical Research Council of Canada.

tory animals including the rat (National Academy of Sciences, 1972), which indicates that it is not yet considered necessary for the optimal performance of biological functions. The recognition that dietary inositol, like choline, can serve as a lipotropic factor for many animal species under various conditions (Gavin and McHenry, 1941; Handler, 1946; Best *et al.*, 1951), and the fact that inositol can be synthesized by intestinal flora (Woolley, 1942) and also in certain tissues (Halliday and Anderson, 1955; Eisenberg and Bolden, 1963) led to a limited nutritional interest in this compound.

It is now known, however, that inositol is an essential growth factor for human cells in tissue culture (Eagle *et al.*, 1957) and promotes the growth of young rats in a manner that is dependent on the diet composition (Karasawa, 1972). Inositol is a required nutrient in the diet of the female gerbil (Hegsted *et al.*, 1974) and other animals under certain conditions. Recent advances in inositol research have resulted in a renewed consideration of its nutritional importance. This is reflected in the thinking of many nutritional scientists as evidenced by experiments using the laboratory rat which have been reported in the *Journal of Nutrition* during the past 3 years (September, 1976 to September, 1979). Diets were supplemented with inositol in 71% of 184 sutdies in which the composition of the experimental diets was listed despite the fact that choline was also present and inositol is not listed as a required nutrient for the rat by the National Research Council. In the field of human nutrition, some infant formulae on the market are now being supplemented with inositol. Recent evidence has indicated that oral doses of inositol may be valuable in the treatment of certain abnormalities in human subjects such as diabetic neuropathy (Salway *et al.*, 1978).

Up until the past decade, much of the nutritionally related work on inositol has been channeled into documenting its role as a lipotropic factor for various animal species. However, a shift in emphasis more recently has led to many exciting developments in our understanding of the metabolism, biochemical function, and potential nutritional importance of *myo*-inositol in both healthy and diseased states. It is now recognized that many of the biological functions of inositol can be attributed at the cellular level to inositol-containing lipids such as phosphatidylinositol. In addition, new chemical forms of inositol have been discovered that are capturing the interest of both nutritionists and biochemists. It is the purpose of this chapter to focus on these recent advancements in the field.

2. Inositol and Its Biological Forms

Myo-inositol occurs most commonly in nature in its free form, as inositol phospholipids, and as phytic acid (inositol hexaphosphate). The cyclitols represent a group of compounds that includes the inositols. Of the nine possible isomers of hexahydroxycyclohexane, *myo*-inositol greatly predominates in biological systems and is the form that has been of primary metabolic and functional interest. The *myo*-inositol-to-*scyllo*-inositol ratio in rabbit tissues ranges from 7 : 1 to 45 : 1 (Sherman *et al.*, 1968). In contrast with the millimolar levels of *myo*-inositol which are found in mammalian tissues, *neo*-inositol is only present in micromolar quantities (Sherman *et al.*, 1971). The nomenclature of the cyclitols has been in a confused state for some time (Angyal and Anderson, 1959), and even an international commission on nomenclature recently published unnatural structures for the inositol-containing phospholipids (IUPAC-IUB Commission on Biochemical Nomenclature, 1977). Thus, erroneous structures for these phospholipids often appear in textbooks and journals. Agranoff (1978) has recently provided a thoughtful and simple approach for arriving at the correct formulae. Figure 1 gives the structures of *myo*-inositol and phosphatidylinositol (1,2-diacyl-*sn*-glycero-3-phosphorylinositol) or monophosphoinositide.

Inositol is a common component of plant foodstuffs, being present mainly as phytate, the chemical designation of which is *myo*-inositol hexakis(dihydrogen phosphate). For example, inositol hexaphosphate can represent up to 75% of the total phosphorus in the seeds of cereals (Schulz and Oslage, 1972a). Numerous data are available on the phytate content of various plant products such as cereals, fruits, and vegetables (Sebrell and Harris, 1967; Oberleas, 1973). Methods for the analysis of phytate in foodstuffs have been reviewed (Oberleas, 1971) and commonly involve the precipitation of phytic acid with ferric ion.

The preponderance of *myo*-inositol in animal sources exists in its free form and as part of the polar headgroup of phosphatidylinositol and, to a lesser extent, the polyphosphoinositides. Data are available on the inositol content of such animal products as meats, fowl, fish, and dairy products (Sebrell and Harris, 1967). It has been known for some time that the organs of the male reproductive tract are rich in free inositol (Eisenberg and Bolden, 1964). More recently, high concentrations have been confirmed in the rat testis, epididymal, vesicular, and prostatic fluids (Voglmayr and Amann, 1973; Lewin and Beer, 1973; Ghafoorunissa, 1976). Mammalian semen is one of the richest sources of free inositol, with the concentration in seminal plasma being severalfold higher than

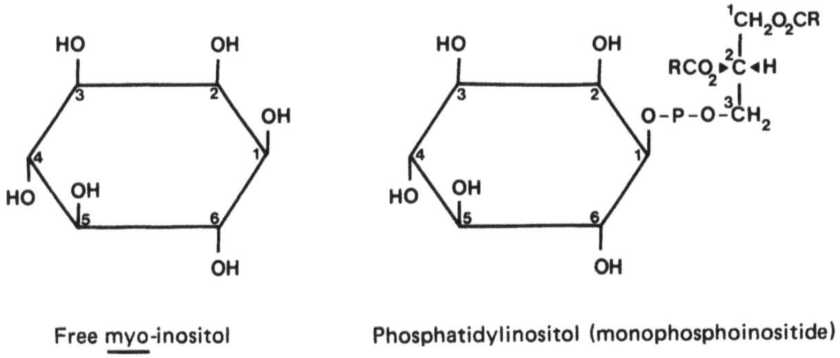

Free *myo*-inositol

Phosphatidylinositol (monophosphoinositide)

Fig. 1. Structures of inositol and phosphatidylinositol.

in blood. The levels of unbound inositol in brain, cerebrospinal fluid, and choroid plexus are also higher than in plasma (Spector and Lorenzo, 1975). The free inositol level in plasma of rats (72 days of age) fed a control diet containing inositol was found to be approximately 50 μM (Burton *et al.*, 1976). In fasting normal human subjects, the plasma concentration is maintained between 10 to 50 μM (Clements and Reynertson, 1977). Human breast milk has been found to contain about 0.6 mM inositol at 3–7 months of lactation (Burton and Wells, 1974). In contrast to the liver, free inositol levels in the small intestine, kidney, and cerebrum from 72-day-old rats were greater than those of lipid-bound inositol (Burton *et al.*, 1976). Inositol levels in natural products, mammalian tissues, single cells, and body fluids have been measured by microbiological, titrimetric, enzymatic, paper chromatographic, and gas chromatographic methods (Yamada and Tsukahara, 1973; Lewin *et al.*, 1974; Pitkänen, 1976) as well as the selected ion-monitoring method of gas chromatography–mass spectrometry and the techniques of quantitative histochemistry (Sherman *et al.*, 1977a).

Phosphatidylinositol is the major inositol-containing phospholipid found in mammalian cells and subcellular membranes. It represents 2–12% of the total phospholipid in various mammalian tissues (White, 1973). The polyphosphoinositides, phosphatidylinositol 4-phosphate (diphosphoinositide) and phosphatidylinositol 4,5-bisphosphate (triphosphoinositide), occur in trace amounts in most tissues, although their concentrations tend to be higher in the nervous sytem (Hawthorne and Pickard, 1979). It is of considerable interest that stearate and arachidonate represent the major fatty acids in phosphatidylinositol isolated from rat liver and other mammalian tissues and are almost exclusively located in the 1 and 2 positions, respectively, of the *sn*-glycero-3-phosphorylinositol backbone (Holub, 1978). This is true also for the polyphosphoinositides from bovine brain (Holub *et al.*, 1970). In contrast, lamb liver phosphatidylinositol contains oleate rather than arachidonate as the major unsaturated fatty acid (Luthra and Sheltawy, 1972). Quantitation of the inositol-containing phospholipids in mammalian tissues has usually involved their isolation by appropriate column and thin-layer chromatographic techniques (Gonzalez-Sastre and Folch-Pi, 1968; Michell *et al.*, 1970; Eichberg and Hauser, 1973; Palmer, 1977; Schacht, 1978).

In addition to the existence of free inositol, phosphatidylinositol, phosphatidylinositol 4-phosphate, and phosphatidylinositol 4,5-bisphosphate, mammalian tissues are also considered to contain low concentrations of the metabolites of these compounds such as their monacyl derivatives (e.g., lysophosphatidylinositol), inositol 1-phosphate, inositol 1:2-cyclic phosphate, inositol diphosphate, inositol triphosphate, and glycerylphosphorylinositol (Dawson *et al.*, 1971; Baker and Thompson, 1973; Koch and Diringer, 1974; Griffin and Hawthorne, 1978). These latter metabolites can be isolated and resolved by the use of ion exchange, thin-layer chromatography, paper chromatography, and silica gel glass-fiber sheets (Hokin-Neaverson and Sadeghian, 1976; Koch-Kallnbach and Diringer, 1977; Diringer *et al.*, 1977). Inositol pentaphosphate has long been recognized as a predominant organic phosphate in the erythrocytes of most avian species (Johnson and Tate, 1969). This compound has also recently been characterized in the red cells of two species of elasmobranch fish—the spiny dogfish and torpedo ray (Borgese and Nagel, 1978). In contrast, the erythrocyte of the adult ostrich contains inositol tetraphosphate as the major organic phosphate (Isaacks *et al.*, 1977). In 1974, Naccarato and Wells reported on the presence of a disaccharide derivative of inositol, 6 O-β-D-galactopyranosyl *myo*-inositol (6-β-galactinol), in human and rat milk as well as rat mammary gland. The sugar represented approximately 17% of the total nonlipid neutral *myo*-inositol in rat milk on the 18th day of lactation and was absent in all other rat tissues examined (Naccarato *et al.*, 1975).

It has been calculated that a 6-month infant weighing 7.5 kg who is being breast fed ingests close to 130 mg of inositol per day (Anderson and Holub, 1980a). It has also been estimated that a mixed North American diet can provide the human adult with approximately 1 g of inositol daily (Goodhart, 1973).

3. Absorption and Metabolism of Inositol and Its Derivatives

3.1. Digestion and Absorption

The inositol present in plant foodstuffs as phytate is hydrolyzed in the gut of monogastric animals by the enzyme phytase. This enzyme is present in plant material (Oberleas, 1973) and also in the intestinal mucosa of various animals (Davies *et al.*, 1970) and catalyzes the release of free inositol, orthophosphate, and intermediary products including the mono-, di-, tri-, tetra-, and pentaphosphate esters of inositol (Van den Berg *et al.*, 1972). Using cereals, milling by-products, and oil-seed residues, the breakdown of native phytate in the pig was found to be within the range of 40–100% (Schulz and Oslage, 1972b). A considerable hydrolysis occurred in the small intestine where most of the released phosphorus was absorbed. A significant influence of phytase present in the ingested plant materials on the degradation of phytate in the gastrointestinal tract of the pig is now recognized (Schulz and Oslage, 1972b; Vemmer and Oslage, 1973). There is also a distinct negative correlation between the extent of intestinal phytate hydrolysis and the level of dietary calcium. Is has not yet been established to what extent small amounts of dietary phytate may be absorbed intact or if inositol polyphosphates, produced by phytase activity within the lumen of the bowel or in the intestinal cells, can enter the circulating blood.

The mode of absorption of free inositol has been studied by Caspary and Crane (1970) using segments of hamster small intestine. By various criteria, inositol was actively transported. Uptake and accumulation occurred against a concentration gradient in an energy- and Na^+-dependent manner and exhibited saturation kinetics with an apparent transport K_m of 0.14 mM. Phlorizin interacted competitively with the inositol binding site with an affinity 10- to 100-fold less than that for the common glucose binding site, suggesting that the pathway by which inositol crosses the brush border membrane is not exactly the same as the D-glucose pathway.

Although many animal tissues and foodstuffs are enriched in the inositol-containing phospholipids, the mode of digestion and absorption of phosphatidylinositol remains to be investigated. It is tempting to speculate, however, that this could be partly mediated by a mechanism analogous to that proposed for phosphatidylcholine (Parthasarathy *et al.*, 1974; LeKim and Betzing, 1976). Thus, dietary phosphatidylinositol might possibly be hydrolyzed by pancreatic phospholipase A_2 in the intestinal lumen to lysophosphatidylinositol which is subsequently reacylated or further hydrolyzed on entering the mucosa. Free inositol is transported in human blood plasma at a concentration of approximately 29 μM in control subjects (Clements and Diethelm, 1979), and phosphatidylinositol is a minor constituent of all serum lipoproteins (Skipski *et al.*, 1967).

3.2. Uptake by Tissues

Lewin *et al.* (1976) have studied the uptake of radioactive inositol following intraperitoneal injection by some organs of sham-operated and nephrectomized male rats. Five hours after injecting [2-^{14}C]inositol to control (sham-operated) rats, the liver and kidney contained 10 and 8%, respectively, of the administered dose, whereas 16% had accumulated in expired CO_2 and $<1\%$ in urine. The liver, spleen, pituitary gland, kidney, and especially the thyroid glands concentrated labeled inositol from the blood quite actively. Although the organs of the male reproductive tract have high concentrations of inositol, the testes did not concentrate radiolabeled inositol from the blood. The vas deferens, epididymis, coagulating gland, seminal vesicle, and prostate had levels of radioactivity that were 8- to 28-fold those in blood serum. The concentration of labeled inositol in the brains of control animals was only moderately higher than in blood serum, indicating a limited rate of transfer for inositol between blood and brain in the rat. The bulk of the radiolabeled inositol in most of the organs was found in trichloroacetic acid-soluble form in contrast to the liver, where the majority of the radioactivity was associated with inositol lipid. These latter patterns are generally consistent with the compositional data discussed earlier (Section 2). In the trichloroacetic acid extracts of all organs studied, the amount of radioactivity found in free inositol greatly predominated over that associated with its metabolites. Radioactive inositol was mainly in the high-speed supernatant fraction isolated from testis, prostate, and seminal vesicles and particulate bound in the microsomes and mitochondria from liver (Lewin and Sulimovici, 1975).

It has been well documented that the uptake of *myo*-inositol by kidney slices occurs against a concentration gradient by means of Na^+- and energy-dependent active transport (Howard and Anderson, 1967; Hauser, 1969; Takenawa and Tsumita, 1974a). The existence of a specific *myo*-inositol transport system has been demonstrated in a plasma membrane preparation from rat kidney which contained brush border membranes (Takenawa and Tsumita, 1974b). The uptake of inositol by the membrane had similar features to that by slices and was temperature dependent, pH sensitive, stereospecific, and inhibited by phlorizin. The results also indicated that *myo*-inositol uptake represented entry into the intravesicular spaces rather than binding to the membrane. Subsequent work revealed that both the binding and transport of inositol in brush border membranes of rat kidney were dependent on Na^+ (Takenawa *et al.*, 1977a).

The uptake of inositol by mammalian brain has also been extensively studied since it resides there in high concentrations relative to plasma. Approximately one-half of the free inositol in rabbit brain has been estimated to be derived from plasma inositol via a saturable transport system based on *in-vivo* experiments (Spector and Lorenzo, 1975; Spector, 1976a). The choroid plexus was implicated as the locus of inositol transport from plasma to the cerebrospinal fluid. Isolated brain slices also contained a saturable uptake system for inositol which was considered an active transport system by various criteria (Spector, 1976b) and capable of explaining the large inositol concentration differential between brain and cerebrospinal fluid. In contrast, the uptake of inositol by rat brain synaptosomes was found to occur via an unsaturable process that did not provide a concentration gradient indicative of active transport (Warfield *et al.*, 1978). These latter workers suggested that the uptake system observed in rabbit brain slices (Spector, 1976b) may reflect a species difference or uptake by a component of the slice other than neurons.

3.3. Biosynthesis

The early work of Eisenberg and Bolden (1963) indicated a very active system for inositol synthesis from [^{14}C]glucose in rat testis as measured per milligram of tissue. Hauser and Finelli (1963) demonstrated a synthetic capacity in slices of rat brain, kidney, and liver. It is of interest that the testis concentrates little inositol, whereas organs with a slight synthetic ability tend to maintain higher concentrations of the compound. The synthesizing capacity of the male reproductive tract has been reported to be of the order of testis > epididymis > seminal vesicles, whereas relative tissue levels of inositol follow a reverse sequence (Eisenberg and Bolden, 1964). Marginal protein deficiency did not alter the biosynthesis of inositol in the testis, epididymis, or seminal vesicles (Ghafoorunissa, 1975). It has been shown that blood glucose and not blood inositol is the major source of inositol in the rete testis fluid of the ram (Middleton and Setchell, 1972) and that prostatic secretion is the source of inositol in human seminal fluid (Lewin and Beer, 1973). In contrast to epididymal or ejaculated spermatozoa, testicular spermatozoa of the ram exhibited a significant capacity for inositol biosynthesis from glucose (Voglmayr and White, 1971), although the activity was considered insufficient to account entirely for the high inositol concentration in rete testis fluid. These latter authors suggested that the cytoplasmic droplet may be the site of inositol synthesis in the testicular spermatozoa.

No definitive *in vivo* studies have yet been conducted on the relative contribution of various organs to the total inositol that is endogenously synthesized in the body. The potential importance of the kidney and liver in this regard is suggested by the work of Hauser (1963) who monitored the incorporation of labeled glucose into free *myo*-inositol. It has been estimated that approximately 50% of the unbound inositol in rabbit brain is synthesized from glucose *in situ* with the remainder being transported into brain from blood (Spector and Lorenzo, 1975). An impressive study by Clements and Diethelm (1979) permitted an *in vivo* measurement of inositol synthesis in human kidney. The rate of endogenous synthesis from one normal human kidney was found to approach 2 g/day, thereby providing 4 g/day in the binephric human which is considerably more than is ingested daily. These experiments also suggested that extrarenal tissues contribute to the endogenous production of inositol as found in the rat.

The enzymatic biosynthesis of *myo*-inositol, as studied extensively in rat testis (Eisenberg, 1967), involves the conversion of glucose 6-phosphate to inositol 1-phosphate by inositol 1-phosphate synthase (EC 5.5.1.4) followed by a dephosphorylation reaction catalyzed via inositol 1-phosphatase activity (EC 3.1.3.25). Chen and Eisenberg (1975) have provided strong evidence for the presence of myoinosose-2 1-phosphate as an intermediate in the synthase reaction, and mechanistic studies on the cyclization of glucose 6-phosphate have been reported recently (Sherman *et al.*, 1977b; Loewus, 1977). Recently, Robinson and Fritz (1979) reported that, among cells in the testis examined, Sertoli cells had the highest levels of enzymes required for inositol biosynthesis and suggested that these cells may comprise the primary source of inositol in the testis. These latter workers also found high levels of synthase and phosphatase in epididymal cells which may synthesize amounts of inositol observed in epididymal fluid provided adequate amounts of glucose 6-phosphate are available. Naccarato *et al.* (1974) have isolated and partially purified the synthase and phosphatase from lactating rat mammary gland. In other work, the activities of both enzymes in the inositol biosynthetic pathway were found to increase in rat mammary gland in close agreement with the inositol content of milk (Burton and Wells, 1974).

3.4. Catabolism

Since [2-^{14}C]inositol was not degraded to respiratory $^{14}CO_2$ by nephrectomized rats, Howard and Anderson (1967) concluded that the kidney is the only organ of importance in inositol catabolism. These workers also conducted experiments with slices from principal zones of the kidney and observed that the cortex, medulla, and isolated tubules were about equally active in catabolizing inositol to CO_2. Additionally, gluconeogenesis from inositol can proceed in kidney via the glucuronic acid pathway and the pentose cycle (Howard and Anderson, 1967; Hankes *et al.*, 1969; Freinkel *et al.*, 1970). The more recent work of Lewin *et al.* (1976) *in vivo* using sham-operated control rats and nephrectomized animals has confirmed earlier work. Bilaterally nephrectomized rats were essentially unable to convert inositol into CO_2, whereas controls catabolized 16% of the injected [2-^{14}C]inositol to $^{14}CO_2$ in 5 hr. The catabolism of inositol by the kidney was of much greater significance than its excretion in urine, sunce $<1\%$ of the administered radioactive inositol was released into the urine over the same time interval. Clements and Diethelm (1979) have observed that urinary excretion accounts for only a small fraction of the disposal of inositol by the kidney in human subjects. The disposed inositol was converted to D-glucose and D-glucuronolactone or completely oxidized to carbon dioxide and water. It is apparent that the kidney is likely the primary regulator of plasma inositol concentrations in man.

3.5. Incorporation into Phospholipid

Tracer studies *in vivo* have revealed that radioactive inositol is readily incorporated into the lipid-bound inositol fraction of all animal organs studied (Lewin *et al.*, 1976). In addition, the entry of inorganic [^{32}P]phosphate into phosphatidylinositol as well as the polyphosphoinositides, phosphatidylinositol 4-phosphate and phosphatidylinositol 4,5-bisphosphate, has been well documented in various tissues (Jungalwala and Dawson, 1971; Friedel and Schanberg, 1971; Gonzalez-Sastre *et al.*, 1971; Holub and Kuksis, 1971; Cohen *et al.*, 1971). Figure 2 indicates the two known biochemical mechanisms by

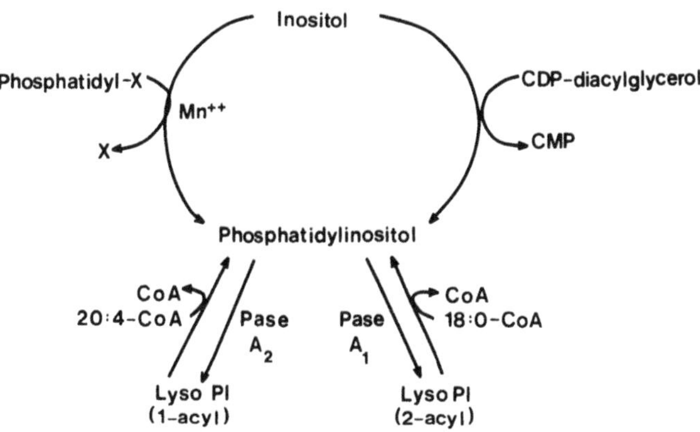

Fig. 2. Pathways of phosphatidylinositol formation.

which radiolabeled free inositol can be incorporated into tissue phosphatidylinositol. Free inositol can provide for the *de novo* biosynthesis of phosphatidylinositol by reaction with the liponucleotide, CDP-diacylglycerol, in the presence of the enzyme CDP-diacylglycerol: inositol phosphatidyltransferase (EC 2.7.8.11) as described (Agranoff *et al.*, 1958; Paulus and Kennedy, 1960; Thompson *et al.*, 1963). The enzyme, which resides mainly in the microsomal fraction (Benjamins and Agranoff, 1969; Bishop and Strickland, 1970; van Golde *et al.*, 1974), has been solubilized and purified from rat brain and liver (Rao and Strickland, 1974; Takenawa and Egawa, 1977). An alternative route for the entry of inositol into phosphatidylinositol involves the Mn^{2+}-activated exchange of free inositol with the base moiety of endogenous microsomal phospholipid (Paulus and Kennedy, 1960; Hubscher, 1962; Broekhuyse, 1971). Chase experiments have indicated that endogenous phosphatidylinositol is the preferred substrate for the Mn^{2+}-stimulated entry of inositol into phospholipid in rat liver microsomes when the exchange reaction is enhanced by the addition of CTP or CDP-choline (Holub, 1975) which would not provide, therefore, for the net synthesis of phosphatidylinositol. The enzyme catalyzing the exchange reaction has been solubilized recently from a rat liver microsomal fraction (Takenawa *et al.*, 1977b). Tracer experiments *in vivo* and *in vitro* using [^{3}H]inositol have suggested that a major proportion of the free inositol that enters rate liver phosphatidylinositol under physiological conditions does so via the CDP-diacylglycerol: inositol phosphatidyltransferase and not by the Mn^{2+}-stimulated exchange reaction (Holub, 1974).

There is considerable evidence accumulating from tracer studies *in vivo* to indicate that the molecular species of phosphatidylinositol formed in rat liver (Holub and Kuksis, 1971, 1972) and brain (MacDonald *et al.*, 1975; Luthra and Sheltawy, 1976) via *de novo* synthesis from the reaction of inositol with CDP-diacylglycerol cannot account for the preponderance of arachidonic acid in tissue phosphatidylinositol. It has been estimated that approximately one-half of the arachidonate in liver phosphatidylinositol is derived via the CDP-diacylglycerol: inositol phosphatidyltransferase in rat liver (Holub, 1978), whereas the other half may originate by the acylation of 1-acyl-*sn*-glycero-3-phosphorylinositol as depicted in Fig. 2. The conversion of phosphatidylinositol to its monoacyl derivatives has been demonstrated in mammalian tissues (White *et al.*, 1971; Strickland *et al.*, 1978; Irvine *et al.*, 1978). Acyl-CoA : 1-acyl-*sn*-glycero-3-phosphorylinositol acyltransferase activity in the microsomal fraction from rat brain (Baker and Thompson, 1973) and liver (Holub, 1976) exhibits a selectivity for arachidonoyl-CoA. The acylation of lysophosphatidylinositol (1-acyl derivative) produces mainly tetraenoic (arachidonoyl) species of phosphatidylinositol when liver homogenate or microsomal preparations are not supplemented with acyl-CoA (Holub, 1976). The existence of acyl-CoA : 2-acyl-*sn*-glycero-3-phosphorylinositol acyltransferase activity has been demonstrated recently in rat liver (Holub and Piekarski, 1979). Thus, the acylation reactions depicted in Fig. 2 could provide for the enrichment of phosphatidylinositol with stearic and arachidonic acids at the 1 and 2 positions, respectively, in keeping with natural tissue compositions.

It is of interest that the phosphatidylinositol pool on the cytoplasmic side of the endoplasmic reticulum is very limited, with about 85% being localized on the luminal side (Guarnieri, 1975; Low and Finean, 1976; Nilsson and Dallner, 1977a,b). The phosphatidylinositol that is synthesized in the endoplasmic reticulum of eukaryotic cells can be transferred to other cellular membranes by phospholipid-exchange proteins residing in the cytoplasm (Helmkamp *et al.*, 1976; Brophy *et al.*, 1978; Brophy and Aitken, 1979). In many tissues, phosphatidylinositol is converted by stepwise phosphorylation to the polyphosphoinositides (Kai and Hawthorne, 1969; Tou *et al.*, 1970; Cooper and Hawthorne, 1976) which occur as very minor constituents of the cellular phospholipid pool. The formation of diphosphoinositide and triphosphoinositide is catalyzed by the ATP : phosphatidylinositol 4-phosphotransferase (EC 2.7.1.67) and ATP : diphosphoinositide 5-phosphotransferase (EC 2.7.1.68), respectively. The enzymes involved in polyphosphoinositide metabolism have been located on the cytoplasmic surface in the case of the human erythrocyte membrane (Garrett and Redman, 1975). The catabolism of tissue phosphatidylinositol can occur by phospholipase-catalyzed deacylation which produces glycerophosphorylinositol or by a phospholipase C type cleavage which releases inositol 1:2-cyclic phosphate or inositol 1-phosphate (Dawson *et al.*, 1971; Irvine *et al.*, 1978). Enzymes exist in mammalian tissues for the release of inositol monophosphate or free inositol from glycerylphosphorylinositol and the hydrolysis of inositol 1:2-cyclic phosphate to inositol 1-phosphate (Dawson and Clarke, 1972; Dawson and Hemington, 1977; Dawson *et al.*, 1979). Tissue phosphomonoesterase hydrolyzes triphosphoinositide and diphosphoinositide to yield diphosphoinositide and phosphatidylinositol, respectively (Lee and Huggins, 1968; Sheltaway *et al.*, 1972; Nijjar and Hawthorne, 1977). In addition, both soluble and particulate forms of phosphodiesterase exist in certain mammalian tissues that degrade triphosphoinositide and diphosphoinositide to free 1,2-diacylglycerols with the release of inositol triphosphate and inositol diphosphosphate, respectively (Keough and Thompson, 1972; Tou *et al.*, 1973; Irvine and Dawson, 1978).

4. Biochemical and Physiological Functions of Inositol

4.1. Function of Free Inositol

There is considerable interest currently in the possible cellular functions of free inositol, since it predominates in concentration over the inositol-containing phospholipids in most tissues. The importance of free inositol in reproduction has been indicated by various workers. Since the epididymis maintains a higher concentration of this cyclitol than the testis, Eisenberg and Bolden (1964) proposed that inositol may play a role in the maturation of spermatozoa as they migrate through the epididymis. Triethylenemelamine administration to rats or surgical cryptorchidism resulting in a decrease in inositol synthesis was concomitant with the disappearance of spermatids and spermatozoa (Morris and Collins, 1971), suggesting the involvement of inositol in spermatogenesis. Robinson and Fritz (1979) have suggested that inositol may be one of the components synthesized by Sertoli cells which are important in establishing the unique microenvironment in the seminiferous tubule required for germinal cell development.

The relatively high intracellular levels of free inositol found in nervous and secretory tissues, which are enriched in microtubules, may reflect a role for this cyclitol in controlling the functional states of microtubules. Kirazov and Lagnado (1977) observed a preferential binding of inositol to assembly-competent tubulin oligomers and a protection by inositol of microtubules against cold- and calcium-induced depolymerization. Thus, inositol served to stabilize both microtubules and the intermediate aggregate species of tubulin with which they are in dynamic equilibrium. Some of these effects could be mediated by an influence on the state of hydration of microtubular protein, since inositol behaves as a water-structuring compound (Suggett, 1975). Pickard and Hawthorne (1978a) have suggested that the effects of free inositol on microtubules may result from nonspecific interactions and be of limited physiological relevance.

4.2. Function of Inositol Phosphates

Although the phosphorylated derivatives of inositol represent minor forms of this cyclitol in animal tissues, there is experimental evidence indicating that certain of these compounds may have important cellular functions. It has been recognized from *in vitro* experiments that various organic phosphates, and particularly inositol hexaphosphate, can decrease the affinity of hemoglobin for oxygen by forming a very tight complex formation with deoxyhemoglobin (Edalji *et al.*, 1976). It is now recognized that the naturally occurring form of inositol polyphosphate in the erythrocytes of most mature species of birds is actually inositol pentaphosphate (Section 2). This latter compound is effective in causing a right shift of the oxygen equilibrium of fetal- and adult-type duck hemoglobins (Borgese and Nagel, 1977). Borgese and Nagel (1978) recently reported that inositol pentaphosphate is also present in the red cells of two species of elasmobranch fish and suggested that it may play a role in regulating the oxygen affinity of dogfish and torpedo hemoglobins. Inositol tetraphosphate is the major organic phosphate in erythrocytes of the adult ostrich (Isaacks *et al.*, 1977). Inositol 1-pyrophosphate has been identified as a coenzyme for thiamine pyrophosphokinase (Okazaki, 1975). Inositol 1 : 2-cyclic phosphate, which is derived from the hydrolysis of phosphatidylinositol, has been suggested (Michell and Lapetine, 1972) as an intracellular "second messenger." This compound can potentiate contraction in the vas deferens of the rat in response to norepinephrine (Lapetina and Zeiher, 1976).

4.3. Function of Phosphatidylinositol

Most of the physiological and biochemical roles for inositol in mammalian tissues have been attributed to phosphatidylinositol. This phospholipid exerts its functions primarily at the membrane level, since it is intimately associated with biological membranes. It is possible that some of the functions that have been attributed to phosphatidylinositol may be imparted by its unique molecular composition—namely, a preponderance of 1-stearoyl 2-arachidonoyl species (Holub, 1978). In this regard, there is experimental evidence to suggest that the triene/tetraene ratio may show more dramatic changes in tissue phosphatidylinositol as compared to other phospholipids under conditions of essential fatty acid deficiency (Andersen, 1977).

Michell (1975, 1979) has reviewed the experimental evidence indicating that membrane phosphatidylinositol may have a special function in the responses of various cells to external stimuli such as hormones and neurotransmitters. Stimuli whose major effects are to produce rapid physiological responses (muscarinic cholinergic, α-adrenergic, etc.) or those that bring about longer-term stimulation of cell proliferation (phytohaemagglutinin and other mitogens, etc.) produce an enhanced phosphatidylinositol metabolism in appropriate target tissues. This "phosphatidylinositol effect" appears to involve an initial degradation of membrane phosphatidylinositol (Hokin-Neaverson *et al.*, 1978; Michell, 1979) and may control cell surface Ca^{2+} permeability which gives rise to an elevation in intracellular Ca^{2+} concentration (Jafferji and Michell, 1976). In very recent work, Kirk *et al.* (1979) have described the relationship between enhanced phosphatidylinositol metabolism and the activation of glycogen phosphorylase in hepatocytes exposed to vasopressin and related peptides. Evidence for a role of phosphatidylinositol turnover in stimulus-secretion coupling has been provided for the Ca^{2+}-mediated histamine secretion in antigen-sensitized rat peritoneal mast cells stimulated with a number of different ligands (Cockcroft and Gomperts, 1979).

Hawthorne and Pickard (1979) have recently reviewed evidence supporting a key role for phosphatidylinositol in synaptic function. The stimulatory effect of acetylcholine on the turnover of phosphatidylinositol in nerve ending fractions of guinea pig cortex has supported a muscarinic and not nicotinic action for the neurotransmitter centrally (Schacht and Agranoff, 1972; Miller, 1977). The cholinergic phosphatidylinositol effect has been postulated as an essential part of the synaptic transmission process (Durrell *et al.*, 1969). Pickard and Hawthorne (1978b) have provided details on the possible role of phosphatidylinositol turnover in transmitter release by showing that electrical stimulation of synaptosomes labeled with ^{32}P *in vivo* caused a loss of radioactivity from the phosphatidylinositol associated with the synaptic vesicles.

The work of Slaby and Bryan (1976) with rat pancreatic tissue *in vitro* has implicated a central role of enhanced phosphatidylinositol turnover in secretagogue-stimulated secretion. Isolated rat pancreatic islets subjected to an increase in the medium glucose concentration have exhibited an increased insulin secretion and ^{32}P-labeling of phosphatidylinositol (Freinkel *et al.*, 1975). Clements and Rhoten (1976) prelabeled the phosphatidylinositol in isolated rat pancreatic islets with [^{3}H]inositol and observed that subsequent incubation with elevated concentrations of carbohydrates that stimulate insulin release (D-glucose and D-mannose) resulted in the release of water-soluble forms of radioactive inositol. Incubation with carbohydrates that do not stimulate insulin secretion did not promote a cleavage of labeled phosphatidylinositol, which supports a role for this phospholipid in the process of insulin secretion from the pancreatic β cell.

A specific role for phosphatidylinositol in providing arachidonic acid for prostaglandin synthesis in the thyroid has been suggested by the work of Haye and colleagues (Haye *et al.*, 1973; Haye and Jacquemin, 1977). In addition to demonstrating a preponderance of arachidonate in pig thyroid phosphatidylinositol, these workers found that this acid is specifically mobilized from phosphatidylinositol by a Ca^{2+}-dependent phospholipase A_2 which is stimulated by thyrotropin without the participation of cyclic AMP. There is also experimental evidence to suggest that phosphatidylinositol in brain might possibly provide the unesterified arachidonic acid for the synthesis of the major prostaglandins and thromboxanes *in vivo* (Marion and Wolfe, 1979).

Evidence for specific interactions between phosphatidylinositol and protein provides the molecular basis of a role for this phospholipid in regulating enzyme activity and transport processes. The early work of Charalampous (1971) indicated that the rate of ATP hydrolysis by the membrane-bound Na^+- and K^+-activated adenosine triphosphatase and the translocation of Na^+ and K^+ in the plasma membrane were dependent on inositol, although the biochemical basis for these phenomena was not elucidated. Recently, Mandersloot *et al.* (1978) have shown that phosphatidylinositol is the endogenous activator of the (Na^++K^+)-adenosine triphosphatase in microsomes isolated from rabbit kidney. Alkaline phosphatase and 5′-nucleotidase also appear to depend specifically on phosphatidylinositol for their association with plasma membranes (Low and Finean, 1978). The formation of a complex between chymotrypsinogen A and phosphatidylinositol has been described which could account for the movement of charged globular protein through nonpolar regions of biological membranes (Rothman, 1978). The results of Heger and Peter (1977) have indicated that acetyl-CoA carboxylase from rat liver is a protein–phosphatidylinositol complex. The kinetic properties of citrate activation of the enzyme were controlled by phosphatidylinositol. Recently, it has been demonstrated that phosphatidylinositol has potent effects on tyrosine hydroxylase, the enzyme that catalyzes the rate-limiting step in the biosynthesis of the catecholamines, dopamine, and norepinephrine (Lloyd, 1979). Phosphatidylinositol could provide a rapid reversible activation of the enzyme and a slower preincubation-dependent irreversible inactivation of the enzyme.

4.4. Function of Polyphosphoinositides

The polyphosphoinositides, phosphatidylinositol 4-phosphate and phosphatidylinositol 4,5-bisphosphate, are considered to have important roles in membrane function despite their occurrence in minor amounts in mammalian tissues. Since the 4- and 5-phosphate groups on the inositol ring undergo a stimulated turnover in nervous tissues under certain conditions, an involvement of the polyphosphoinositides in membrane excitability, permeability, and nervous conduction has been implicated (Salway and Hughes, 1972; Hitzemann *et al.*, 1978; Hawthorne and Pickard, 1979). A rise in intracellular Ca^{2+} concentrations which occurs in nerve fibers during passage of the action potential may activate catabolism and turnover of the polyphosphoinositides (Griffin and Hawthorne, 1978). An increased metabolism of the polyphosphoinositides in polymorphonuclear leukocytes has been reported during phagocytosis (Tou and Stjernholm, 1974). The strongly ionic nature of the polyphosphoinositides has led to the suggestion that they may be directly involved in the active transport of cations (Kai and Hawthorne, 1969), although limited support for this concept has been forthcoming. These phospholipids have been implicated in the regulation of Ca^{2+} binding to the erythrocyte membrane and in the regulation of intracellular Ca^{2+} levels (Buckley and Hawthorne, 1972). The polyphosphoinositides may also regulate the adenylate energy charge in the cell through a futile cycle (Talwalkar and Lester, 1973; Buckley, 1977).

5. Effects of Dietary Inositol

5.1. Effect of Feeding Inositol-Deficient Diets

The role of dietary inositol as a lipotropic factor has been the subject of renewed interest over the past few years. Many of the nutritional conditions used in early work to produce an inositol-dependent response in the rat have employed feeding a low-protein depletion diet free from B vitamins and fat before administering the B vitamins with or without inositol (Handler, 1946; Kotaki *et al.*, 1968). In more recent work, Hayashi *et al.*, (1974a) reported an inositol-deficient experimental diet containing phthalylsulfathiazole that resulted in an inositol-responsive accumulation of triglyceride in liver when highly saturated fats such as hydrogenated soybean oil, hydrogenated cottonseed oil, and coconut oil were fed, whereas the highly unsaturated oil, natural cottonseed oil, was ineffective in producing the syndrome. The drug was presumably added to inhibit the growth of intestinal bacteria which can synthesize inositol, although its essentiality in this regard was not demonstrated by these authors. Wells and Burton (1978) have reported that the incorporation of 0.5% phthalylsulfathiazole into the inositol-deficient diet of the lactating rat was necessary for fatty liver development. Hayashi *et al.* (1974a) found that after only 1 week of dietary treatment, rats fed a basal diet containing 10% by weight of hydrogenated cottonseed oil without inositol had liver triglyceride concentrations that were 158% higher than those in animals fed the control diet with inositol at the 0.5% level. Hepatic cholesterol levels were higher by only 15% in the deficient animals.

In a 14-day feeding experiment using choline-deficient diets with corn oil as the dietary lipid, the inclusion of inositol lowered hepatic lipid levels by only 20% (Shepherd and Taylor, 1974a). Andersen and Holub (1976) observed little influence of inositol on hepatic triglyceride levels in male rats fed diets containing corn oil or partially hydrogenated soybean oil as the fat, whereas inositol deficiency produced triglyceride concentrations that were two- and four-fold higher when a low-erucate rapeseed oil and tallow, respectively, were present. The degree of saturation of the dietary fats or the levels of linoleic acid in them could not readily provide a simple explanation for the dependency of the magnitude of the inositol response on the type of dietary fat. The relative abundance of individual saturated and unsaturated fatty acids in the dietary lipid appears to be of significance in this regard. Liver phospholipid levels were slightly lower in all cases when inositol-deficient diets were fed. It has recently been shown that inositol deficiency in the male gerbil could produce an accumulation of inositol deficiency in the male gerbil could produce an accumulation of hepatic lipids, similar to the situation in the rat, when coconut oil or safflower oil comprised the dietary lipid (Hoover *et al.,* 1978).

Experimental conditions have been devised (Andersen and Holub, 1980a) for studying inositol deficiency in the young rat that are nutritionally more acceptable and relevant than those employing diets low in protein, deficient in B vitamins and choline, almost devoid of essential fatty acids, containing a sulfathiazole drug, or a combination of the latter. The exclusion of inositol from these diets that met or exceeded the NRC requirements for all nutrients, including choline, gave rise to mean hepatic triglyceride concentrations that were higher by 230% and 90% in males and females, respectively, as compared to corresponding rats fed control diets supplemented with 0.1% inositol. The greater sensitivity of younger rats to dietary inositol (Andersen and Holub, 1980b) may reflect a greater nutritional requirement or a lower capacity for endogenous biosynthesis and/or higher capacity for catabolism of this cyclitol.

In this regard, Hauser (1963) concluded that organs of the young rat appear to be better able to convert glucose to inositol than those of the fully mature animal. Andersen and Holub (1980b) have also studied the relative response of hepatic lipids in the rat to graded levels of dietary inositol and other lipotropes. In these experiments, inositol and choline were found to have an equivalent lipotropic potency when compared at a dietary level equal to the NRC requirement for choline. Fatty acid analyses have revealed that the weight percentage of 18:2 is reduced in the phospholipid, and the relative abundance of 16:1 tends to be elevated in the triglyceride when inositol-deficient diets producing an accumulation of rat liver triglyceride are fed (Andersen and Holub, 1976, 1980b).

Burton *et al.* (1976) have studied the effect on tissue inositol levels of feeding inositol-depleted and -supplemented diets to the neonatal and developing rat. Dietary inositol deprivation produced a significant lowering of free inositol levels in all tissues studied (testis, liver, plasma, lung, heart, lens, kidney, small intestine) with the exception of the cerebrum and cerebellum. Liver was the only tissue studied in which the lipid-bound inositol was also lowered along with the free inositol levels. Inositol-deficient diets have also been shown to reduce the level of free inositol in the urine of male rats (Shepherd and Taylor, 1974b).

Burton and Wells (1976) have also examined the effect of inositol deprivation in pregnant and lactating rats on inositol metabolism in fetal and postnatal offspring. The pups were fed the corresponding diets after weaning until 3 months of age. A strong correlation between the free inositol content in the diet and in the milk was observed. At day 8 of lactation, the concentrations of free inositol in mammary gland and milk and of 6-β-galactinol in milk for animals receiving the supplemented diet were severalfold higher than for those fed the deficient diet. Supplementation with inositol significantly increased the levels of free inositol in plasma, liver, kidney, and intestine of pups at all ages examined. During lactation, the inositol-deprived dams developed severe fatty livers which were alleviated by dietary inositol supplementation or by termination of lactation. After 14 days of lactation, triglyceride and cholesterol ester levels were greatly elevated in inositol-deficient dams, whereas liver free cholesterol and phospholipid levels, particularly phosphatidylinositol, were significantly depressed (Burton and Wells, 1977). Electron microscopy revealed an increase in the size and number of fat droplets in the livers of the deficient dams.

The biochemical mechanisms responsible for the accumulation of liver triglyceride under conditions of inositol deficiency have recently been the subject of intense investigation. There is considerable support for the concept that the transport of lipoproteins from liver into plasma is impeded when inositol-deficient diets are consumed. The experiments of Hasan *et al.* (1970, 1971) have indicated that inositol promotes the synthesis of phosphatidylinositol which increases the synthesis of β-lipoprotein in liver and its secretion. Nicolosi *et al.* (1976) have measured hepatic triglyceride secretion rates in *in vivo* following the injection of Triton which coats very-low-density lipoproteins, thereby preventing their catabolism by lipoprotein lipase. With this technique, secretion rates were found to be significantly reduced in inositol-deficient gerbils (Hoover *et al.*, 1978), which implicates an inadequate release of lipoprotein as a causative factor in the hepatic lipid accumulation.

Burton and Wells (1977) have observed a depression in the levels of total plasma lipoprotein lipid, very-low-density lipoprotein, high-density lipoprotein, total phospholipid, and plasma phosphatidylinositol in inositol-deprived dams during lactation, thereby suggesting a block in hepatic lipoprotein secretion. This latter conclusion has been further supported by recent work demonstrating that lactating rats supplemented with inositol exhibited a greater loss of isotope from liver triglycerides with a more rapid appearance in serum triglyceride following the intravenous injection of [^{14}C]palmitate as compared to those given a deficient diet (Burton and Wells, 1979). Evidence for an impaired release of triglycerides from the livers of inositol-deficient lactating rats was provided by the use of intravenously injected Triton. No difference in the *in vivo* incorporation of [^{14}C]arginine into total liver protein between the dietary groups was revealed, although the specific radioactivity of the serum protein was decreased in deficient rats compared to those given supplementary inositol. It was suggested that inositol deficiency inhibited or delayed the hepatic release of lipoproteins.

An interest in the possibility that other mechanisms might contribute to an elevation of hepatic triglycerides in inositol deficiency was stimulated by experiments with eukaryotic cells not actively involved in lipoprotein secretion. The effects of inositol deprivation have been examined in the cells of the yeast *S. carlsbergensis* and involve a marked accumulation of total neutral lipids including triglyceride (Paultauf and Johnston, 1970; Hayashi *et al.*, 1976). It has been suggested that the lipid that accumulates in inositol-deficient yeast may result in part from an enhancement of acetyl-CoA carboxylase activity (Hayashi *et al.*, 1978a; Tomita *et al.*, 1979). However, Daum *et al.* (1979) have provided data to indicate that increased fatty acid synthesis is not primarily responsible for triglyceride accumulation.

In contrast to the experimental results discussed above, Hayashi *et al.*, (1974b) have concluded that an elevation of hepatic triglycerides in the inositol-deficient rat is primarily the result of an increased rate of fatty acid mobilization from adipose depots to the liver. In support of this concept, these workers found that the elevation in serum nonesterified fatty acid and liver triglyceride levels observed in inositol deficiency was inhibited by reserpine treatment. Furthermore, the incorporation of radioactivity from the epididymal fat pads labeled with [^{14}C]palmitate into the liver lipids of the inositol-deficient animals was almost threefold that of the control rats. Andersen and Holub (1980b) have observed a marked increase in the concentration of nonesterified fatty acids in the livers of rats consuming inositol-deficient diets. Recent evidence suggests that the increased lipolysis associated with inositol deficiency results from an activation of hormone-sensitive lipase in adipose tissue (Hayashi *et al.*, 1978b). The level of plasma epinephrine, a potential activator, was higher in inositol-deficient rats concomitant with an increase in blood pressure. Since adrenalectomy did not influence the liver lipid accumulation caused by inositol deficiency, the stimulated lipolysis appeared to be caused by an excitation of sympathetic nerve terminals innervating the adipose tissues and not by an elevation of serum epinephrine released from the adrenals. The elevation of serum free fatty acids and hepatic triglyceride levels in inositol deficiency was impeded by treatment with sympathetic nervous blockers such as hexamethonium and bupranolol, suggesting that the central autonomic discharge to the adipose tissue in the deficient rat may be increased. These same workers (Hayashi *et al.*, 1978b) have suggested that the decrease in the inositol level of the brain, especially of the hypothalamus, in the deficient animal may link an excitation of this region to certain metabolic alternations associated with inositol deficiency.

No definitive studies have been conducted to determine if dietary inositol influences the catabolism and oxidation of lipid in liver. Andersen and Holub (1976) have observed that liver homogenates prepared from inositol-deficient animals have a moderately higher capacity for glyceride synthesis from glycerol-3-phosphate than do controls receiving an inositol-supplemented diet when saturating levels of this substrate are added to the assay medium. However, the actual rate of acylation of glycerol-3-phosphate may not be higher *in vivo,* since the level of glycerol-3-phosphate in liver is decreased in rats consuming an inositol-deficient diet (Hayashi *et al.*, 1974b). By incubating rat liver slices with [^{14}C]oleic acid or [^{14}C]acetate, Hoover *et al.* (1978) found that oleate entry into lipid was not influenced by dietary inositol but that incorporation of acetate into triglyceride was markedly increased in the inositol-deficient animals, suggesting an enhancement in fatty acid synthesis.

Hegsted *et al.* (1973) were the first to document the development of an intestinal lipodystrophy in female gerbils fed a diet containing coconut oil which was prevented by the inclusion of inositol in the diet. The syndrome was characterized by the accumulation of fat in the intestinal mucosal cells which resulted in eventual debilitation and death. In the chronic condition, a progressive loss of body weight was associated with alopecia which became further complicated by an exudative dermatitis and inanition. At necroscopy, the small intestine of these animals was greatly enlarged, and the serosal surface was unusually white with the exposed mucosa appearing swollen, corrugated, and equally whitened. Subsequent investigation revealed that feeding inositol-deficient diets containing triglycerides rich in lauric, capric, or myristic acids to female gerbils produced the maximum accumulation of gut lipid (Kroes *et al.*, 1973). Recent work supports the concept that dietary inositol deprivation limits the transport of saturated fat by the mucosal cells more than that of unsaturated lipid (Watkins and Hegsted, 1979).

Castration of male gerbils eliminated their resistance to the lipodystrophy, suggesting that testicular synthesis of inositol might account for the different response between the sexes (Kroes *et al.*, 1973). Hegsted *et al.* (1974) have observed that male animals are able to maintain a higher content of inositol in intestinal tissue than females. These latter workers have estimated the inositol requirement of the female gerbil fed a purified diet containing 20% coconut oil to be 70 to 120 mg/kg of diet. It has been suggested that the intestinal lipodystrophy in inositol-deficient gerbils may result from insufficient amounts of inositol to provide an adequate supply of phosphatidylinositol for the synthesis and transport of intestinal lipoproteins (Hegsted *et al.*, 1973). Recently, Chu and Hegsted (1979) found the content of phosphatidylinositol in microsomes of intestinal mucosal cells to be greatly depleted in inositol deficiency, which may relate to the ability of inositol-deficient gerbils to transport fat rich in saturated fatty acids.

5.2. Effect of Inositol Supplementation of Practical Diets

A number of experiments have been conducted to determine whether inositol supplementation of practical poultry rations containing endogenous inositol and choline might provide any beneficial result. Inositol has been investigated as a lipotropic agent in the laying hen in relation to the fatty liver syndrome. It has been reported that the addition of inositol decreased the liver fat content and increased the egg production of affected birds (Reed *et al.*, 1968; Bull, 1968). In contrast, inositol addition to a cereal-based ration for the laying hen was found to have no significant effect on liveweight, egg production, egg size, food consumption, or the efficiency of food utilization (Pearce, 1972). The incidence of the fatty liver and kidney syndrome in broiler chickens fed diets high in wheat and low in protein was also not alleviated by inositol supplementation (Pearce, 1975). Other work has failed to demonstrate an influence of supplementary inositol on the fatty liver syndrome in laying hens (Hamilton and Garlich, 1972; Wolford and Murphy, 1972; Schexnailder and Griffith, 1973; Wolford and Polin, 1975). The quantity of abdominal fat in broilers was also found not to be affected by the addition of inositol to practical rations (Kubena *et al.*, 1974). Smith *et al.* (1974) have suggested that small but frequent doses of inositol or other lipotropes to beef cattle fed high-concentrate low-roughage diets may produce leaner carcasses with less external fat.

5.3. Effect of Dietary Phytate

Much attention has been given in the past to the rachitogenic effect of dietary phytate (inositol hexaphosphate), and appropriate review articles are available (Oberleas, 1973). Phytate has the capacity to form insoluble salts with calcium, thereby decreasing the absorption of this cation in the gut. Van den Berg *et al.* (1972) have shown that different polyphosphate esters of inositol, which may enter the circulation, are potent inhibitors of the calcification *in vitro* of rachitic rat cartilage, whereas phytate itself is inert. Phytate can also decrease zinc availability by complexing with it to form a very insoluble salt at pH values found in the upper small intestine in a manner that is also affected by the phytate–calcium synergism (Oberleas, 1973). An effect of phytate on calcium and zinc balances in human subjects has been demonstrated (Reinhold *et al.*, 1973).

6. Nutritional Significance and Metabolism of Inositol in Disease States

6.1. Inositol and Diabetes

There has recently been a great surge in interest in the relationship of inositol to various diseases and associated abnormalities since this topic was reviewed by Milhorat (1971). Diabetics are known to exhibit decreased peripheral motor and sensory nerve conduction velocities with or without evidence of polyneuropathy. Lowered concentrations of *myo*-inositol have been found in rabbit and rat nerves during the first stage of Wallerian degeneration when axonal disintegration was complete and before Schwann cells had proliferated (Kusama and Stewart, 1970). Green *et al.* (1975) have studied the relationship of inositol to impaired functioning of the peripheral nervous system in rats with acute streptozotocin diabetes. These workers demonstrated that experimental diabetes resulted in an impaired ability of these animals to maintain normal concentrations of free inositol in peripheral nerve which was related to a decreased motor nerve conduction velocity. This decrease in nerve free inositol occurred despite the fact that free inositol levels in plasma were similar in normal and diabetic rats. Dietary supplements of 1% inositol increased plasma and nerve levels of free inositol and significantly improved motor nerve conduction velocities in the diabetic rats. Insulin treatment was found to prevent the decrease in nerve inositol levels and the impaired nerve conduction velocity in the diabetic animals. Electron microscopy has revealed structural changes in nerve membranes in diabetes and their reversal by inositol and insulin administration (Fukuma *et al.*, 1978).

It is noteworthy that an excessive elevation in plasma inositol levels induced by feeding a diet containing 3% rather than 1% inositol decreased the motor nerve conduction velocity in both normals and diabetics. Palmano *et al.* (1977) have confirmed the observed decrease of free inositol in sciatic nerve of streptozotocin diabetic rats and also found a lowered concentration of lipid-bound inositol in the nerve of acutely diabetic animals. Whiting *et al.* (1979) have recently reported that the specific activity of the inositol 1-phosphate synthase involved in inositol biosynthesis was lower in the testis but not in the sciatic nerve of diabetic rats relative to controls. However, the specific activity of the enzymes responsible for the synthesis of phosphatidylinositol and phosphatidylinositol 4,5-bisphosphate was significantly lower in the sciatic nerve and brain of diabetic animals, which suggests that an altered metabolism of the inositol lipids in nerve tissue may contribute to the human diabetic neuropathy and to a decrease in the nerve conduction velocity. Hothersall and McLean (1979) observed a fall in the rate of [^{3}H]inositol incorporation into phosphatidylinositol in intact nerve segments from diabetic rats but not in broken cell preparations and suggested, therefore, that a depression in inositol transport occurs in diabetes.

The activity of the kidney inositol oxygenase which catabolizes inositol was markedly decreased in experimental diabetes, which may explain the elevated concentration of inositol in diabetic kidney and the increased clearance of inositol (Palmano *et al.* 1977; Whiting *et al.*, 1979). It has been suggested that the inositoluria that is observed in human diabetes may arise mainly from the inhibitory effect of glucose on renal tubular reabsorption of inositol (Clements and Reynertson, 1977). Thus, urinary inositol excretion can account for a significant fraction of the dietary inositol intake of the untreated diabetic. The urinary excretion is lowered toward normal levels with insulin treatment.

Clements and Reynertson (1977) have proposed that hyperglycemia in the untreated human diabetic may impair inositol transport, resulting in a widespread relative intracellular deficiency in man. A 3-g oral load of inositol was found to significantly elevate plasma inositol concentrations in human subjects, with diabetics showing a greater response. These workers suggested, therefore, that oral inositol supplementation might possibly be of benefit in the prevention and treatment of certain complications associated with human diabetes mellitus. In this regard, Salway *et al.* (1978) have investigated the effect of inositol on neurophysiological measurements in diabetic patients. Giving a 500-mg oral dose of inositol twice a day for 2 weeks increased the amplitude of the evoked action potentials of the median, sural, and popliteal nerves by an average of 76, 160, and 40%, respectively. These results indicate that inositol may provide a therapeutic role in diabetic neuropathy.

6.2. Inositol and Chronic Kidney Disorders

A dramatic elevation in serum levels of free inositol has been documented in human subjects with chronic renal failure (Clements *et al.*, 1973; Lewin *et al.*, 1974; Pitkänen, 1976). In normal subjects, inositol can be filtered through the renal glomeruli and reabsorbed by tubular cells in the kidney. Since the kidney is the major site for inositol catabolism in the body, it has been suggested that an impaired renal oxidation of inositol to D-glucuronate may be responsible for the abnormally elevated plasma inositol levels in uremic patients (Clements *et al.*, 1973). In advanced forms of glomerulonephritis, a decreased glomerular filtration rate and a disturbed inositol reabsorption are also present (Pitkänen, 1976). This has led to the suggestion that estimation of serum and urinary inositol has advantages in the evaluation of kidney function. Plasma inositol levels decrease during hemodialysis but to a lesser extent than the plasma urea nitrogen. By injecting [^{3}H]inositol into an antecubital vein of fasting subjects, Clements and Diethelm (1979) have observed that the half-time of inositol disappearance, which was prolonged in patients with chronic renal failure, was obviated following successful renal transplantation.

The potential toxic effects of abnormally evelvated plasma inositol levels have been studied in both experimental animals and human subjects. In addition to increasing plasma and tissue levels of free inositol, large doses of inositol were found to increase phosphatidylinositol levels in the endoplasmic reticulum of rat liver (Yagi and Kotaki, 1969) with no obvious deleterious effect. When studied morphologically by light microscopy, massive doses of inositol did not produce an appreciable change in the liver or kidney of male and female rats (Hasan *et al.*, 1974). The adverse effects of raised plasma inositol levels on peripheral nervous function have been examined in normal male rats by placing them on a diet enriched in inositol for 1 week (Clements *et al.*, 1973). A striking decrease in the sciatic nerve motor neuron conduction velocity developed in these animals which improved when they were restored to a normal diet. These latter experiments have been extended in the rat and support the possibility that hyperinositolemia may contribute to the pathogenesis of uremic polyneuropathy in subjects with chronic renal failure (De Jesus *et al.*, 1974). Based on experiments with uremic patients, Reznek *et al.* (1977) have shown that rises in plasma inositol concentration were related to a depression of sural nerve conduction velocity, but a relationship with clinically evident neuropathy was not established. It is of interest that Liveson *et al.* (1977) reported the development of cytoplasmic abnormalities within several days following exposure of dorsal root ganglion cells to levels of inositol analogous to the serum concentration found in uremic patients. It remains to be established whether a reduction in plasma inositol levels by dietary modification might prove to be beneficial in patients with chronic renal failure.

6.3. Inositol and Cardiovascular Disease

The potential effect of dietary inositol on serum cholesterol levels has been studied in the rat and other animals. Hayashi *et al.* (1974a) as well as Basarkar and Hatwalne (1975) found that the addition of inositol to semipurified diets did not significantly influence serum cholesterol levels. In contrast, it has been observed that dietary inositol deprivation produces a reduction in the concentration of serum free cholesterol in lactating dams (Burton and Wells, 1977). The effect of supplementing inositol-deficient diets with 0.1% inositol on cholesterol metabolism has been studied in the male gerbil, since this animal model appears suitable for studying the effect of dietary lipid on plasma cholesterol levels (W. Clark, D. B. Anderson, N. J. Mercer, and B. J. Holub, unpublished observations, 1979; Mercer and Holub, 1979). Inositol supplementation was found to elevate mean plasma cholesterol levels by 28% without influencing liver cholesterol concentrations. The aforementioned studies suggest that further research into the potential hypercholesterolemic effect of dietary inositol is warranted. It is of interest to note that a rapid metabolism of phosphatidylinositol, that appears to be influenced by cholesterol feeding, has been reported to occur in the pig and monkey aorta (Borensztajn *et al.*, 1973; Day *et al.*, 1974). Furthermore, the effect of different dietary fats on the recalcification (platelet-rich) clotting time in rabbits has been significantly correlated with the fatty acid composition of a platelet phospholipid fraction rich in phosphatidylinositol (Renaud and Gautheron, 1975).

6.4. Inositol and Other Diseases

An alteration in inositol metabolism has also been implicated in assorted other diseases, although, in most cases, no definitive interrelationships have been established. As would be expected from the discussion in Section 5.1, dietary inositol has the potential for exerting a lipotropic action in patients with fatty infiltration of the liver under certain conditions (Milhorat, 1971). It has been reported that an elevation in galactitol concentration and a depression in free and lipid-bound inositol levels develop in the brains of galactosemic infants or animals subjected to experimental galactose toxicity (Wells and Wells, 1967). Recent evidence indicates that the phosphatidylinositol response to acetylcholine is impaired in synaptosomes from galactose-fed rats, which suggests that these animals may be deficient in number of acetylcholine recpetors or have a defect in a step between receptor–neurotransmitter interaction and phosphatidylinositol breakdown (Warfield and Segal, 1978). Lipid analysis of a biopsy specimen from the liver of an adult patient with hepatosplenomegaly and hyperlipidemia has revealed a marked elevation of phosphatidylinositol (Yamamoto *et al.*, 1970). The level and fatty acid composition of phosphatidylinositol in the plasma membranes of fibroblasts from patients with cystic fibrosis and matched controls were found not to differ significantly (Riordan *et al.*, 1979).

Wood (1975) has compared the content and fatty acid composition of phosphatidylinositol in hepatoma, normal rat liver, and host liver of animals maintained on normal and fat-free diets. In general, phosphatidylinositol was much lower in concentration and richer in octadecanoic acid in hepatoma relative to normal liver, which may be related to the differential functioning of neoplastic cells as compared to normal cells. Diringer *et al.* (1977) have observed that conditions causing a cessation of growth by normal but not by tumor cells were accompanied by changed levels for glycerylphosphorylinositol and free inositol, suggesting, therefore, that these compounds may be involved in the regulation of cell growth. It is of interest that the lymphocytes from multiple sclerosis patients have exhibited a lower incorporation of [^{3}H]inositol into phosphatidylinositol than those from control patients when stimulated by phytohemagglutinin (Offner *et al.*, 1974), since an immunologic abnormality appears to be involved in the pathogenesis of multiple sclerosis.

7. Summary

Recent advances in nutritional and biochemical research have substantiated the importance of inositol as a dietary and cellular constituent. The processes involved in the metabolism of inositol and its derivatives in mammalian tissues have been characterized both *in vivo* and at the enzyme level. Biochemical functions elucidated for phosphatidylinositol in biological membranes include the mediation of cellular responses to external stimuli, nerve transmission, and the regulation of enzyme activity through specific interactions with various proteins. Inositol deficiency in animals has been shown to produce an accumulation of triglyceride in liver, intestinal lipodystrophy, and other abnormalities. The metabolic mechanisms giving rise to these latter phenomena have been extensively studied as a function of dietary inositol. Altered metabolism of inositol has been documented in patients with diabetes mellitus, chronic renal failure, galactosemia, and multiple sclerosis. A moderate increase in plasma and nerve inositol levels by dietary supplementation has been suggested as a means of treating diabetic neuropathy, although excessively high levels, such as are found in uremic patients, may be neurotoxic. A thorough consideration of the biochemical functions of inositol and a further characterization of various diseases with the aid of appropriate animal models may suggest a possible role for inositol and other dietary components in their prevention and treatment.

References

Agranoff, B. W., 1978, Cyclitol confusion, *Trends Biochem. Sci.* **3**:283.

Agranoff, B. W., Bradley, R. M., and Brady, R. O., 1958, The enzymatic synthesis of inositol phosphatide, *J. Biol. Chem.* **233**:1077.

Andersen, D. B., 1977, *Inositol—A Lipotrophic Factor,* M.Sc. Thesis, University of Guelph, Guelph, Ontario.

Andersen, D. B., and Holub, B. J., 1976, The influence of dietary inositol on glyceride composition and synthesis in livers of rats fed different fats, *J. Nutr.* **106**:529.

Andersen, D. B., and Holub, B. J., 1980a, *myo*-Inositol-responsive liver lipid accumulation in the rat, *J. Nutr.* **110**:488.

Andersen, D. B., and Holub, B. J., 1980b, The relative response of hepatic lipids in the rat to graded levels of dietary *myo*-inositol and other lipotropes, *J. Nutr.* **110**:496.

Angyal, S. J., and Anderson, L., 1959, The cyclitols, *Advances in Carbohydrate Chemistry,* Vol. 14 (M. L. Wolfrom, ed.), pp. 135–212, Academic Press, New York.

Baker, R. R., and Thompson, W., 1973, Selective acylation of 1-acylglycerophosphorylinositol by rat brain microsomes: Comparison with 1-acylglycerophosphoryl choline, *J. Biol. Chem.* **248**:7060.

Basarkar, P. W., and Hatwalne, V. G., 1975, Studies on hypocholesterolemic action of *m*-inositol, quercetin and epicatechin, *Baroda J. Nutr.* **2**:99.

Benjamins, J. A., and Agranoff, B. W., 1969, Distribution and properties of CDP-diglyceride: Inositol transferase from brain, *J. Neurochem.* **16**:513.

Best, C. H., Lucas, C. C., Patterson, J. M., and Ridout, J. H., 1951, The rates of lipotropic action of choline and inositol under special dietary conditions, *Biochem. J.* **48**:452.

Bishop, H. H., and Strickland, K. P., 1970, On the specificity of cytidine diphosphate diglycerides in monophosphoinositide biosynthesis by rat brain preparations, *Can. J. Biochem.* **48**:269.

Borensztajn, J., Getz, G. S., and Wissler, R. W., 1973, The *in vitro* incorporation of [^{3}H]thymidine into DNA and ^{32}P into phospholipids and RNA in the aorta of rhesus monkeys during early atherogenesis, *Atherosclerosis* **17**:269.

Borgese, T. A., and Nagel, R. L., 1977, Differential effects of 2,3-DPG, ATP and inositol pentaphosphate (IP5) on the oxygen equilibria of duck embryonic fetal and adult hemoglobins, *Comp. Biochem. Physiol.* **56A**:539.

Borgese, T. A., and Nagel, R. L., 1978, Inositol pentaphosphate in fish red blood cells, *J. Exp. Zool.* **205**:133.

Broekhuyse, R. M., 1971, Lipids in tissues of the eye. V. Phospholipid metabolism in normal and cataractous eyes, *Biochim. Biophys. Acta* **231**:360.

Brophy, P. J., and Aitken, J. W., 1979, Phosphatidylinositol transfer activity in rat cerebral hemispheres during development, *J. Neurochem.* **33**:355.

Brophy, B. J., Burbach, P., Nelemans, S. A., Westerman, J., Wirtz, K. W. A., and Van Deenen, L. L. M., 1978, The distribution of phosphatidylinositol in microsomal membranes from rat liver after biosynthesis *de novo*, *Biochem. J.* **174**:413.

Buckley, J. T., 1977, Properties of human erythrocyte phosphatidylinositol kinase and inhibition by adenosine, ADP, and related compounds, *Biochim. Biophys. Acta* **498**:1.

Buckley, J. T., and Hawthorne, J. N., 1972, Erythrocyte membrane polyphosphoinositide metabolism and the regulation of calcium binding. *J. Biol. Chem.* **247**:7218.

Bull, H. S., 1968, Fatty liver syndrome in laying hens, in *Proceedings 23rd Annual Texas Nutrition Conference,* pp. 219–225.

Burton, L. E., and Wells, W. W., 1974, Studies on the developmental pattern of the enzymes converting glucose-6-phosphate to *myo*-inositol in the rat, *Dev. Biol.* **37**:35.

Burton, L. E., and Wells, W. W., 1976, *myo*-Inositol metabolism during lactation and development in the rat. The prevention of lactation-induced fatty liver by dietary *myo*-inositol, *J. Nutr.* **106**:1617.

Burton, L. E., and Wells, W. W., 1977, Characterization of the lactation dependent fatty liver in *myo*-inositol deficient rats, *J. Nutr.* **107**:1871.

Burton, L. E., and Wells, W. W., 1979, *myo*-Inositol deficiency: Studies on the mechanism of lactation-dependent fatty liver formation in the rat, *J. Nutr.* **109**:1483.

Burton, L. E., Ray, R. E., Bradford, J. R., Orr, J. P., Nickerson, J. A., and Wells, W. W., 1976, *myo*-Inositol metabolism in the neonatal and developing rat fed a *myo*-inositol-free diet, *J. Nutr.* **106**:1610.

Caspary, W. F., and Crane, R. K., 1970, Active transport of *myo*-inositol and its relation to the sugar transport system in hamster small intestine, *Biochim. Biophys. Acta* **203**:308.

Charalampous, F. C., 1971, Metabolic functions of *myo*-inositol, VIII. Role of inositol in Na^+–K^+transport and in Na^+- and K^+-activated adenosine triphosphatase of KB cells, *J. Biol. Chem.* **246**:455.

Chen, C. H., and Eisenberg, F., Jr., 1975, Myoinosose-2 1-phosphate: An intermediate in the myoinositol 1-phosphate synthetase reaction, *J. Biol. Chem.* **250**:2963.

Chu, S. W., and Hegsted, D. M., 1979, Changes in the intestinal composition and transport of lipids in inositol deficient gerbils, *Fed. Proc.* **38**:279.

Clements, R. S., Jr., and Diethelm, A. G., 1979. The metabolism of *myo*-inositol by the human kidney, *J. Lab. Clin. Med.* **93**:210.

Clements, R. S., Jr., and Reynertson, R., 1977, *myo*-Inositol metabolism in diabetes mellitus: Effect of insulin treatment, *Diabetes* **26**:215.

Clements, R. S., Jr., and Rhoten, W. B., 1976, Phosphoinositide metabolism and insulin secretion from isolated rat pancreatic islets, *J. Clin. Invest.* **57**:684.

Clements, R. S., Jr., De Jesus, P. V., Jr., and Winegrad, A. I., 1973, Raised plasma-myoinositol levels in uremia and experimental neuropathy, *Lancet* **1**:1377.

Cockcroft, S., and Gomperts, B. D., 1979, Evidence for a role of phosphatidylinositol turnover in stimulus–secretion coupling, *Biochem. J.* **178**:681.

Cohen, P., Broekman, M. J., Verkley, A., Lisman, W. W., and Derksen, A., 1971, Quantification of human platelet inositides and the influence of ionic environment on their incorporation of orthophosphate-^{32}P, *J. Clin. Invest.* **50**:762.

Cooper, P. H., and Hawthorne, J. N., 1976, Phosphatidylinositol kinase and diphosphoinositide kinase of rat kidney cortex, *Biochem. J.* **160**:97.

Daum, G., Gamerith, G., and Paltauf, F., 1979, The effect of cerulenin and exogenous fatty acids on triacylglycerol accumulation in an inositol-deficient yeast, *Saccharomyces carlsbergensis, Biochim. Biophys. Acta* **573**:413.

Davies, M. I., Ritcey, G. M., and Motzok, I., 1970, Intestinal phytase and alkaline phosphatase of chicks: Influence of dietary calcium, inorganic and phytate phosphorus and vitamin D_3, *Poult. Sci.* **49**:1280.

Dawson, R. M. C., and Clarke, N., 1972, D-*myo*-Inositol 1:2-cyclic-phosphate 2-phosphohydrolase, *Biochem. J.* **127**:113.

Dawson, R. M. C., and Hemington, N., 1977, A phosphodiesterase in rat kidney cortex that hydrolyses glycerylphosphorylinositol, *Biochem. J.* **162**:241.

Dawson, R. M. C., Freinkel, N., Jungalwala, F. B., and Clarke, N., 1971, The enzymatic formation of *myo*-inositol 1:2-cyclic phosphate from phosphatidylinositol, *Biochem. J.* **122**:605.

Dawson, R. M. C., Hemington, N., Richards, D. E., and Irvine, R. F., 1979, *sn*-Glycero(3)phosphoinositol glycerophosphohydrolase: A new phosphodiesterase in rat tisssues, *Biochem. J.* **182**:39.

Day, A. J., Bell, F. P., and Schwartz, C. J., 1974, Lipid metabolism in focal areas of normal-fed and cholesterol-fed pig aortas, *Exp. Mol. Pathol.* **21**:179.

De Jesus, P. V., Jr., Clements, R. S., Jr., and Winegrad, A. I., 1974, Hypermyoinositolemic polyneuropathy in rats: A possible mechanism for uremic polyneuropathy, *J. Neurol. Sci.* **21**:237.

Diringer, H., Koch-Kallnbach, M. E., and Friis, R. R., 1977, Quantitative determination of *myo*-inositol, inositol 1-phosphate, inositol cyclic 1:2-phosphate and glycerylphosphoinositol in normal and Rous-sarcoma-virus-transformed quail fibroblasts under different growth conditions, *Eur. J. Biochem.* **81**:551.

Durell, J., Garland, J. T., and Friedel, R. O., 1969, Acetylcholine action: Biochemical aspects, *Science* **165**:862.

Eagle, H., Oyama, V. I., Levy, M., and Freeman, A. E., 1957. *myo*-Inositol as an essential growth factor for normal and malignant human cells in tissue culture, *J. Biol. Chem.* **266**:191.

Edalji, R., Benesch, R. E., and Benesch, R., 1976, Binding of inositol hexaphosphate to deoxyhemoglobin, *J. Biol. Chem.* **251**:7720.

Eichberg, J., and Hauser, G., 1973, The subcellular distribution of polyphosphoinositides in myelinated and unmyelinated rat brain, *Biochim. Biophys. Acta* **326**:210.

Eisenberg, F., Jr., 1967, D-*myo*-Inositol 1-phosphate as product of cyclization of glucose-6-phosphate and substrate for a specific phosphatase in rat testis, *J. Biol. Chem.* **242**:1375.

Eisenberg, F., Jr., and Bolden, A. H., 1963, Biosynthesis of inositol in rat testis homogenate, *Biochem. Biophys. Res. Commun.* **12**:72.

Eisenberg, F., Jr., and Bolden, A. H., 1964, Reproductive tract as site of synthesis and secretion of inositol in the male rat, *Nature* **202**:599.

Freinkel, N., Antony, G., Williams, H., and Landau, B. R., 1970, Metabolism of *myo*-inositol in rabbit kidney and in man, *Biochim. Biophys. Acta* **201**:425.

Freinkel, N., El Younsi, C., and Dawson, R. M. C., 1975, Inter-relations between the phospholipids of rat pancreatic islets during glucose stimulation and their response to medium inositol and tetracaine, *Eur. J. Biochem.* **59**:245.

Friedel, R. O., and Schanberg, S. M., 1971, Incorporation *in vivo* of intracisternally injected ^{33}Pi into phospholipids of rat brain, *J. Neurochem.* **18**:2191.

Fukuma, M., Carpentier, J. L., Orci, L., Greene, D. A., and Winegrad, A. I., 1978, An alteration in internodal myelin membrane structure in large sciatic nerve fibers in rats with acute streptozotocin diabetes and impaired nerve conduction velocity, *Diabetologia* **15**:65.

Garrett, R. J., and Redman, C. M., 1975, Localization of enzymes involved in polyphosphoinositide metabolism on the cytoplasmic surface of the human erythrocyte membrane, *Biochim. Biophys. Acta* **382**:58.

Gavin, G., and McHenry, E. W., 1941, Inositol: A lipotropic factor, *J. Biol. Chem.* **139**:485.

Ghafoorunissa, 1975, Effect of dietary protein on the biosynthesis of inositol in rat testis, *J. Reprod. Fertil.* **42**:233.

Ghafoorunissa, 1976, Effect of dietary protein and inositol on sperm metabolism and fructose content of male accessory sex organs of rat, *Indian J. Exp. Biol.* **14**:564.

Gonzalez-Sastre, F., and Folchi-Pi, J., 1968, Thin-layer chromatography of phosphoinositides, *J. Lipid Res.* **9**:532.

Gonzalez-Sastre, F., Eichberg, J., and Hauser, G., 1971, Metabolic pools of polyphosphoinositides in rat brain, *Biochim. Biophys. Acta* **248**:96.

Goodhart, R. S., 1973, Bioflavonoids, in: *Modern Nutrition in Health and Disease* (R. Goodhart and M. Shils, eds.), pp. 259–267, Lea and Febiger, Philadelphia.

Greene, D. A., De Jesus, P. V., and Winegrad, A. I., 1975, Effects of insulin and dietary myoinositol on impaired peripheral motor nerve conduction velocity in acute streptozotocin diabetes, *J. Clin. Invest.* **55**:1326.

Griffin, H. D., and Hawthorne, J. N., 1978, Calcium-activated hydrolysis of phosphatidyl-*myo*-inositol 4-phosphate and phosphatidyl-*myo*-inositol 4,5-bisphosphate in guinea-pig synaptosomes, *Biochem. J.* **176**:541.

Guarnieri, M., 1975, Reaction of anti-phosphatidyl inositol antisera with neutral membranes, *Lipids* **10**:294.

Halliday, J. W., and Anderson, L., 1955, The synthesis of *myo*-inositol in the rat, *J. Biol. Chem.* **217**:797.

Hamilton, P. B., and Garlich, J. D., 1972, Failure of vitamin supplementation to alter the fatty liver syndrome caused by aflatoxin, *Poul. Sci.* **51**:688.

Handler, P., 1946, Dietary factors in the regulation of liver lipid concentration, *J. Biol. Chem.* **162**:77.

Hankes, L. V., Politzer, W. M., Touster, O., and Anderson, L., 1969, *myo*-Inositol catabolism in human pentosurics: The predominant role of the glucuronate–xylulose–pentose phosphate pathway, *Ann. N.Y. Acad. Sci.* **165**:564.

Hasan, S. H., Kotaki, A., and Yagi, K., 1970, Studies on myoinositol. VI. Effect of myoinositol on plasma lipoprotein metabolism of rats suffering from fatty liver, *J. Vitaminol.* **16**:144.

Hasan, S. H., Nakagawa, Y., Nishigaki, I., and Yagi, K., 1971, Studies on myoinositol VIII. The incorporation of ^{3}H-myoinositol into phosphatidylinositol of fatty liver, *J. Vitaminol.* **17**:159.

Hasan, S. H., Nishigaki, I., Tsutsui, Y., and Yaki, K., 1974, Studies on myoinositol. IX. Morphological examination of the effect of massive doses of myoinositol on liver and kidney of rat, *J. Nutr. Sci. Vitaminol. (Tokyo)* **20**:55.

Hauser, G., 1963, The formation of free and lipid *myo*-inositol in the intact rat, *Biochim Biophys. Acta* **70**:278.

Hauser, G., 1969, *myo*-Inositol transport in slices of rat kidney cortex, II. Effect of the ionic composition of the medium, *Biochim. Biophys. Acta* **173**:267.

Hauser, G., and Finelli, V. M., 1963, The biosynthesis of free and phosphatide *myo*-inositol from glucose by mammalian tissue slices, *J. Biol. Chem.* **238**:3224.

Hawthorne, J. N., and Pickard, M. R., 1979, Phospholipids in synaptic function, *J. Neurochem.* **32**:5.

Hayashi, E., Maeda, T., and Tomita, T., 1974a, The effects of *myo*-inositol deficiency on lipid metabolism in rats I. The alteration of lipid metabolism in *myo*-inositol deficient rats, *Biochim. Biophys. Acta* **360**:134.

Hayashi, E., Maeda, T., and Tomita, T., 1974b, The effect of *myo*-inositol deficiency on lipid metabolism in rats. II. The mechanism of triacylglycerol accumulation in the liver of *myo*-inositol-deficient rats, *Biochim. Biophys. Acta* **360**:146.

Hayashi, E., Hasegawa, R., and Tomita, T., 1976, Accumulation of neutral lipids in *Sacchromyces carlsbergensis* by *myo*-inositol deficiency and its mechanism, *J. Biol. Chem.* **251**:5759.

Hayashi, E., Hasegawa, R., and Tomita, T., 1978a, The fluctuation of various enzyme activities due to *myo*-inositol deficiency in *Saccharomyces carlsbergensis, Biochim. Biophys. Acta* **540**:231.

Hayashi, E., Maeda, T., Hasegawa, R., and Tomita, T., 1978b, The effect of *myo*-inositol deficiency on lipid metabolism in rats. III. The mechanism of an enhancement in lipolysis due to *myo*-inositol deficiency in rats, *Biochim. Biophys. Acta* **531**:197.

Haye, B., and Jacquemin, C., 1977, Incorporation of [^{14}C] arachidonate in pig thyroid lipids and prostaglandins, *Biochim. Biophys. Acta* **487**:231.

Haye, B., Champion, S., and Jacquemin, C., 1973, Control by TSH of a phospholipase A_2 activity, a limiting factor in the biosynthesis of prostaglandins in the thyroid, *FEBS Lett.* **30**:253.

Heger, H. W., and Peter, H. W., 1977, Phosphatidylinositol as essential constituent of the acetyl-CoA carboxylase from rat liver, *Int. J. Biochem.* **8**:841.

Hegsted, D. M., Hayes, K. C., Gallagher, A., and Hanford, H., 1973, Inositol deficiency: An intestinal lipodystrophy in the gerbil, *J. Nutr.* **103**:302.

Hegsted, D. M., Gallagher, A., and Hanford, H., 1974, Inositol requirement of the gerbil, *J. Nutr.* **104**:588.

Helmkamp, G. M., Wirtz, K. W. A., and van Deenen, L. L. M., 1976, Phosphatidylinositol exchange protein: Effects of membrane structure on activity and evidence for a ping-pong mechanism, *Arch. Biochem. Biophys.* **174**:592.

Hitzemann, R. J., Natsuki, R., and Loh, H. H., 1978, Effects of nicotine on brain 1-phosphatidylinositol-4-phosphate and 1-phosphatidylinositol-3,4-biphosphate synthesis and metabolism—possible relationship to nicotine-induced behaviors, *Biochem. Pharmacol.* **27**:2519.

Hokin-Neaverson, M., and Sadeghian, K., 1976, Separation of [^{3}H]inositol monophosphates and [^{3}H]inositol on silica gel glass fiber sheets, *J. Chromatog.* **120**:502.

Hokin-Neaverson, M., Sadeghian, K., Harris, D. W., and Merrin, J. S., 1978, The mechanism of stimulated phosphatidylinositol breakdown, in: *Cyclitols and Phosphoinositides* (W. W. Wells and F. Eisenberg, Jr., eds.), pp. 349–360, Academic Press, New York.

Holub, B. J., 1974, The Mn^{2+}-activated incorporation of inositol into molecular species of phosphatidylinositol in rat liver microsomes, *Biochim. Biophys. Acta* **369**:111.

Holub, B. J., 1975, Role of cytidine triphosphate and cytidine diphosphate choline in promoting inositol entry into microsomal phosphatidylinositol, *Lipids* **10**:483.

Holub, B. J., 1976, Specific formation of arachidonoyl phosphatidylinositol from 1-acyl-*sn*-glycero-3-phosphorylinositol in rat liver, *Lipids* **11**:1.

Holub, B. J., 1978, Studies on the metabolic heterogeneity of different molecular species of phosphatidylinositols, in *Cyclitols and Phosphoinositides* (W. Wells and F. Eisenberg, Jr., eds.), pp. 523–534, Academic Press, New York.

Holub, B. J., and Kuksis, A., 1971, Differential distribution of orthophosphate-^{32}P and glycerol-^{14}C among molecular species of phosphatidylinositols of rat liver *in vivo, J. Lipid Res.* **12**:699.

Holub, B. J., and Kuksis, A., 1972, Further evidence for the interconversion of monophosphoinositides *in vivo, Lipids* **7**:78.

Holub, B. J., and Piekarski, J., 1979, The formation of phosphatidylinositol by acylation of 2-acyl-*sn*-glycero-3-phosphorylinositol in rat liver microsomes, *Lipids* **14**:529.

Holub, B. J., Kuksis, A., and Thompson, W., 1970, Molecular species of mono-, di, and triphosphoinositides of bovine brain, *J. Lipid Res.* **11**:558.

Hoover, G. A., Nicolosi, R. J., Corey, J. E., El Losy, M., and Hayes, K. C., 1978, Inositol deficiency in the gerbil: Altered hepatic lipid metabolism and triglyceride secretion, *J. Nutr.* **108**:1588.

Hothersall, J. S., and McLean, P., 1979, Effect of experimental diabetes and insulin on phosphatidylinositol synthesis in rat sciatic nerve, *Biochem. Biophys. Res. Commun.* **88**:477.

Howard, C. F., Jr., and Anderson, L., 1967, Metabolism of *myo*-inositol in animals, II. Complete catabolism of *myo*-inositol-^{14}C by rat kidney slices, *Arch. Biochem. Biophys.* **118**:332.

Hubscher, C., 1962, Metabolism of phospholipids: VI. The effect of metal ions on the incorporation of L-serine into phosphatidylserine, *Biochim. Biophys. Acta* **57**:555.

Irvine, R. F., and Dawson, R. M. C., 1978, The distribution of calcium-dependent phosphatidylinositol-specific phosphodiesterase in rat brain, *J. Neurochem.* **31**:1427.

Irvine, R. F., Hemington, N., and Dawson, R. M. C., 1978, The hydrolysis of phosphatidylinositol by lysosomal enzymes of rat liver and brain, *Biochem. J.* **176**:475.

Isaacks, R., Harkness, D., Sampsell, R., Adler, J., Roth, S., Kim, C., and Goldman, P., 1977, Studies on avian erythrocyte metabolism: Inositol tetrakisphosphate: The major phosphate compound in the erythrocytes, *Eur. J. Biochem.* **77**:567.

IUPAC-IUB Commission on Biochemical Nomenclature, 1977, Nomenclature of phosphorus-containing compounds of biochemical importance (recommendations, 1976), *Proc. Natl. Acad. Sci. U.S.A.* **74**:2222.

Jafferji, S. S., and Michell, R. H., 1976, Effects of calcium antagonistic drugs on the stimulation by carbamylcholine and histamine of phosphatidylinositol turnover in longitudinal smooth muscle of guinea-pig ileum, *Biochem. J.* **160**:163.

Johnson, L. F., and Tate, M. E., 1969, Structure of phytic acids, *Can. J. Chem.* **47**:63.

Jungalwala, F. B., and Dawson, R. M. C., 1971, The turnover of myelin phospholipids in the adult and developing rat brain, *Biochem. J.* **123**:683.

Kai, M., and Hawthorne, J. N., 1969, Physiological significance of polyphosphoinositides in brain, *Ann. N.Y. Acad. Sci.* **165**:761.

Karasawa, K., 1972, The effect of carbohydrate and inositol on the growth of rats, *Jpn. J. Nutr.* **30**:3.

Keough, K. M. W., and Thompson, W., 1972, Soluble and particulate forms of phosphoinositide phosphodiesterase in ox brain, *Biochim. Biophys. Acta* **270**:324.

Kirazov, E. P., and Lagnado, J. R., 1977, Interaction of *myo*-inositol with brain microtubules, *FEBS Lett.* **81**:173.

Kirk, C. J., Rodrigues, L. M., and Hems, D. A., 1979, The influence of vasopressin and related peptides on glycogen phosphorylase activity and phosphatidylinositol metabolism in hepatocytes, *Biochem. J.* **178**:493.

Koch, M. A., and Diringer, H., 1974, Isolation of cyclic inositol-1,2-phosphate from mammalian cells and a probable function of phosphatidylinositol turnover, *Biochem. Biophys. Res. Commun.* **58**:361.

Koch-Kallnbach, M. E., and Diringer, H., 1977, Isolation and separation of inositol 1-phosphate, cyclic inositol 1,2-phosphate, and glycerylphosphoinositol from tissue culture cells labelled with [^{3}H]inositol, *Hoppe-Seylers Z. Physiol. Chem.* **358**:367.

Kotaki, A., Sakurai, T., Kobayashi, M., and Yagi, K., 1968, Studies on *myo*inositol. IV: Effects of myoinositol on the cholesterol metabolism of rats suffering from experimental fatty liver, *J. Vitaminol.* **14**:87.

Kroes, J. F., Hegsted, D. M., and Hayes, K. C., 1973, Inositol deficiency in gerbils: Dietary effects on the intestinal lipodystrophy, *J. Nutr.* **103**:1448.

Kubena, L. F., Deaton, J. W., Chen, T. C., and Reece, F. N., 1974, Factors influencing the quality of abdominal fat in broilers. 1. Rearing temperature, sex, age or weight, and dietary choline chloride and inositol supplementation, *Poult. Sci.* **53**:211.

Kusama, H., and Stewart, M. A., 1970, Levels of *Myo*-inositol in normal and degenerating periopheral nerve, *J. Neurochem.* **17**:317.

Lapetina, E. G., and Zeiher, L. M., 1976, Phosphatidylinositol metabolism and *myo*-inositol 1,2-cyclic phospate action in smooth muscle, *Adv. Exp. Med. Biol.* **72**:257.

Lee, T. C., and Huggins, C. G., 1968, Triphosphoinositide phosphomonoesterase in rat kidney cortex. 1. General properties and subcellular locations, *Arch. Biochem. Biophys.* **126**:206.

LeKim, D., and Betzing, H., 1976, Intestinal absorption of polyunsaturated phosphatidylcholine in the rat, *Hoppe-Seylers Z. Physiol. Chem.* **357**:1321.

Lewin, L. M., and Beer, R., 1973, Prostatic secretion as the source of *myo*-inositol in human seminal fluid, *Fertil. Steril.* **24**:666.

Lewin, L. M., and Sulimovici, S., 1975, The distribution of radioactive *myo*-inositol in the reproductive tract of the male rat, *J. Reprod. Fertil.* **43**:355.

Lewin, L. M., Melmed, S., and Bank, H., 1974, Rapid screening test for detection of elevated *myo*-inositol levels in human blood serum, *Clin. Chim. Acta* **54**:377.

Lewin, L. M., Yannai, Y., Sulimovici, S., and Kraicer, P. F., 1976, Studies on the metabolic role of *myo*-inositol, *Biochem. J.* **156**:375.

Liveson, J. A., Gardner, J., and Bornstein, M. B., 1977, Tissue culture studies of possible uremic neurotoxins: *myo*Inositol, *Kidney Int.* **12**:131.

Lloyd, T., 1979, The effects of phosphatidylinositol on tyrosine hydroxylase: Stimulation and inactivation, *J. Biol. Chem.* **254**:7247.

Loewus, M. W., 1977, Hydrogen isotope effects in the cyclization of D-glucose 6-phosphate by *myo*-inositol-1-phosphate synthase, *J. Biol. Chem.* **252**:7221.

Low, M. G., and Finean, J. B., 1976, The action of phosphatidylinositol-specific phospholipases C on membranes, *Biochem. J.* **154**:203.

Low, M. G., and Finean, B. J., 1978, Specific release of plasma membrane enzymes by a phosphatidylinositol-specific phospholipase C, *Biochim. Biophys. Acta* **508**:564.

Luthra, M. G., and Sheltawy, A., 1972, The fractionation of phosphatidylinositol into molecular species by thin-layer chromatography on silver nitrate-impregnated silica gel, *Biochem. J.* **126**:1231.

Luthra, M. G.,and Sheltawy, A., 1976, The metabolic turnover of molecular species of phosphatidylinositol and its precursor phosphatidic acid in guinea-pig cerebral hemispheres, *J. Neurochem.* **27**:1503.

MacDonald, G., Baker, R. R., and Thompson, W., 1975, Selective synthesis of molecular classes of phosphatidic acid, diacylglycerol and phosphatidylinositol in rat brain, *J. Neurochem.* **24**:655.

Mandersloot, J. G., Roelofsen, B., and De Grier, J., 1978, Phosphatidylinositol as the endogenous activator of the (Na^+ + K^+)-ATPase in microsomes of rabbit kidney, *Biochim. Biophys. Acta* **508**:478.

Marion, J., and Wolfe, L. S., 1979, Origin of the arachidonic acid released post-mortem in rat forebrain, *Biochim. Biophys. Acta* **574**:25.

Martin, G. J., 1941, The mouse antialopecia factor, *Science* **93**:422.

Mercer, N. J. H., and Holub, B. J., 1979, Response of free and esterified plasma cholesterol levels in the mongolian gerbil to the fatty acid composition of dietary lipid, *Lipids*, **14**:1009.

Michell, R. H., 1975, Inositol phospholipids and cell surface receptor function, *Biochim. Biophys. Acta* **415**:81.

Michell, R. H., 1979, Inositol phospholipids in membrane function, *Trends Biol. Sci.* **4**:128.

Michell, R. H., and Lapetina, E. G., 1972, Production of cyclic inositol phosphate in stimulated tissues, *Nature [New Biol.]* **240**:258.

Michell, R. H., Hawthorne, J. N., Coleman, R., and Karnovsky, M. L., 1970, Extraction of polyphosphoinositides with neutral and acidified solvents: A comparison of guinea-pig brain and liver, and measurements of rat liver inositol compounds which are resistant to extraction, *Biochim. Biophys. Acta* **210**:86.

Middleton, A., and Setchell, B. P., 1972, The origin of inositol in the rete testis fluid of the ram, *J. Reprod. Fertil.* **30**:473.

Milhorat, A. I., 1971, Inositols XI. Deficiency effects in human beings in *The Vitamins* Vol. III. (W. H. Sebrell, Jr. and R. S. Harris, eds.), pp. 398–405, Academic Press, New York.

Miller, J. C., 1977, A study of the kinetics of the muscarinic effect on phosphatidylinositol and phosphatidic acid metabolism in rat brain synaptosomes, *Biochem. J.* **168**:549.

Morris, R. N., and Collins, A. C., 1971, Biosynthesis of *myo*inositol by rat testis following surgically induced cryptorchidism or treatment with triethylenemelamine, *J. Reprod. Fertil.* **27**:201.

Naccarato, W. F., and Wells, W. W., 1974, Identification of 6-O-β-D-galactopyranosyl *myo*-inositol: A new form of *myo*-inositol in mammals. *Biochem. Biophys. Res. Commun.* **57**:1026.

Naccarato, W. F., Ray, R. E., and Wells, W. W., 1974, Biosynthesis of *myo*-inositol in rat mammary gland. Isolation and properties of the enzymes, *Arch. Biochem. Biophys.* **164**:194.

Naccarato, W. F., Ray, R. E., and Wells, W. W., 1975, Characterization and tissue distribution of 6-O-β-D-galactopyranosyl *myo*-inositol in the rat, *J. Biol. Chem.* **250**:1872.

National Academy of Sciences, 1972, *Nutrient Requirements of Domestic Animals, 10. Nutrient Requirements of Laboratory Animals, Second Edition,* National Academy of Sciences, Washington.

Nicolosi, R. J., Herrera, M. G., El Lozy, M., and Hayes, K. C., 1976, Effect of dietary fat on hepatic metabolism of ^{14}C-oleic acid and very low density lipoprotein triglyceride in the gerbil, *J. Nutr.* **106**:1279.

Nijjar, M. S., and Hawthorne, H. N., 1977, Purification and properties of polyphosphoinositide phosphomonoesterase from rat brain, *Biochim. Biophys. Acta* **480**:390.

Nilsson, O. S., and Dallner, G., 1977a, Enzyme and phospholipid asymmetry in liver microsomal membranes, *J. Cell Biol.* **72**:568.

Nilsson, O. S., and Dallner, G., 1977b, Transverse asymmetry of phospholipids in subcellular membranes of rat liver, *Biochim. Biophys. Acta* **464**:453.

Oberleas, D., 1971, The determination of phytate and inositol phosphates, in *Methods of Biochemical Analysis,* Vol. 20 (D. Glick, ed.) pp. 87–101, Wiley, New York.

Oberleas, D., 1973, Phytates, in *Toxicants Occurring Naturally in Foods,* pp. 363–371, National Academy of Sciences, Washington.

Offner, H., Konat, G., and Clausen, J., 1974, Effect of phytohemagglutinin, basic protein and measles antigen on myo-(2-^{3}H)inositol incorporation into phosphatidylinositol of lymphocytes from patients with multiple sclerosis, *Acta Neurol. Scand.* **50**:791.

Okazaki, K., 1975, Evidence for existence and a tentative identification of coenzyme in yeast thiamine pyrophosphokinase, *Biochem. Biophys. Res. Commun.* **64**:20.

Palmano, K. P., Whiting P. H., and Hawthorne, J. N., 1977, Free and lipid *myo*-inositol in tissues from rats with acute and less severe streptozotocin-induced diabetes, *Biochem. J.* **167**:229.

Palmer, F. B. St. C., 1977, The enzymatic preparation of diphosphoinositides, *Prepar. Biochem.* **7**:457.

Parthasarathy, S., Subbaiah, P. V., and Ganguly, J., 1974, The mechanism of intestinal absorption of phosphatidylcholine in rats, *Biochem. J.* **140**:503.

Paultauf, F., and Johnston, J. M., 1970, Lipid metabolism in inositol-deficient yeast, *Saccharomyces carlsbergensis.* 1. Influence of different carbon sources on the lipid composition of deficient cells, *Biochim. Biophys. Acta* **218**:424.

Paulus, H., and Kennedy, E. P., 1960, The enzymatic synthesis of inositol monophosphatide, *J. Biol. Chem.* **235**:1303.

Pearce, J., 1972, The lack of effect of dietary inositol supplementation on egg production and liver lipid metabolism in the laying hen, *Poul. Sci.* **51**:1998.

Pearce, J., 1975, The effects of choline and inositol on hepatic lipid metabolism and the incidence of the fatty liver and kidney syndrome in broilers, *Br. Poult. Sci.* **16**:1975.

Pickard, M. R., and Hawthorne, J. N., 1978a, Does *myo*-inositol specifically interact with brain microtubules? *FEBS Lett.* **93**:78.

Pickard, M. R., and Hawthorne, J. N., 1978b, The labelling of nerve ending phospholipids in guinea-pig brain *in vivo* and the effect of electrical stimulation on phosphatidylinositol metabolism in prelabelled synaptosomes, *J. Neurochem.* **30**:145.

Pitkänen, E., 1976, Changes in serum and urinary *myo*-inositol levels in chronic glomerulonephritis, *Clin. Chim. Acta* **71**:461.

Rao, R. H., and Strickland, K. P., 1974, On the solubility, stability and partial purification of CDP diacyl-*sn*-glycerol: inositol transferase from rat brain, *Biochim. Biophys. Acta* **348**:306.

Reed, J. R., Deacon, L. E., Farr, F., Gouch, J. R., 1968, Inositol and the fatty liver syndrome, in: *Proceedings, 23rd Annual Texas Nutrition Conference,* pp. 204–218.

Reinhold, J. G., Nasr, K., Lahimgarzadeh, A., and Hedayati, H., 1973, Effects of purified phytate and phytate-rich bread upon metabolism of zinc, calcium, phosphorous and nitrogen in man, *Lancet* **1**:283.

Renaud, S., and Gautheron, P., 1975, Influence of dietary fats on atherosclerosis, coagulation and platelet phospholipids in rabbits, *Atherosclerosis* **21**:115.

Reznek, R. H., Salway, J. G., and Thomas, P. K., 1977, Plasma-*myo*inositol concentrations in uraemic neuropathy, *Lancet* **1**:675.

Riordan, J. R., Alon, N., and Buchurald, M., 1979, Plasma membrane lipids of human diploid fibroblasts from normal individuals and patients with cystic fibrosis, *Biochim. Biophys. Acta* **574**:39.

Robinson, R.,and Fritz, I. B., 1979, *Myo*inositol biosynthesis by Sertoli cells, and levels of *myo*inositol biosynthetic enzymes in testis and epididymis, *Can. J. Biochem.* **57**:962.

Rothman, S. S., 1978, Chymotrypsinogen inositol phosphatide complexes and the transport of digestive enzyme across membranes, *Biochim. Biophys. Acta* **509**:374.

Salway, J. G., and Hughes, I. E., 1972, An investigation of the possible role of phosphoinositides as regulators of action potentials by studying the effect of electrical stimulation, tetrodotoxin and cinchocaine on phosphoinositide labelling by ^{32}P in rabbit vagus, *J. Neurochem.* **19**:1233.

Salway, J. G., Finnegan, J. A., Barnett, D., Whitehead, L., Kacununayaka, A., and Payne, R. B., 1978, Effect of *myo*-inositol on peripheral-nerve function in diabetes, *Lancet* **2**:1282.

Schacht, J., 1978, Purification of polyphosphoinositides by chromatography on immobilized neomycin, *J. Lipid Res.* **19**:1063.

Schacht, J., and Agranoff, B. W., 1972, Effects of acetylcholine on labeling of phosphatidate and phosphoinositides by [^{32}P]orthophosphate in nerve ending fractions of guinea-pig cortex, *J. Biol. Chem.* **247**:771.

Schexnailder, R., and Griffith, M., 1973, Liver fat and egg production of laying hens as influenced by choline and other nutrients, *Poul. Sci.* **52**:1188.

Schulz, V. E., and Oslage, H. J., 1972a, Untersuchugen zur intestinalen Hydrolyse von Inositphosphorsäureester und zur Absorption von Phytinphosphor beim Schwein: Problemstellung der Untersuchungen und analytische Methodik, *Z. Tierphysiol.* **30**:55.

Schulz, V. E., and Oslage, H. J., 1972b, Untersuchungen zur intestinalen Hydrolyse von Inositphosphorsäureester und zur Absorption von Phytinphosphor beim schwein: Untersuchungen zur Hydrolyse der Inositphosphorsäureester im Verdauungstrakt des Schweines, *Z. Tierphysiol.* **30**:76.

Sebrell, W. H., Jr., and Harris, R. S., 1967, *The Vitamins,* Vol. III, Academic Press, New York.

Sheltawy, A., Brammer, M., and Borrill, D., 1972, The subcellular distribution of triphosphoinositide phosphomonoesterase in guinea-pig brain, *Biochem. J.* **128**:579.

Shepherd, N. D., and Taylor, T. G., 1974a, The lipotropic action of *myo*-inositol, *Proc. Nutr. Soc.* **33**:64A.

Shepherd, N. D., and Taylor, T. G., 1974b, A re-assessment of the status of *myo*-inositol as a vitamin, *Proc. Nutr. Soc.* **33**:63A.

Sherman, W. R., Stewart, M. A., Kurien, M. M., and Goodwin, S. L., 1968, The measurement of *myo*-inositol, *myo*-inosose-2 and *scyllo*-inositol in mammalian tissues, *Biochim. Biophys. Acta* **158**:197.

Sherman, W. R., Goodwin, S. L., and Gunnell, K. D., 1971, *neo*-Inositol in mammalian tissues. Identification, measurement and enzymatic synthesis from mannose-6-phosphate, *Biochemistry* **10**:3491.

Sherman, W. R., Packman, P. M., Laird, M. H.,and Boshans, R. L., 1977a, Measurement of *myo*-inositol in single cells and defined areas of the nervous system by selected ion monitoring, *Anal. Biochem.* **78**:119.

Sherman, W. R., Rasheed, A., Mauck, L., and Wiecko, J., 1977b, Incubations of testis *myo*-inositol-1-phosphate synthase with D-[5-^{18}O]glucose 6-phosphate and with $H_2{}^{18}O$ show no evidence of Schiff base formation, *J. Biol. Chem.* **252**:5672.

Skipski, V. P., Barclay, M., Barclay, R. K., Fetzer, V. A., Good, J. J., and Archibald, F. M., 1967, Lipid composition of human serum lipoproteins, *Biochem. J.* **104**:340.

Slaby, F., and Bryan, J., 1976, High uptake of *myo*-inositol by rat pancreatic tissue *in vitro* stimulates secretion, *J. Biol. Chem.* **251**:5078.

Smith, G. S., Chambers, J. W., Neumann, A. L., Ray, E. E., and Nelson, A. B., 1974, Lipotropic factors for beef cattle fed high-concentrate diets, *J. Anim. Sci.* **38**:627.

Spector, R., 1976a, The specificity and sulfhydryl sensitivity of the inositol transport system of the central nervous system, *J. Neurochem.* **27**:229.

Spector, R., 1976b, Inositol accumulation by brain slices *in vitro*. *J. Neurochem.* **27**:1273.

Spector, R., and Lorenzo, A. V., 1975, The origin of *myo*-inositol in brain, cerebrospinal fluid and choroid plexus, *J. Neurochem.* **25**:353.

Strickland, L. P., Shum, P., and Rao, R. H., 1978, Phosphatidylinositol metabolism in rat brain preparations, in *Cyclitols and Phosphoinositides* (W. W. Wells and F. Eisenberg, Jr., eds.), pp. 201–214, Academic Press, New York.

Sugget, A., 1975, Polysaccharides, in: *Water: A Comprehensive Treatise,* Vol. 4 (F. Franks, ed.), pp. 519–567, Plenum Press, New York.

Takenawa, T., and Egawa, K., 1977, CDP-diglyceride : inositol transferase from rat liver, *J. Biol. Chem.* **252**:5419.

Takenawa, T., and Tsumita, T., 1974a, Properties of scyllitol transport in rat kidney slices, *Biochim. Biophys. Acta* **373**:490.

Takenawa, T., and Tsumita, T., 1974b, *myo*-Inositol transport in plasma membrane of rat kidney, *Biochim. Biophys. Acta* **373**:106.

Takenawa, T., Wada, E., and Tsumita, T., 1977a, *myo*-Inositol binding and transport in brush border membranes of rat kidney, *Biochim. Biophys. Acta* **464**:108.

Takenawa, T., Saito, M., Nagai, Y., and Egawa, K., 1977b, Solubilization of the enzyme catalyzing CDP-diglyceride-independent incorporation of *myo*-inositol into phosphatidyl inositol and its comparison to CDP-diglyceride : inositol transferase, *Arch Biochem. Biophys.* **182**:244.

Talkwalkar, R. T., and Lester, R. L., 1973, The response of diphosphoinositide and triphosphoinositide to perturbations of the adenylate energy charge in cells of *Saccharomyces cerevisiae, Biochim. Biophys. Acta* **306**:412.

Thompson, W., Strickland, K. P., and Rossiter, R. J., 1963, Biosynthesis of phosphatidylinositol in rat brain, *Biochem. J.* **87**:136.

Tomita, T., Hasegawa, R., and Hayashi, E., 1979, Neutral lipid accumulation in yeast due to inositol deficiency: Kinetic studies on the reciprocal regulation by fructose bisphosphate and citrate of yeast acetyl CoA carboxylase, *J. Nutr. Sci. Vitaminol.* (*Tokyo*) **25**:59.

Tou, J. S., and Stjernholm, R. L., 1974, Stimulation of the incorporation of ^{32}Pi and *myo*-[2-^{3}H]inositol into the phosphoinositides in polymorphonuclear leukocytes during phagocytosis, *Arch. Biochem. Biophys.* **160**:487.

Tou, J. S., Hurst, M. W., Huggins, C. G., and Foor, W. E., 1970, Biosynthesis of triphosphoinositide in rat kidney cortex, *Arch. Biochem. Biophys.* **140**:492.

Tou, J. S., Hurst, M. W., Baricos, W. H., and Huggins, C. G., 1973, The hydrolysis of triphosphoinositide by a phosphodiesterase in rat kidney cortex, *Arch. Biochem. Biophys.* **154**:593.

van den Berg, C. J., Hill, L. F., and Stanbury, S. W., 1972, Inositol phosphates and phytic acid as inhibitors of biological calcification in the rat, *Clin. Sci.* **43**:377.

van Golde, L. M. G., Raben, J., Batenburg, J. J., Fleisher, B., Zambrano, F., and Fleischer, S., 1974, Biosynthesis of lipids in golgi complex and other subcellular fractions from rat liver, *Biochim. Biophys. Acta* **360**:179.

Vemmer, V. H., and Oslage, H. J., 1973, Untersuchungen zur intestinalen Hydrolyse von Inositphosphorsäureester und zur Absorption von Phytinphosphor beim Schwein: Untersuchungen zur intestinalen Absorption von Gesamtphosphor und Phytinphosphor an wachsenden Schweinen bei niedriger Phosphor- und Calciumversorgung, *Z. Tierphysiol.* **31**:129.

Voglmayr, J. K., and Amann, R. P., 1973, The distribution of free *myo*-inositol in fluids, spermatozoa, and tissues of the bull genital tract and observations on its uptake by the rabbit epididymis, *Biol. Reprod.* **8**:504.

Voglmayr, J. K., and White, I. G., 1971, Synthesis and metabolism of myoinositol in testicular and ejaculated spermatozoa of the ram, *J. Reprod. Fertil.* **24**:29.

Warfield, A. S., and Segal, S., 1978, *Myo*inositol and phosphatidylinositol metabolism in synaptosomes from galactose-fed rats, *Proc. Natl. Acad. Sci. U.S.A.* **75**:4568.

Warfield, A., Hurang, S. H., and Segal, S., 1978, On the uptake of inositol by rat brain synaptosomes, *J. Neurochem.* **31**:957.

Watkins, T. R., and Hegsted, D. M., 1979, Tissue inositol metabolism in gerbil lipodystrophy, *Fed. Proc.* **38**:279.

Wells, H. J., and Wells, W. W., 1967, Galactose toxicity and *myo*-inositol metabolism in developing rat brain, *Biochemistry* **6**:1168.

Wells, W. W., and Burton, L. E., 1978, Requirement for dietary *myo*-inositol in the lactating rat, in *Cyclitols and Phosphoinositides* (W. W. Wells and F. Eisenberg Jr., eds.), pp. 471–485, Academic Press, New York.

White, D. A., 1973, The phospholipid composition of mammalian tissues, in: *Form and Function of Phospholipids* (G. Ansell, R. Dawson, and J. Hawthorne, eds.), pp. 441–482, Elsevier, Amsterdam.

White, D. A., Pounder, D. J., and Hawthorne, J. N., 1971, Phospholipase A_1 activity of guinea pig pancreas, *Biochim. Biophys. Acta* **242**:99.

Whiting, P. H., Palmano, K. P., and Hawthorne, J. N., 1979, Enzymes of *myo*-inositol lipid metabolism in rats with streptozotocin-indiced diabetes, *Biochem. J.* **179**:549.

Wolford, J. H., and Murphy, D., 1972, Effect of diet on fatty liver-hemorrhagic syndrome incidence in laying chickens, *Poult. Sci.* **51**:2087.

Wolford, J. H., and Polin, D., 1975, Effect of inositol, lecithin, vitamins (B_{12} with choline and E), and iodinated casein on induced fatty liver-hemorrhagic syndrome in laying chickens, *Poul. Sci.* **54**:981.

Wood, R., 1975, Hepatoma, host liver, and normal rat liver phospholipids as affected by diet, *Lipids* **10**:736.

Woolley, D. W., 1941, Identification of the mouse antialopecia factor, *J. Biol. Chem.* **139**:29.

Woolley, D. W., 1942, Synthesis of inositol in mice, *J. Exp. Med.* **75**:277.

Yagi, K., and Kotaki, A., 1969, The effect of massive doses of *myo*-inositol on hepatic phospholipid metabolism, *Ann. N.Y. Acad. Sci.* **165**:710.

Yamada, M., and Tsukahara, T., 1973, A microbiological assay of inositol by cup-plate method with *Kloeckera apiculata, J. Nutr. Sci. Vitaminol.* (*Tokyo*) **19**:205.

Yamamto, A., Adachi, S., Takeshi, I., Yoshitake, S., Kaki-Uchi, Y., Seki, K., and Kitani, T., 1970, Accumulation of acidic phospholipids in a case of hyperlipidemia with hepatosplenomegaly, *Lipids* **5**:566.

Chapter 6

Neurobiology of Pyridoxine

Krishnamurti Dakshinamurti

1. Introduction

There has been a continuing interest in the role of pyridoxine (vitamin B_6) in maintenance of the structure and function of the nervous system. At the International Symposium (Harris *et al.*, 1964) on vitamin B_6 in honor of Professor Paul György, Roberts *et al.* (1964) mentioned that "γ-amino butyric acid (GABA) plays an inhibitory or modulatory role in the central nervous sytem (CNS)." Since then, more direct evidence has accumulated to confirm both a neurotransmitter role for GABA in the vertebrate CNS and the regulation of its synthesis by pyridoxal phosphate (PALP).

Recent investigations unraveling the function of pyridoxine in the metabolism of a number of putative neurotransmitters have focused our attention on other areas including the function of the anterior pituitary gland. The possibility that PALP might be involved in the regulation of steroid hormone activity has been indicated. The areas of suggested involvement of pyridoxine in the function of the nervous system have increased since the conference on vitamin B_6 in metabolism of the nervous system (Kelsal, 1969). Some of these aspects have been reviewed by Ebadi and Costa (1972) and Dakshinamurti (1977). Pyridoxine has a role in the metabolism of protein, lipid, and carbohydrate in all tissues; however, it has a prominent position among the vitamins affecting the nervous system.

The deficiency of vitamin B_6 is a problem of nutritional concern in that a

---

Krishnamurti Dakshinamurti • Department of Biochemistry, Faculty of Medicine, University of Manitoba, Winnipeg, Manitoba R3E OW3, Canada. This work was supported by a grant from the Medical Research Council of Canada.

number of factors not only related to its dietary content may contribute to a deficiency or to an increased human requirement for this vitamin. Apart from some selected studies on pregnant women or those using the anovulatory steroids, there has been no assessment of the vitamin B_6 nutrition of various population groups, some of them at high risk. The determination of the vitamin B_6 status of people is complicated by the presence of this vitamin in tissues in a number of related chemical forms. Any assessment should take into consideration this diverse distribution.

Current work on the neurobiology of putative neurotransmitters, polyamines, and hormones as related to pyridoxine is reviewed in this chapter. Examination of experimental pyridoxine deficiency, particularly in the perinatal period, is followed by a consideration of human pyridoxine deficiency and dependence. The concluding section attempts to integrate the experimental and clinical observations.

2. Nutritional Aspects of Pyridoxine

2.1. Biochemical Reactions

Pyridoxal phosphate is the major coenzymatic form of pyridoxine. Pyridoxamine phosphate can function in transamination reactions by the cyclic regeneration of its two active phosphate forms. The principal biochemical reaction is between the carbonyl group of pyridoxal with a primary amine to form a Schiff base (Fig. 1). As a result, the carbon atom of the substrate that is adjacent to the aldimino group is destabilized. The further reactions can be classified into three groups depending on the site of elimination and replacement of the substituents. Reactions occurring at the α-carbon atom include those catalyzed by transaminases, racemases of α-amino acid, α-amino acid decarboxylases, the condensation reactions of glycine, the α-β cleavage of β-hydroxy amino acids such as that catalyzed by δ-amino levulinic acid synthetase, serine hydroxymethylase, and sphingosine synthetase. Reactions occurring at the β-carbon atom include those catalyzed by serine and threonine dehydrases, cystathionine synthetase, tryptophanase, and kynureniase. Reactions catalyzed by homoserine dehydrase and γ-cystathionase occur at the γ-carbon atom.

Pyridoxine-dependent enzymes investigated in the nervous system are in-

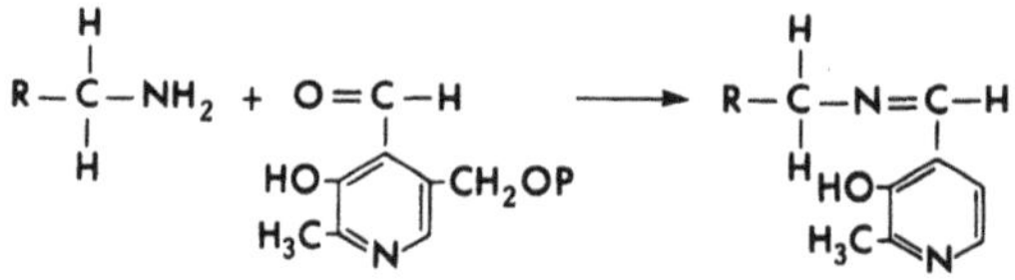

Fig. 1. Formation of Schiff base.

volved in catabolic reactions of various amino acids. They include the following: glutamic-oxaloacetic aminotransferase, glutamic-pyruvic aminotransferase, amino acid (valine, leucine, isoleucine, cysteine, methionine, ornithine)-α-ketoglutarate aminotransferase, γ-aminobutyrate transaminase, cystathionine synthetase, cystathionase, kynureniase, glutamic acid decarboxylase (GAD), aromatic amino acid (tyrosine, dihydroxyphenylalanine, 5-hydroxytryptophan, phenylalanine, histidine) decarboxylases, ornithine decarboxylase, and cysteine sulfinic acid (cysteic acid) decarboxylase. An enzyme not related to amino acid metabolism but of significance to nervous tissue is dihydrosphingosine synthetase.

The crucial role played by pyridoxine in the nervous system is evident from the fact that the various putative neurotransmitters—dopamine, norephinephrine, tyramine, tryptamine, serotonin, histamine, γ-aminobutyric acid, and taurine—are synthesized and/or metabolized by pyridoxine-dependent enzymes. The role of pyridoxine in the synthesis of sphingolipids and polyamines highlights its importance in the development as well as in the maintenance of the integrity of the nervous system.

2.2. Interconversion of Pyridoxine Vitamers

The term "vitamin B_6" is used to refer to the group of naturally occurring pyridine derivatives that includes pyridoxine (pyridoxol), pyridoxal, pyridoxamine, and their phosphorylated derivatives. Pyridoxal phosphate generally, and pyridoxamine phosphate in the case of many of the aminotransferases are the coenzyme forms. The conversion of pyridoxine to pyridoxal is catalyzed by an NADPH-dependent pyridoxine oxidase. Pyridoxal is oxidized to pyridoxic acid by an aldehyde oxidase. Pyridoxic acid, the major catabolite of pyridoxine is excreted in urine. In most tissues including brain, the major forms of vitamin B_6 are pyridoxal phosphate and pyridoxamine phosphate.

The enzyme pyridoxal kinase increases during brain maturation in most species. Antimetabolites such as 4-deoxypyridoxine act by inhibiting the kinase. The enzyme seems to be regulated by several factors. An inverse relationship between the tissue concentration of pyridoxal phosphate and the activity of pyridoxal kinase has been reported by Ebadi *et al.* (1970). Ebadi *et al.* (1968) and Ebadi and Govitrapong (1979) also claim a similar inverse relationship between the activity of this enzyme and the concentration of monoamines in the rabbit brain. Neary *et al.* (1972) report that serotonin inhibits pyridoxal kinase and could be a regulator of brain pyridoxine metabolism. The kinase is activated by phospholipids extracted from brain cortex (Maniero *et al.*, 1973). In contrast, phospholipids from other tissues have no effect on the enzyme.

Although the acute administration of levo-DOPA reduced the concentration of pyridoxal phosphate, the chronic oral administration of levo-DOPA (100 mg/kg per day) did not alter the concentration of pyridoxal phosphate in rat basal ganglia (Ebadi *et al.*, 1978), suggesting an adaptive increase in the activity of pyridoxal kinase. Pyridoxine phosphate oxidase is inhibited by its product, pyridoxal phosphate. This could be another mechanism controlling cellular pyridoxal phosphate concentration. The phosphorylated coenzymes in the free form are hydrolyzed by phosphatases (Saraswathi and Bachhawat, 1963; Li and Lumeng, 1974). A large part of the cellular pyridoxal phosphate is protein-bound. Thus, reactions that form pyridoxal phosphate are balanced by its sequestration as the protein-bound form and by its degradation by phosphatases.

2.3. Uptake of Vitamin B_6 by Brain

The cerebral capillaries are the sites of the blood–brain barrier through which molecules diffuse from blood to enter the extracellular space of the brain. The blood–brain barrier is a major feature of the mature brain. Lipophilic compounds are better able to cross this barrier than hydrophobic ones. Material from blood can pass into the cerebrospinal fluid (CSF) which bathes the brain. The site of this blood–CSF transfer is the choroid plexus. In addition to passive diffusion, facilitated diffusion is involved in the transport of material from blood to the CSF. In the reverse direction, products of catabolism in brain find their way into CSF and from there to blood by bulk flow of CSF through the arachnoid granulations. The concentration ratio of various vitamins between the CSF and plasma varies considerably, with the CSF concentration generally being higher (Spector, 1977).

Pyridoxine is transported to brain and CSF from plasma by a saturable transport process (Spector, 1978a,b). The mechanism of this transfer, which can maintain high brain levels even at very low plasma concentrations, is not known. Concentrating mechanisms at the level of blood–CSF and at the level of CSF–brain cells should be operative. However, increasing the plasma concentration of a vitamin such as ascorbic acid over 100-fold results in only a tenfold increase in its concentration in CSF and barely a twofold increase in its concentration in brain (Spector, 1977). Thus, the mechanisms that help in building up high levels of vitamins in the brain also operate to maintain vitamin homeostasis. Pyridoxine transport systems in brain and the choroid plexus are present in the newborn as well as developing animals (Spector, 1979).

It is possible that in certain conditions the transport of some vitamins to the brain is inadequate even when dietary intake and plasma levels are quite high. Spector *et al.* (1979) have hypothesized that dementia of as yet unknown origin might be the result of this inadequate cerebral uptake of certain vitamins. Carney (1967) has reported the presence of low levels of CSF folate in some demented patients. Treatment in these cases would require development of lipid-soluble forms of these vitamins or direct injection of the vitamin into ventricular CSF. As the majority of cases of dementia are classified as "idiopathic" and are not amenable to current modes of treatment, this attractive hypothesis (Spector *et al.*, 1979) needs to be tested not only in the case of folate but also for pyridoxine.

2.4. Pyridoxine Depletion—Physiological and Pathological

The prevalence of a relative deficiency of pyridoxine in a significant percentage of pregnant women is suggested by the work of Karlin and Dumont (1963) and Karlin *et al.* (1968), who found an increased excretion of xanthurenic acid by pregnant women following a load of tryptophan. The tryptophan load test is not a reliable criterion of pyridoxine deficiency in view of the hormonal effects on the enzymes of tryptophan metabolism (Rose and Braidman, 1971; Brin, 1971). However, the use of other criteria such as a methionine load test (Krishnaswamy, 1974) or direct measurement of pyridoxine vitamers (Contractor and Shane, 1970; Hamfelt and Tuvemo, 1972; Heller *et al.*, 1973; Cleary *et al.*, 1975) has substantiated this observation.

Even a moderate deficiency of pyridoxine in the mother could have deleterious effects on the growth and development of the fetus and infant as seen from experimental studies (Dakshinamurti and Stephens, 1969; Stephens *et al.*, 1971). Klieger *et al.*, (1969), in their study of the eclamptogenic syndrome which affects 6–7% of all obstetric cases, rerported that the toxemic placenta is markedly deficient in pyridoxine when compared to normal placenta. The concentration of pyridoxal phosphate is reduced to almost a fifth of normal levels in cord blood of infants of preeclamptic mothers (Brophy and Siiteri, 1975). Choriodecidual blood flow is reduced by at least a third in preeclampsia (Dixon *et al.*, 1963). The reduced PALP in cord plasma from toxemic patients may be a reflection of reduced blood flow to the placenta. Decreased transport of pyridoxine to the toxemic placenta is excluded, as the low levels of PALP are increased by loading the toxemic mother with pyridoxine. Pyridoxine phosphate oxidase, a key enzyme in the formation of pyridoxal phosphate, is reduced in the toxemic placenta (Gaynor and Dempsey, 1972). Retardation of maturation and development of the brain of small-for-gestational-age infants of toxemic mothers have been reported (Schulte *et al.*, 1971). However, the role of pyridoxine in the development of the eclamptogenic syndrome is not known.

Many of the clinical side effects seen in users of oral contraceptive steroids are those associated with pregnancy. In one study, about 20% of women who have used oral contraceptives for at least 6 months had lower than normal concentrations of pyridoxal phosphate in plasma (Lumeng *et al.*, 1974; Driskell *et al.*, 1976). The majority showed abnormal tryptophan metabolism (Lubby *et al.*, 1971; Leklem *et al.*, 1975). The altered metabolism of tryptophan in both pregnant women and users of oral contraceptive steroids (OCS) is corrected by the administration of 10–50 mg pyridoxine per day (Brown *et al.*, 1968; Ahmed and Bamji, 1976). Perioral dermatosis and neuropsychiatric disorders such as changes in sleep pattern and mood seen in a small group of women using OCS are reported to be corrected by high doses of pyridoxine (Baumblatt and Winston, 1970). Pyridoxine deficiency produces a decrease in the content of brain serotonin (Dakshinamurti *et al.*, 1976). Green *et al.* (1970) suggest that the induction of tryptophan 2,3-dioxygenase diverts tryptophan metabolism predominantly toward the nicotinic acid pathway, decreasing the availability of both tryptophan and pyridoxal phosphate for the serotonin pathway. The conjugates formed in the liver from the increased estrogenic steroids of endogenous origin in pregnancy and of exogenous origin in OCS users could compete with pyridoxal phosphate for the apoenzyme and thus cause its inhibition.

Evidence to suggest a functional deficiency of pyridoxine in uremic patients has been reported. Stone *et al.* (1975) have suggested that pyridoxine deficiency might contribute to the symptomatology of renal failure based on observations of low plasma pyridoxal phosphate as well as low plasma and erythrocyte aspartate aminotransferase in undialyzed and dialyzed uremic patients. Dobbelstein *et al.* (1974) have reported similar findings.

Serum pyridoxal has been reported to be significantly decreased in patients with diabetic neuropathy when compared with randomly selected diabetic patients matched for age and sex but without any evidence of neuropathy (McCann and Davis, 1978). The deficiency of pyridoxine in these conditions could have various causes such as impaired intestinal absorption or phosphorylation, increased dephosphorylation, or other mechanisms inactivating pyridoxal phosphate.

Pyridoxal phosphate is chemically a very reactive compound and forms Schiff bases with compounds having an amino group. Formation of such complexes would reduce the concentration of the biologically active form of pyridoxine. Various therapeutic drugs thus have an antipyridoxine effect. Included in this group are isonicotinic acid hydrazide, used in the treatment of pulmonary tuberculosis (Levy, 1969; Coyer and Nicholson, 1976); cycloserine, also used in the treatment of human tuberculosis resistant to treatment with a streptomycin-paraaminosalicylic acid-isonicotinic acid hydrazide regimen (Cohen, 1968); D-penicillamine, used in the treatment of Wilson's disease, lead intoxication, and rheumatoid arthritis (Walshe, 1956; Smith and Gallagher, 1970; Tomono *et al.*, 1973); and peripheral DOPA decarboxylase inhibitors such as benserazide and carbidopa, which are used clinically to reduce the amount of DOPA needed to treat patients with Parkinson's disease (Bartholini *et al.*, 1967; Airoldi *et al.*, 1978).

2.5. Determination of Pyridoxine Status

Any assay for vitamin B_6 should include all pyridoxine vitamers. Several chemical, microbiological, enzymatic, and fluorometric methods are available, each with its own advantages and limitations. Thiele and Brin (1968) adapted Toepfer and Lehmann's (1961) chromatographic separation for the microbiological assay of these compounds. Microbiological procedures give values with a wide scatter in view of the lack of specificity of the assay organism (Storvick and Peters, 1964). The enzymatic methods use tyrosine decarboxylase apoenzyme (Dakshinamurti and Stephens, 1969; Chabner and Livingston, 1970) or apotryptophanase (Haskell and Snell, 1972). They are very sensitive and specific but can be applied only to pyridoxal phosphate. Fluorometric methods are available for the assay of pyridoxine vitamers alone or in any combination (Loo and Badger, 1969; Durko *et al.*, 1973; Chauhan and Dakshinamurti, 1979a,b). They are sensitive and specific and applicable to tissues and body fluids.

Blood level or urinary excretion of a vitamin generally reflects the vitamin status. Pyridoxine is excreted in the urine as 4-pyridoxic acid. Variations in the excretion of pyridoxic acid unrelated to the intake of pyridoxine have been reported (Baysal *et al.*, 1966; Kelsay *et al.*, 1968). As pyridoxal phosphate is involved extensively in the metabolism of tryptophan and methionine, load tests using these amino acids have been used. Canham *et al.* (1968) have pointed out that the effects of protein intake, stress, and hormonal imbalances on the metabolism of tryptophan have to be taken into consideration.

The extent of the saturation of erythrocyte aspartate aminotransaminase *in vitro* by added pyridoxal phosphate is related to the saturation of the enzyme with the coenzyme *in vivo* (Kishi *et al.*, 1975) and has been used to assess the pyridoxine status on the assumption that the concentration of the apoenzyme is invariant. Such assumptions have not been rigorously tested.

3. Pyridoxine and the Nervous System

3.1. The γ-Aminobutyric Acid System

3.1.1. Role of γ-Aminobutyric Acid

The unique localization of GABA in the nervous system led to investigations about its function. γ-Aminobutyric acid has been shown to have inhibitory physiological effects in experiments with a wide variety of vertebrate and invertebrate test systems ranging from the crayfish stretch receptor to the monkey cortex (Roberts and Eidelberg, 1960). Kravitz and Potter (1965) have identified GABA as an inhibitory transmitter in the lobster neuromuscular junction. Intracellular recordings combined with iontophoresis have shown that GABA can hyperpolarize CNS neurons in vertebrates and that intracellular injection of Cl^- reverses the GABA effect. The neurophysiological action of GABA studied by iontophoretic application resembles that observed in postsynaptic inhibition produced by electrical stimulation (Krnjević and Schwartz, 1967). The inhibitory effect following electrical stimulation of affected tracts to the cuneate nucleus is abolished by bicuculline, a convulsant, which also blocks the inhibitory action of iontophoretically applied GABA (Curtis *et al.*, 1971a,b; Kelly and Renaud, 1971). Specific GABA receptors have been demonstrated in cerebral cortex, cerebellar cortex, vestibular cortex, hippocampus, thalamus, olfactory bulb, and spinal cord (Bloom, 1972). Regional variations in the concentration of GABA correlate with distribution of glutamic acid decarboxylase (GAD) and high-affinity GABA uptake (Enna *et al.*, 1975; Hökfelt *et al.*, 1970). Immunocytochemical methods used to study the distribution of GABA-ergic neurons indicate that the cerebellar Purkinje, Golgi II, basket, and stellate cells use GABA as the inhibitory neurotransmitter (Wood *et al.*, 1976).

When GABA was injected into young chicks, it produced abolition of photically evoked responses. Jasper *et al.* (1965) have shown that during sleep GABA was present in cerebral cortex perfusates with a decrease in glutamic acid release. γ-Aminobutyric acid was also released into the fourth ventricle after stimulation of the cerebellar cortex (Curtis and Johnston, 1970). There are suggestions that GABA might act through alteration of cerebellar cyclic GMP. The cerebellar levels of cyclic nucleotides increases concomitantly with increased neuronal activity induced by various convulsants. The effect on cyclic GMP, however, is reversed by chemicals that elevate cerebellar GABA (Mao *et al.*, 1975). Similar effects have been observed in animals subject to electroconvulsive shock and protected by diphenylhydantoin, phenobarbital, or dipropyl acetate (Lust *et al.*, 1975). Dipropyl acetate, whose side effects are minimal compared with aminooxyacetic acid, has been used in the treatment of patients with petit mal and grand mal seizures (Meinardi, 1971).

γ-Aminobutyric acid is generally considered to be involved in the etiology of convulsive seizures (Sze, 1979). Most consideration has been given to the possibility that a decrease in GABA levels at a vital location would be the primary lesion leading to a decrease in inhibition. The other possibility of an impairment at the postsynaptic membrane has not been considered. Wood (1975) compiled the changes in GABA levels in brains of animals at the onset of seizures induced by a variety of convulsive agents, e.g., glutamic acid-γ-hydrazide, hydrazine, aminooxyacetic acid, semicarbazide, thiosemicarbazide, isonicotinic acid hydrazide, and allylglycine. The changes in GABA levels ranged from an increase of 256% to a decrease of 56%. The change in GAD activity at the beginning of convulsions correlated better. Wood and Peeskar (1974) empirically related the excitable state of the brain to a function containing the change in GAD as the major factor and the change in GABA as the minor factor:

$$RE_{\mathrm{GABA}} = 100 + \frac{\Delta\mathrm{GAD} + 0.4\,\Delta\,(\sqrt{\mathrm{GABA}})}{1.4} \qquad (1)$$

where RE_{GABA} is the effectiveness of the GABA system to modulate the excitability of the brain, and ΔGAD and $\Delta(\sqrt{\mathrm{GABA}})$ are the percent changes in GAD activity and in the square root of GABA content brought about by the convulsive agent. Thus, depending on the changes in these paremeters, the excitability function could vary between 100% (normal) and 0%. However, at or below a critical RE_{GABA} value, which is different for different species, convulsive seizures result regardless of the chemical nature of the anticonvulsant. The possibility that this equation might represent the GABA concentration in the synaptic cleft is speculative. Wood (1975) pointed out the obvious limitation in this attempted correlation: a convulsive agent that has no, or only a partial, effect on the GABA system would yield meaningless data for RE_{GABA}.

Apart from the involvement of GABA in the etiology of certain convulsive seizures, abnormalities in GABA-ergic neuronal pathways contributing to other CNS disorders have been recognized. In Huntington's chorea, a disease marked by the onset of dementia and choreiform movements, there is a marked decrease in the concentration of GABA as well as in the activity of GAD in the basal ganglia (McGeer *et al.*, 1973). In Parkinson's disease, the main neurochemical lesion is associated with degeneration of the dopaminergic neurons in the substantia nigra. In addition to this, the observed decreases in both GABA and GAD (Lloyd and Hornykiewicz, 1973) in the basal ganglia are of interest, suggesting an interrelationship between the two neuronal systems (Hornykiewicz *et al.*, 1976). In view of the involvement of GABA in a variety of clinical abnormalities, there is considerable interest in the possibility of modifying GABA-ergic neuronal activity through use of specific drug intervention.

3.1.2. Regulation of γ-Aminobutyric Acid

γ-Aminobutyric acid is formed from glutamic acid through the action of glutamic acid decarboxylase (GAD) and is catabolized by transamination catalyzed by GABA transaminase (GABA-T) to yield succinic semialdehyde (SSA). Both GAD and GABA-T are pyridoxal phosphate (PALP)-dependent enzymes. Succinic semialdehyde is oxidized by succinic semialdehyde dehydrogenase to succinic acid. The concentration of "effective" GABA is controlled by regulation of its synthesis, catabolism, and uptake or through alteration of the GABA receptors. Alteration of enzyme activity is achieved by various factors including effector binding, cofactor availability, and covalent modification of the enzyme, as well as by regulation of enzyme synthesis and its posttranslational modification or degradation. Metabolic events within the cell, which are themselves affected by extracellular signals such as neurotransmitters or hormones, control the factors listed above.

The increase in activity of brain GAD during development results from increased enzyme synthesis (Matsuda *et al.*, 1973). γ-Aminobutyric acid itself depresses the level of GAD activity (Haber *et al.*, 1970; Sze, 1970), perhaps through feedback repression. Sze (1979) has reported that acetylcholine affects the activity of GAD in the cerebellar cortex *in vivo*. Transsynaptic mechanisms have been suggested to regulate GAD. Thus, there is a decrease in GAD activity in cervical spinal cord following section of the dorsal roots (Kelly *et al.*, 1973). An increase in GAD activity has been reported following administration of sedatives such as morphine (Kuriyama and Yoneda, 1978) or phenobarbital (Tzeng and Ho, 1977). The synthesis of GABA is also regulated through alterations in the saturation of GAD with PALP. Brain GAD is not saturated with PALP even when its concentration is high. Thus, factors that affect the endogenous binding of PALP to the apoenzyme regulate the increased formation and release of GABA (Miller and Walters, 1979). Glutamic acid decarboxylase activity (GADII) is present in nonneural tissues. In contrast to the brain enzyme (GADI), which is inhibited by anions and carbonyl-trapping agents (Chase and Walters, 1976) and stimulated by PALP, the nonneural GAD requires high concentration of anions for maximal activity and is unaffected by added PALP and carbonyl-trapping agents (Haber *et al.*, 1970).

An increase in CNS GABA could also be achieved through inhibition of the catabolic enzyme GABA-T or by blocking the reuptake of GABA; GABA-T is inhibited by "K Cat" inhibitors such as γ-vinyl-GABA which are metabolized by GABA-T to form an acylating molecules that act at the active site (Rando, 1974). Even more significant is the inhibition of GABA-T by branched-chain fatty acids such as di-*n*-propyl acetate (Simler *et al.*, 1973) which are used in the treatment of epilepsy (Simon and Penry, 1975).

3.2. Taurine

3.2.1. Distribution and Synthesis

Taurine (2-aminoethanesulfonic acid) is ubiquitous in distribution. It is present in high concentration in many body tissues including brain and heart. Significant amounts are present in plasma, and in most species large amounts of taurine are excreted in urine. In the vertebrate brain there is a distinct nonuniform distribution of taurine in various areas. The cerebral and cerebellar cortex, olfactory bulb, and striatum contain the highest concentration of taurine, whereas the hypothalamus, medulla, and spinal cord contain the lowest. In the visual system, retina contains a very high concentration of taurine. The perinatal brain of most species contains a higher concentration of taurine than the mature brain (Sturman and Gaull, 1975). Within the nervous system, a high concentration of taurine is found in the synaptic vesicles (DeBelleroche and Bradford, 1973).

The major pathway of biosynthesis of taurine is by the transsulfuration pathway (Fig. 2). In addition to the enzymes forming and degrading cystathionine, cysteine sulfinic acid decarboxylase (CSD), the rate limiting enzyme of this pathway, is dependent on pyridoxal phosphate. The brain enzyme (CSD) is present both in the cytosol and in the particulate fraction (synaptosomes). During postnatal development, the particulate form of the enzyme, which reflects the increase in the synaptosomal enzyme, increases almost tenfold (Rassin and Sturman, 1975; Pasantes-Morales *et al.*, 1976). The soluble enzyme of brain is similar to the liver enzyme and is only slightly activated by added PALP, as this CSD apoenzyme is bound tightly to PALP. The particulate enzyme, however, is activated more than 90% by free PALP. Further confirmation of the dependence

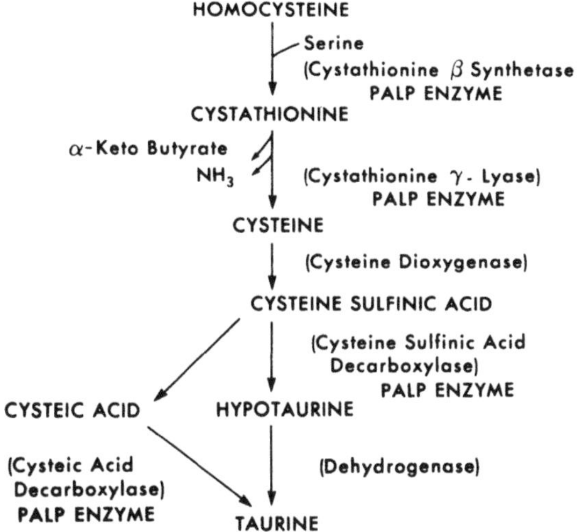

Fig. 2. Biosynthesis of taurine.

on PALP is obtained from the inhibition of the particulate rather than the soluble enzyme by pyridoxal phosphate oxime-O-acetic acid. This inhibitor binds only to those B_6-dependent enzymes that have a low-affinity site for pyridoxal phosphate. Thus, enzyme activity in the subcellular fraction is dependent on the concentration of free PALP in the synaptic terminal (Pasantes-Morales *et al.*, 1976) as has been suggested for glutamate decarboxylase (Bergamini *et al.*, 1974). The CSD activity increases in the synaptic terminals during maturation involving synaptogenesis and the development of electrical activity (Myslivecek, 1970).

Other pathways of taurine formation are known. Cysteine sulfinic acid is oxidized to cysteic acid, which forms taurine on decarboxylation, by the same enzyme, cysteine sulfinic acid decarboxylase. Another pathway, referred to as the inorganic pathway, involves transfer of sulfate from phosphoadenosine-5′-phosphosulfate to dehydrated serine to form cysteic acid which is decarboxylated to form taurine. The significance of this pathway is uncertain. Yet another route, the cysteamine pathway, involves the synthesis of pantothenyl cysteine, its decarboxylation to pantetheine, degradation to cysteamine, and oxidation to taurine. However, this pathway is of limited significance in the nervous tissue (Novelli *et al.*, 1954). It is to be noted that both the "transsulfuration" and "inorganic sulfate" pathways of taurine biosynthesis utilize cysteine sulfinic acid (cysteic acid) decarboxylase which is pyridoxal phosphate dependent.

High levels of taurine are maintained in the nervous tissue, with a heterogeneous distribution. The synthesis in the synaptic terminals seems to be regulated by the local concentration of pyridoxal phosphate. There is very little catabolic breakdown of taurine to isethionic acid or to sulfate. These degradative reactions do not seem to have a role in the control of the taurine pool in the central nervous system. Brain taurine is remarkably stable metabolically, with a half-life of over 24 days (Sturman, 1973). It is excreted as such in urine in appreciable amounts. Increased excretion of urinary taurine is observed following intraperitoneal injection of β-alanine, β-aminobutyric, and β-aminoisobutyric acids to mice, suggesting that these amino acids decrease renal tubular reabsorption of taurine by a competitive transport mechanism (Gilbert *et al.*, 1960).

Growing animals of most species have the capacity to synthesize taurine in the nervous system. Two possible exceptions to this are the kitten and the human infant (Sturman *et al.*, 1978). Cats and kittens become taurine-deficient with low concentration of taurine in various brain areas and in the retina if fed a taurine-deficient diet (Schmidt *et al.*, 1976). A dietary supplement of taurine corrects this deficiency. The human infant probably requires a dietary source of taurine, although specific taurine deficiency symptoms in human infants have not been reported. The evidence is only circumstantial. Human preterm infants fed synthetic formulas have low concentrations of taurine in plasma and urine. Human milk contains 30-fold more taurine than cow's milk (Sturman *et al.*, 1978; Gaull *et al.*, 1977). If [^{35}S]taurine is injected into a pregnant rat, it is taken up by the brain of the pup as rapidly as by its liver. The uptake in young animals depends on the absence of an effective blood–brain barrier in the developing animal. Following intraperitoneal injection of massive doses of taurine, there is no significant increase in the concentration of taurine in most brain areas of the adult animal, suggesting the existence of a strong blood–brain barrier to the entry of taurine into the adult brain from blood (Levi *et al.*, 1967; Battistini *et al.*, 1969).

Available information would suggest that in most species, since taurine is synthesized in brain through the action of a pyridoxal phosphate-dependent decarboxylase, a dietary deficiency of pyridoxine should lead to a decrease in the concentration of taurine in various brain areas. The synaptosomal enzyme is activated by added pyridoxal phosphate and, *in vivo,* seems to be regulated by its concentration locally (Pasantes-Morales *et al.*, 1976). However, various reports (Hope, 1957; Sturman *et al.*, 1969; Sturman, 1973) indicate that even in severe pyridoxine deficiency, the brain levels of taurine remain unchanged. Cysteine sulfinic acid decarboxylase is not detectable in brain of pyridoxine-deficient rats when assayed in the absence of exogenous PALP. However *in-vitro* addition of pyridoxal phosphate restores the activity of CSD to the normal level. In addition, the CSD activity in crude particulate fractions of rat brain reflects the distribution of endogenous pyridoxine. It has been suggested (Martin *et al.*, 1974) that in pyridoxine deficiency there is a switchover in the pathway of taurine biosynthesis from the "transsulfuration" to the "phosphoadenosyl-phosphosulfate" (inorganic sulfate) pathway. However, even in this pathway, the crucial decarboxylase reaction is catalyzed by a pyridoxal phosphate-dependent decarboxylase. It is questionable whether the renal taurine-conserving mechanism

(Sturman, 1973) would be adequate, in the absence of synthesis, to maintain the large concentration of taurine in brain. The decreased urinary excretion of taurine might only be the result of decreased body stores.

3.2.2. Physiological Effects of Taurine

There is increasing acceptance of a role for taurine as a neurotransmitter (Phillis, 1978). It is present in nerve terminals of the central nervous system which also contains the synthetic enzymes. Release from and reuptake by neuronal preparations have been reported. Much evidence has accumulated to demonstrate the inhibiting effect of taurine on neuronal function. In this regard, its action is analogous to those of glycine and γ-aminobutyric acid.

Iontophoretic administration of taurine in the cerebral cortex and cerebellum inhibits spontaneous or induced firing of neurons (Krnjević and Phillis, 1963; Krnjević and Puil, 1976). Similar effects in the spinal cord and brain stem have been reported (Curtis and Watkins, 1960; Haas and Hösli, 1973). Depression of neuronal activity by taurine is associated with hyperpolarization, a decrease in membrane conductance, and an increased permeability to chloride and potassium ions (Krnjević, 1974). Studies using specific inhibitors indicate that taurine may be the transmitter at synapses possessing strychnine- and bicuculline-sensitive receptors (Curtis *et al.*, 1968; Curtis and Tebēcis, 1972). Taurine can suppress the stimulated release of acetylcholine from cholinergic neurons, presumably by stabilizing excitable membranes. The protection of cholinergic neurons against the neurotoxic action of kainic acid (a rigid glutamic acid analogue with potent excitatory action on neurons) is consistent with the membrane-stabilizing properties of taurine (Sanberg *et al.*, 1979).

The membrane potential changes mediated by taurine are transient, and recovery is rapid. This indicates rapid removal of taurine from the synaptic cleft. In view of the metabolic stability of taurine, the reuptake mechanism should be involved in this. Both high- and low-affinity transport of taurine into nerve cells have been reported. The high-affinity system has a K_m for taurine of 10^{-5} M, exhibits high specificity, and is sodium and energy dependent and is suggested to be involved in terminating taurine action at the synapse (Kaczmarek and Davison, 1972). An efficient uptake by glial cells (Borg *et al.*, 1977) could also be involved in the removal of extraneuronal taurine. The release of taurine from nervous tissue by *in vitro* electrical stimulation (Pasantes-Morales *et al.*, 1974) or depolarization by exposure to high potassium concentration (Mandel and Pasantes-Morales, 1978) is similar to the normal release of neurotransmitters. Calcium seems to be involved in the release of taurine (Salceda and Pasantes-Morales, 1975). Available evidence supports the hypothesis that taurine is an inhibitory transmitter in the spinal cord, the brainstem, and the cerebellar cortex.

3.2.3. Anticonvulsant Action of Taurine

Decreased levels of taurine have been reported in epileptogenic foci of human brain (Van Gelder *et al.*, 1972) and in cobalt-induced epileptogenic foci in experimental animals (Craig and Hartman, 1973). The anticonvulsant action of taurine against seizures induced by colbalt lesion (Van Gelder, 1972) or by convulsants such as strychnine, ouabain, and pentylenetetrazol (Izumi *et al.*, 1973, 1974) have been reported. It has been suggested (Van Gelder, 1978) that the anticonvulsant and antiepileptic effects of taurine are related to the effect of taurine in maintaining the tissue contents of glutamic acid, glutamine, GABA, and aspartic acid by regulating glutamic acid retention by neuronal structures.

3.2.4. Taurine and the Visual System

Structures in the visual system (retina, geniculate bodies, and the optic tract) contain large amounts of taurine (Casola and Di Matteo, 1972; Mandel and Pasantes-Morales, 1978). The concentration of taurine and the activity of cysteine sulfinic acid decarboxylase in the retina of various species increase during development. In the retinal cells, the photoreceptors contain the highest concentration of taurine (Orr *et al.*, 1976).

Taurine has an inhibitory effect on bioelectric response of the isolated, perfused retina (Urban *et al.*, 1976). The addition of taurine at a concentration less than that present in retina to the perfusion medium produces a rapid depressant effect on the b-wave amplitude of the electroretinogram. Washing the retina with a taurine-free medium reverses this effect. Intravitreal injection of taurine results in decreased tectal evoked potential (Pasantes-Morales *et al.*, 1973a,b) consistent with the suggestion that taurine may block the transmission of visual information to higher centers. Electrical stimulation of chick retina results in efflux of taurine (Pasantes-Morales *et al.*, 1974). A similar light-stimulated release specific for taurine with no release of glycine or GABA has also been reported (Pasantes-Morales, 1973b). This release of taurine is calcium dependent and conforms to the pattern of neurotransmitter release.

Cats and kittens develop taurine deficiency when fed a diet containing partially purified casein as the source of protein. When the taurine concentration of retina decreases by 50%, they develop retinal degeneration and eventually blindness. This can be prevented or reversed by taurine (Berson *et al.*, 1976). Taurine might have a role in maintaining the structural integrity of the photoreceptors (Hayes *et al.*, 1975) in addition to its role as a neurotransmitter.

3.3. Polyamines

Putrescine, spermidine, and spermine are aliphatic polyamines exhibiting a wide range of biological activity. They are normal constituents of all cell types in both pro- and eukaryotes. Four enzymes are involved in the synthesis of these polycations (Fig. 3). Of the two decarboxylases, only ornithine decarboxylase (ODC) requires pyridoxal phosphate as a coenzyme (Pegg and Williams-Ashman, 1968). The lack of requirement of PALP for *S*-adenosylmethionine decarboxylase has been clearly established (Pegg, 1977).

In most eukaryotic cells, ODC is present in low concentration and is considered to be rate limiting in the synthesis of the polyamines. It has a remarkably short half-life. ODC is inhibited by putrescine, spermidine, and spermine. Putrescine, which is incorporated directly into spermidine, regulates spermidine synthesis through precursor availability. The effectiveness of polyamines follows a cationic progression, with spermine, the strongest base, being the most active. These organic cations offer an advantage over inorganic cations in that they are synthesized intracellularly. Spermidine has been shown to be required for optimal activity of DNA replicase. In the reaction sequence DNA–RNA–protein, polyamines have been implicated in transcription, strand selection, chain initiation, extension, or termination, amino acylation, and ribosomal subunit assembly (Tabor and Tabor, 1976; Maudsley, 1979). An increase in cellular ornithine decarboxylase is the first detectable response of tissues to stimulation of the differentiated function of the cell. This is followed by an increase in cellular putrescine and in the accentuation of the protein-synthetic reaction sequences.

Polyamines are present in the brain of all species examined (Shaw, 1979). The concentration of putrescine in brain is 2–3% of the concentration of spermidine which is of the order of 100 nmole/g. Ornithine decarboxylase, which reflects putrescine concentration, seems to be a sensitive index of brain maturation. In the developing brain, protein-synthetic reactions are central to develop-

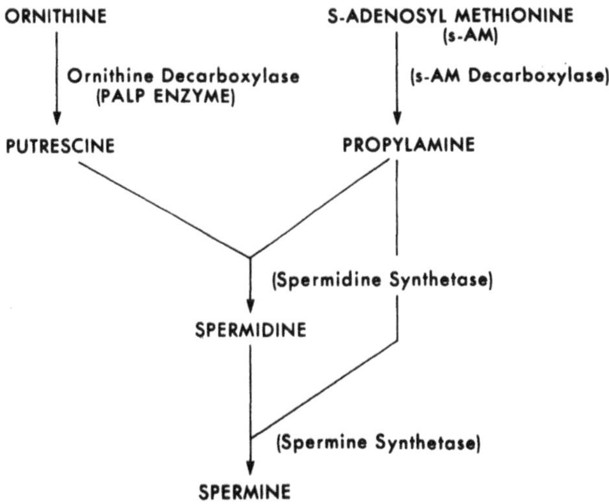

Fig. 3. Biosynthesis of polyamines.

mental maturation. In view of the role of the polyamines in stimulation of protein synthesis, their relationship to maturation is clear.

However, further investigation reveals a wide variation in putrescine content of different brain areas, with higher concentration in the spinal cord, cerebellum, hypothalamus, and cerebral cortex (Harik and Snyder, 1974; Seiler and Schmidt-Glenwinkel, 1975). The concentration of spermidine seems to parallel the white matter content of any brain area. Polyamines, being cationic, bind to acidic macromolecules. The most striking relationship is between spermidine and myelin. Thus, a deficiency of spermidine has been reported in the myelin-deficient "quaking" or "jimpy" mutant mouse (Russel and Meier, 1975). Subfractionation of endogenous polyamine content of crude mitochondrial fraction of brain reveals substantial binding to both myelin and synaptosomes (Shaw, 1979). Whereas the polyamines bound to myelin could be displaced by the addition of excess of polyamine, binding to synaptosomal membrane is resistant to displacement. The binding of spermine to synaptic vesicles has been shown to inhibit the binding of acetylcholine but not that of GABA to synaptic vesicles (Kuriyama *et al.*, 1968).

Intraventricular injection of microgram amounts of spermine or spermidine leads to either sedation or convulsions (Anderson *et al.*, 1975). Polyamines activate or inhibit cholinesterase activity depending on the concentration of acetylcholine (Anand *et al.*, 1976). The observation that doses of spermine or spermidine sufficient to produce central effects are without effect on the brain concentration of the putative neurotransmitters (Shaw, 1977) has been reported with the suggestion that spermine and spermidine may modulate central synaptic activity. In the brain, ODC can be induced by electrical stimulation (Pajunen *et al.*, 1978) and also transsynaptically by depletion of amine stores with reserpine (Deckardt *et al.*, 1978). The release of endogenous spermidine and spermine from the motor cortex of the Rhesus monkey by electrical stimulation has been reported by Russel *et al.* (1974). The relationship between increased concentration of GABA in mouse brain and the concomitant increase of putrescine has been suggested to be the result of the stabilization of ODC by GABA (Seiler *et al.*, 1979). It has been suggested that the modulation of ODC by GABA may be important in tissues such as brain where the concentration of GABA is compartmentalized.

The central feature in the relationship between the polyamines and pyridoxine is the enzyme ornithine decarboxylase which has an absolute requirement for pyridoxal phosphate. Thus, in pyridoxine deficiency, the synthesis not only of putrescine but also those of spermidine and spermine are decreased. This decrease seems to have a very drastic effect on the growing animal with very active cell proliferation where polyamine action is mediated through interaction with nucleic acids. The growth retardation of the pyridoxine-deficient animal is in part caused by this deficiency of polyamines. The suggested role for spermine as a modulator of neurotransmission is yet to be established. Specific uptake and transport (Halliday and Shaw, 1978), axonal transport (Ingoglia *et al.*, 1977), nonuniform distribution in brain (Russel *et al.*, 1974), changes following distribution of specific brain areas (Russel *et al.*, 1974), and the reported correlation between behavior and brain content of spermidine (Sakurada *et al.*, 1976) all suggest the need for further investigation of a specific role of polyamines in modulation of neurotransmission.

3.4. Secretion of Pituitary Hormones

The secretion of pituitary hormones is under the influence of the hypothalamus. The concentrations of hypothalamic dopamine and serotonin seem to be the controlling factors with essentially opposing effects on pituitary secretions.

The secretion of prolactin is regulated by the prolactin-inhibiting factor secreted by the hypothalamus. It is suggested that dopamine itself might be the prolactin-inhibiting factor (Weiner and Ganong, 1978). Treatment of patients with nonpuerperal galactorrhea with L-dihydroxyphenylalanine (L-DOPA) results in a decrease in the plasma concentration of prolactin (Kleinberg *et al.*, 1977). Serotonin, on the other hand, stimulates prolactin secretion (Kamberi *et al.*, 1971b). Serotonergic neurons seem to be involved in the suckling-induced or estrogen-induced surges in prolactin secretion. Chlorophenylalanine treatment, which inhibits serotonin synthesis, prevents the suckling-induced release of prolactin in rats (Kordon *et al.*, 1973–1974). Lancranjan *et al.* (1977) inhibited peripheral aromatic amino acid decarboxylase with benserazide and administered 1,5-dihydroxytryptophan intravenously. Under these conditions, the concentration of serotonin in brain was increased. The significant increase in plasma prolactin in these animals has been suggested by these authors to be the result of the elevation of brain serotonin.

Women with hyperprolactinemia and the galactorrhea–amenorrhea syndrome respond to treatment with massive doses (200–600 mg) of pyridoxine (Foukas, 1973). The increase in brain dopamine following pyridoxine treatment has been implicated in the response to pyridoxine. This pharmacological effect of pyridoxine has been misinterpreted in some recent reports (Greentree, 1979; Lande, 1979). The deletion of pyridoxine from multivitamin preparations has been recommended in the belief that pyridoxine inhibits milk production by the nursing mother. There is clearly no basis for this suggestion.

The release of thyrotropin-releasing hormone (TRH) from the hypothalamus is also controlled by the concentrations of dopamine and serotonin in the hypothalamus. Again, these neurotransmitters have opposing effects on TRH release (Chen and Meites, 1975; Krulich, 1979). The injection of pyridoxine (300 mg) in hypothyroid individuals suppressed both plasma thyroid-stimulating hormone (TSH) and prolactin levels. This response was not elicited in euthyroid individuals (Delitala *et al.*, 1976, 1977). Robbins *et al.* (1967) have shown that

5-hydroxytryptophan decarboxylase activity is very low in human brain and have suggested that this enzyme rather than tryptophan hydroxylase might be rate limiting in the formation of serotonin in the human brain. In view of our observation (Dakshinamurti *et al.*, 1976) of nonparallel changes in the concentration of catecholamines and serotonin in the brain of the pyridoxine-deficient rat, it is possible that the decarboxylations of the precursors of dopamine and serotonin are catalyzed by distinct enzymes with differing affinities for pyridoxal phosphate. This would allow a better regulation of the synthesis of these neurotransmitters with opposing actions.

The synthesis and secretion of growth hormone by the pituitary are controlled by the opposing actions of growth hormone-releasing factor (GH-RF) and growth hormone-release inhibiting factor (GH-RIF). The hypothalamic secretion of these factors is regulated by dopaminergic activity originating in the median eminence of the hypothalamus (Kamberi *et al.*, 1970, 1971a). Growth hormone release is stimulated by treatment with L-DOPA (Boyd *et al.*, 1970). Contradictory reports have appeared on the effect of 5-hydroxytryptophan in mediating the rise in plasma growth hormone. Lancranjan *et al.* (1977) have reported an increase in plasma growth hormone following injection of 1,5-dihydroxytryptophan and benzserazide, a peripheral decarboxylase inhibitor. Pyridoxine administration has been shown to increase plasma growth hormone (Delitala *et al.*, 1976). Makris and Gershoff (1973) have shown that both pituitary and serum levels of growth hormone are low in pyridoxine-deficient young rats.

3.5. The Modulation of Steroid-Receptor Complex by Pyridoxal Phosphate

Glucocorticoid binds to cytosolic receptors in various tissues. This complex undergoes an energy-dependent conformational change to an activated form able to translocate to the nucleus to bind to chromatin or DNA or both. This transition of the complex results in exposure of positive charges on the surface of the receptor (Parchman and Litwack, 1977). Various authors have reported a cytosolic translocation inhibitor, a nonpeptide, small-molecular-weight compound (Milogram and Atger, 1975; Simons *et al.*, 1976; Cake *et al.*, 1976). Cake *et al.* (1978) have presented evidence to show that the endogenous inhibitor could be pyridoxal phosphate. The activated receptor is prevented from binding to DNA-cellulose or isolated nuclei by the addition of PALP to the incubation medium. The mechanism appears to involve the formation of a Schiff base with lysine residues located at the DNA binding site of the receptor. DiSorbo *et al.* (1980) have further shown that pyridoxine deficiency results in an increase in the number of activatable glucocorticoid-receptor complexes and an increase in the rate of translocation of the complex to the nuclei. The deficient animals take up and retain less labeled steroid than control animals and yet translocate to the nucleus the same number of steroid–receptor complexes as do control animals.

Administration of pyridoxine to deficient rats restored their ability to concentrate steroid in liver. Thus, there are two components in this system—uptake and retention of steroid by tissue and the translocation of the steroid-receptor complex. Some component of the plasma membrane or the cytoplasm involved in the transport or storage of the glucocorticoid may require pyridoxal phosphate for its synthesis or function. Various reports indicate an increase in the concentration of apoproteins of pyridoxal phosphate enzymes such as glutamic acid decarboxylase (Stephens *et al.*, 1971) and tyrosine amino transferase (Puskar and Tryfiates, 1974) in pyridoxine-deficient rat livers. The induction of the enzyme (Puskar and Tryfiates, 1974) by glucocorticoid is greater in the pyridoxine-deficient rat than in the normal animal.

4. Experimental Pyridoxine Deficiency

4.1. Neurotransmitters

Congenital deficiency of pyridoxine in the neonatal rat was produced by feeding pregnant rats a pyridoxine-deficient diet during the last 2 weeks of gestation (Dakshinamurti and Stephens, 1969). There was a significant (50%) reduction in the concentration of pyridoxal phosphate in the deficient brain at birth. The activity of glutamic acid decarboxylase was significantly reduced (by approximately 75%). However, there was no difference between deficient and normal groups in the concentration of the apoenzyme. In a study of oxidative reactions, no difference was observed between mitochondria prepared from brains of pyridoxine-deficient and normal neonates in terms of oxygen uptake, ADP/oxygen, and respiratory control ratios, as well as the concentration of the respiratory carriers (Bhuvaneswaran and Dakshinamurti, 1972). In a further study pyridoxine deficiency was produced in rats during the period of nervous system development (Stephens *et al.*, 1971). The lowered concentrations of pyridoxal phosphate and GABA in whole brain of these rats confirmed the existence of deficiency. The activity of GAD holoenzyme was decreased significantly in the deficient rat brain, whereas the apoenzyme level in deficient rat brain was 150% that in brain from pyridoxine-supplemented rats. The specific activity of brain GAD apoenzyme in a group of normal rats on a restricted diet complete in all nutrients was significantly lower than in the group on the unrestricted diet. Thus, the pyridoxine specificity of the observed differences was established.

Concomitantly, some electrophysiological parameters such as EEG and auditory evoked potentials were analyzed. The EEG of pyridoxine-deficient rats showed spike activity, presumably indicative of the existence of seizures. Evoked potentials presented abnormalities in their latency, waveform, and response to repetitive stimuli. The extent to which they were affected correlated with the intensity of the deficiency. It was suggested that the changes observed in the auditory evoked potentials in the deficient rats were the result of retardation of normal ontogenetic development of the CNS of these animals. In the adult rat subjected to pyridoxine deficiency, there was an increased latency of visually evoked potentials (Stewart *et al.*, 1973) which was reversed by supplementation with pyridoxine. However, there was no alteration in the shape of the evoked potentials, since the deficiency was imposed after CNS development was completed.

Decarboxylation of the precursor acid is a necessary step in the formation of the putative neurotransmitters such as dopamine (3,4-dihydroxyphenethylamine), norepinephrine, 5-hydroxytryptamine, and γ-aminobutyric acid. As pyridoxal phosphate is the coenzyme of these decarboxylases, it was decided to determine the brain concentrations of the various neurotransmitters in pyridoxine deficiency. The affinity of the coenzyme for the different apocarboxylases could vary. The activity of the decarboxylase with the most tightly bound coenzyme would be higher than those with lesser affinities between the apoenzyme and the coenzyme. In addition, deficiency of pyridoxine might alter the concentrations of the precursors of the various neurotransmitters to different extents. This would also result in nonparallel alterations in the brain concentrations of the various neurotransmitters. We have examined the effect of pyridoxine deficiency caused by dietary deprivation or the use of antagonists such as deoxypyridoxine and penicillamine on the brain concentrations of the putative neurotransmitters. The nature of the results was the same regardless of the method used for pyridoxine depletion, either dietary or by administration of the antagonists (Dakshinamurti *et al.*, 1976). Deficient animals had significantly lower concentrations of brain γ-aminobutyric acid than did controls. This correlated well with brain concentrations of pyridoxal phosphate and the activity of glutamate decarboxylase. Brain concentrations of dopamine and norepinephrine were unaffected by pyridoxine deficiency.

Our results on brain catecholamine levels are in agreement with the report of Sourkes (1972) that there is no change in the steady-state concentration of norepinephrine and dopamine in pyridoxine deficiency. The normal levels of both catecholamines in the brain of pyridoxine-deficient rats are in striking contrast to the decrease of these amines seen by Shoemaker and Wurtman (1971) in rats subjected to undernutrition perinatally. In our study (Dakshinamurti *et al.*, 1976), we found a significant decrease in the concentration of serotonin in the brain of pyridoxine-deficient rats. We ruled out the possibility that this decrease was the result of inanition and the generalized malnutrition. Increased catabolism of 5-hydroxytryptamine or transport to cerebrospinal fluid did not contribute to this. The levels of brain tryptophan were not affected by dietary deficiency of pyridoxine or treatment of rats with pyridoxine antagonists. The possibility of a decrease in the activity of brain tryptophan hydroxylase was also excluded by assay of this enzyme as well as by 5-hydroxytryptophan loading experiments. Our observations of nonparallel changes in brain concentration of catecholamines and serotonin, respectively, indicate that the synthesis of various amines is regulated separately.

The hypothermia seen in pyridoxine-deficient rats (Dakshinamurti *et al.*, 1976) might be related to the decrease in brain serotonin. Modigh (1974) has presented evidence to suggest that the antagonizing effects of *p*-chlorophenylalanine (PCPA) on hyperthermia and behavioral stimulation induced by nialamide in mice are caused by the inhibitory effect of PCPA on the synthesis of serotonin. Additional evidence has been presented by Myers (1975) to support the view that a serotonergic mechanism in the hypothalamus is involved in thermoregulation in the rat.

Pyridoxine-deficient young rats showed a significant decrease in their motility. Tunnicliff *et al.* (1972) have also shown that locomotor activity in the open field is affected by dietary pyridoxine deficiency in two inbred strains of mice. They suggest that PALP-requiring systems are more important for locomotor activity and are more sensitive to dietary pyridoxine levels than are those systems directly involved in active and passive learning.

We find that pyridoxine deficiency in rats affects sleep in two ways (V. Kamaya, V. Havlicek, and K. Dakshinamurti, unpublished observation, 1980). The duration of deep slow-wave sleep 2 (SWS 2) is shortened, and in some instances this stage of sleep is completely abolished; REM sleep is also affected in the same manner. These animals are in shallow slow-wave sleep (SWS 1). The effects of pyridoxine deficiency on sleep parallel the effects of experimental serotonergic deficit in animals and man in keeping with the view of Jouvet (1972) that serotonergic neurons play a major role in maintenance of slow-wave sleep 2 and REM (paradoxical sleep) events. Thus, treatment of men with *p*-chlorophenylalanine, an inhibitor of serotonin synthesis, was shown to decrease REM sleep. This returned to normal on administration of 5-hydroxytryptophan (Wyatt, 1972). In another human study, methylsergide, a blocker of serotonergic receptors, was administered to adults at a dose of 8 mg per 24 hr. The REM sleep time was significantly reduced in these people, although total sleep time was not changed (Mendelson *et al.*, 1975). In more controlled animal studies, Kiianma and Fuxe (1977) injected 5,7-dihydroxytryptophan bilaterally into the rat dorsomedial mesencephalic tegmentum close to serotonergic pathways and recorded EEG and EMG continuously for 2 to 4 postoperative days in order to determine the time the animals spent awake and in different stages of sleep. They also analyzed brain serotonin. A significant positive correlation was observed between cortical serotonin stores and the time spent in SWS 2 and paradoxical sleep (REM), and a significant negative correlation was seen between cortical serotonin stores and time spent in SWS 1.

4.2. Myelination

Myelin is formed in the CNS by the oligodendrocytes. Myelination is generally studied as a parameter of development (Dobbing, 1974). Glial cell multiplication and rapid myelination accomplished by the glial cells occupy the first and second halves of the period of postnatal brain growth spurts. The period of rapid myelination of various animals is related to the extent of CNS maturation at birth. In the rat, this period extends from about 10 to 45 days postnatally. In humans, the period of very active myelination is around the perinatal period. This phase extends until the end of the second year of life. However, the human neocortex is myelinated at a very slow rate to the end of the second decade of life. The concentration of cerebroside, a myelin-typical lipid, is directly proportional to the extent of maturation.

During this period, the growing animal is vulnerable to many kinds of insults including generalized malnutrition (Fishman *et al.*, 1971; Nakhasi *et al.*, 1975; Wiggins *et al.*, 1976), specific deficiencies (Clausen, 1969; Trapp and Bernsohn, 1978), or hypothyroidism (Rosman *et al.*, 1972; Malone *et al.*, 1975). In addition, certain genetic errors such as phenylketonuria (Gerstl *et al.*, 1967), Down syndrome (Banik *et al.*, 1975), and the "jimpy" (Herschkowitz *et al.*, 1971) and "quaking" (Bauman *et al.*, 1968) mutants in mice also result in defective myelination. Myelination is a complex process and could be affected at various steps. For example, there is a lack of myelinating glia in the "jimpy" mutant, whereas in the "quaking" mutant glial cells multiply normally prior to myelination, and the oligodendrocytes seem to be qualitatively abnormal (Bauman *et al.* 1972). In vitamin A deficiency, the decreased sulfatide synthesis (Clausen, 1969) seems to be related to the decreased formation of active sulfate.

Although a role for pyridoxal phosphate as a cofactor in one of the steps leading to the synthesis of sphingosine was established quite early (Brady and Koval, 1958; Braun and Snell, 1968), the consequences of pyridoxine deficiency during the critical period of development of the rat on cerebral lipids have been investigated only recently. The incorporation of [^{14}C]acetate into all major lipid classes in brain was significantly decreased (Dakshinamurti and Stephens, 1971; Stephens and Dakshinamurti, 1975). The specific radioactivities of purified cerebrosides and sulfatides from pyridoxine-deficient rat brain were only one-fifth those found in pyridoxine-treated controls. Less myelin seems to be synthesized by pyridoxine-deficient rats (Kurtz and Kanfer, 1973). Morre and Kirksey (1978) have shown that the specific activity of 2′,3′-cyclic nucleotide 3′-phosphohydrolase, a marker enzyme for myelin, is decreased significantly in pyridoxine-deficient neonatal rat brain. Williamson and Coniglio (1971) have reported diminution in the sphingomyelin content in 3-week-old pyridoxine-deficient rats. They have also shown that restriction of caloric intake alone did not significantly affect the content of phospholipids, cerebrosides, and sulfatides when calcualted as a percentage of total lipid.

We have examined the fatty acid composition of the galactolipids of the brain of 4- to 6-week-old rats subjected to pyridoxine deficiency since birth (Dakshinamurti *et al.*, 1973; Stephens and Dakshinamurti, 1976). Fatty acids constitute a major portion of brain lipids and appear to have a crucial role in determining the properties of membranes. The hydrophobic interior region of myelin is made up largely of apolar hydrocarbon chains. Cerebrosides and sulfatides of myelin contain fatty acids with chain lengths 25% longer than the chains of the fatty acid component of cerebroside and sulfatide of gray matter (O'Brien, 1965). There are suggestions that the interdigitation of the longer chain saturated fatty acids of cerebroside and sulfatides contributes to the structural stability of myelin. Hence, conditions leading to any substantial derangement of neural tissues might be related to changes in the fatty acid composition. An analysis of the nonhydroxy fatty acids of the galactolipid fraction of brain lipids of pyridoxine-deficient rats indicated an accumulation of stearic acid and, correspondingly, a significant decrease in the content of lignoceric and nervonic acids (Stephens and Dakshinamurti, 1976). Thus, the CNS myelin of pyridoxine-deficient rat is qualitatively and quantitatively different from that in normal rats.

The biosynthesis of the long-chain fatty acids has been shown to involve *de novo* synthesis of palmitic acid by cytoplasmic enzymes (Volpe and Kishimoto, 1972; Cantrill and Carey, 1975) and chain elongation by microsomal (Bauman *et al.*, 1970; Goldberg *et al.*, 1973) and mitochondrial (Boone and Wakil, 1970) enzymes. Our studies (Chauhan and Dakshinamurti, 1977, 1979a) indicate the presence of a complex elongation system in the rat brain microsomes. It seems very likely that the same enzyme is responsible for the elongation of $C_{16:0}$ and $C_{18:0}$ fatty acyl-CoAs. Brophy and Vance (1975) have also reported on this enzyme from rat brain microsomes. Another enzyme that elongates behenyl-CoA seems to be distinct developmentally and is impaired considerably in pyridoxine deficiency. The enzyme elongating arachidyl-CoA, in view of its impairment in pyridoxine deficiency, might be distinct from the enzyme elongating $C_{16:0}$ and $C_{18:0}$ fatty acyl-CoAs which is not affected in deficiency. The specificity of the decrease in brain microsomal elongation enzymes in pyridoxine deficiency needs to be investigated further. A cofactor role for a pyridoxine derivative in the elongation reaction is not indicated. Various pathological conditions associated with impaired myelination seem to share the common defect of a reduced ability to synthesize long-chain fatty acids. The evidence for this has been obtained by the analysis of the profiles of the galactolipid fatty acids. A study of the microsomal elongation enzymes in all of these conditions might uncover the underlying cause of the impaired myelination.

5. Human Pyridoxine Deficiency and Dependency

5.1. Deficiency

Impairment of somatic growth, a pellagralike dermatitis, and ataxia have been reported in all species of pyridoxine-deficient animals (Gries and Scott, 1972). Anemia occurs in all species except the rat (Harris and Horrigan, 1964).

Among the most outstanding symptoms of deficiency are those that affect the nervous system. Thus, besides ataxia, hyperacousis, and hyperirritability, altered mobility and alertness, abnormal head movement, and convulsions are observed in the chicken, duck, turkey, rat, guinea pig, pig, cow, and human (Dakshinamurti, 1977).

Snyderman *et al.* (1953) reported on the production of pyridoxine deficiency in a 2-month-old hydrocephalic fed a deficient diet for 76 days. The biochemical correlates of pyridoxine deficiency were present, and the child had convulsive seizures that were relieved by intravenous administration of pyridoxine. This report was followed by others of widespread occurrence of convulsive seizures in infants receiving a proprietary milk formula inadvertently rendered pyridoxine-deficient during manufacture. Prompt relief was obtained following intramuscular injection of 100 mg pyridoxine (Coursin, 1954; Molony and Parmalee, 1954). Electroencephalographic techniques were used to monitor the effectiveness of treatment. Within minutes after the administration of pyridoxine, marked improvement in the wave forms and normalization of the amplitude and frequency were seen. In more extensive studies, Bessey *et al.* (1957) correlated dietary intake of pyridoxine of less than 100 μg/liter in infants with biochemical correlates of deficiency and convulsive seizures. They found that convulsive seizures could be corrected with a lower dose of pyridoxine than was needed to reverse the abnormal metabolism of a load of tryptophan.

Of the iatrogenic causes of pyridoxine deficiency, particularly in adults, that related to the use of oral contraceptive steroids has been discussed earlier (Section 2.4). Of significance are neuropsychiatric disorders such as changes in sleep pattern and mood that are related to the altered serotonin metabolism in some OCS users (Wynn *et al.*, 1975). Diabetics and patients with renal failure also seem to be at risk in terms of pyridoxine depletion.

5.2. Dependency

Two infant sibs with seizures uncontrolled by anticonvulsants were described by Hunt *et al.* (1954). They had to be maintained on a large dose of pyridoxine (5–25 mg) to control the seizures. Hunt applied the term "pyridoxine dependency" to describe this condition. Since then, there have been more than 40 reports of such patients. "Vitamin-dependency" states are characterized by a biochemical abnormality affecting one or more reactions catalyzed by the vitamin-containing holoenzyme. There is no deficiency of the vitamin in question. The biochemical lesion in pyridoxine dependency appears not to be an inability of the body to retain pyridoxine vitamers but, rather, a structural abnormality of one or more apoenzymes resulting in a low affinity for PALP.

An autosomal recessive mode of inheritance is indicated. The trait is variable in intensity of expression and time of first appearance. Some probably have intrauterine convulsions (Bejsovec *et al.*, 1967), and in many others tonic-clonic seizures progressing to status epilepticus are seen soon after birth. If the condition is not diagnosed and treated, early severe mental retardation results.

It is to be noted that in patients with pyridoxine dependency, the seizures and EEG abnormalities can be quickly abolished not only by intravenous (i.v.) administration of pyridoxine but also by i.v. administration of GABA (Marie *et al.*, 1961). It is not clear how GABA crosses the blood-brain barrier. Based on Roberts' hypothesis (1964) on the role of GABA in neuronal function, Scriver (1964) proposed that the defect was caused by a mutation in the gene locus for glutamic acid decarboxylase with the result that the apoenzyme had very poor affinity for pyridoxal phosphate. No abnormality in any of the other pyridoxal phosphate-requiring reactions was seen in these patients. Following Scriver and Whelan's observation (1969) that this enzyme was also present in rat kidney, Yoshida *et al.* (1971) established the presence of the defective enzyme in the kidney of a patient with pyridoxine-dependency seizures. The *in vitro* activity was fully restored by addition of pyridoxal phosphate, thus establishing the rationale for treatment with large doses of pyridoxine.

Recently, two cases of pyridoxine-responsive infantile convulsions were investigated fully. In one instance, three consecutive siblings in one family were diagnosed to be pyridoxine dependent (Miyasaki *et al.*, 1978). Electroencephalogram abnormalities and the response to pyridoxine were as expected in the dependency syndrome. Brain lesions at autopsy of one of these children were similar to those seen in cases of cryptogenic epilepsy. Lott *et al.* (1978) have to date presented the most complete neuropathologic and biochemical findings in the brain of a child dying at the age of 13½ years with clinically proven pyridoxine-dependent convulsions. Biochemically, the concentration of glutamic acid was increased and that of GABA decreased in both the frontal and occipital lobes, whereas cystathionine was elevated in the occipital cortex only. The concentration of PALP was reduced by one-third to one-half the values for controls in the frontal cortex. There was no significant decrease in PALP in the occipital cortex. Neuropathologic findings included a striking degree of neuronal loss bilaterally in the thalamus. The sparseness of central white matter between gray-matter structures was the most striking abnormality. Hypoplasia of myelin could explain the symptoms of spastic quadriparesis, optic atrophy, and dementia. These findings could be explained on the basis of the known involvement of PALP as cofactor of GAD and cystathionase as well as in the synthesis of myelin-typical lipids.

Other pyridoxine-dependent aminoacidopathies have been reported, e.g., homocystinuria and xanthurenic aciduria. In neither of these situations is mental retardation always the normal sequela of the untreated condition. The accumulation of homocysteine and xanthurenic acid has not been shown to be specifically related to a neurological impairment.

6. Conclusions

Most of the pyridoxal phosphate-dependent enzymes are involved in the catabolism of amino acids. Apart from this, the presence of covalently bound PALP in glycogen phosphorylase has been recognized for a long time. A structural role for PALP in this enzyme has been demonstrated (Hedrick, 1972). The glycogen reserves of brain are modest, and brain depends mainly on blood glucose for its energy requirement. Hence, the decrease in phosphorylase activity seen in pyridoxine deficiency will not have any adverse effect on cerebral glucose metabolism. The synthesis of all sphingolipids—sphingomyelin, cerebroside, sulfatide, and ganglioside—would be affected in pyridoxine deficiency, as pyridoxal phosphate is involved in the synthesis of sphingosine. Of the sphingolipids, cerebrosides and sulfatides are considered to be myelin-specific. In pyridoxine deficiency, myelin formation is impaired. In addition to the effect on the synthesis of the base sphinosine, myelin lipids exhibit another defect, a deficiency of the very-long-chain fatty acids that are considered to contribute to the structural integrity and stability of myelin. This defect in fatty acid chain elongation is shared by a number of conditions, inherited as well as of dietary origin. There are suggestions that these conditions also share hypothyroid activity.

Of the pyridoxal phosphate-dependent enzymes involved in the catabolism of amino acids four—glutamic acid decarboxylase, cysteine sulfinic acid decarboxylase, 5-hydroxytryptophan decarboxylase, and ornithine decarboxylase—seem to have crucial roles (Fig. 4). The clinical effects of pyridoxine deficiency can be explained on the basis of the known decrease in activity of these enzymes. The effects of deficiency are devastating in the growing animal during the period of maturation of the nervous system. Thus, even a moderate deficiency of pyridoxine during gestation in the rat results in impairment of the maturation process and a very high degree of mortality. Decreases in polyamines because of decreased ODC activity might be responsible for the impairment in protein synthetic machinery through effects on gene expression. The effects on the nervous system are drastic during this period of growth. In the adult animal, the effects of deficiency are much less pronounced, as the neuronal structures are formed already. Of the effects on the neurotransmitters, those on the formation of GABA, taurine, and serotonin seem to be significant. Decreases in GABA and in GAD are correlated. The decrease in serotonin seems to be functionally significant in view of the correlation between pyridoxine deficiency and specific serotonin depletion as far as effects on behavior, thermoregulation, and sleep are concerned. The effects on serotonin as against dopamine are of particular significance, as these two neurotransmitters have opposing effects in the secretion of the various pituitary hormones. Although a specific decrease in brain taurine has not been shown, the known properties of CSD would seem to indicate such a depletion at specific loci.

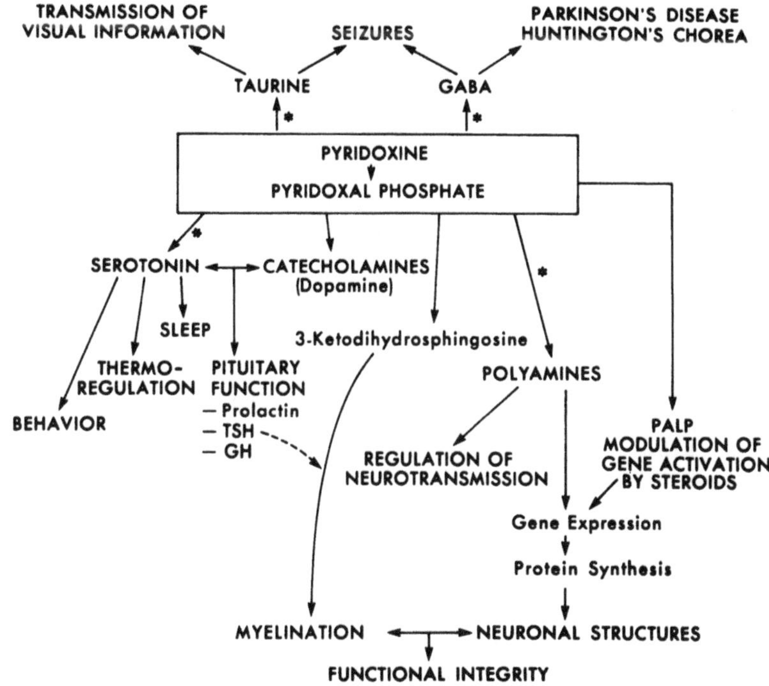

Fig. 4. Involvement of pyridoxine of CNS.

The author has attempted to correlate the effects of pyridoxine in the whole animal with the properties of some of the pyridoxal phosphate-dependent enzymes. It is recognized that there is an element of conjecture in this integration. It is hoped that this would lead to further investigations to understand the neurobiology of pyridoxine at the molecular level.

References

Ahmed, F., and Bamji, M. S., 1976, Vitamin supplements to women using oral contraceptives, *Contraception* **14**:309.

Airoldi, L., Watkins, C. J., Wiggins, J. F., and Wurtman, R. J., 1978, Effect of pyridoxine depletion of tissue pyridoxal phosphate by carbidopa, *Metabolism* **27**:771.

Anand, R., Gore, M. G., and Kerkut, G. A., 1976, The effect of spermine and spermidine on the hydrolysis of acetylcholine in the presence of rat caudate nucleus homogenate or acetylcholinesterase from *Electrophorus electricus*, *J. Neurochem.* **27**:381.

Anderson, D. J., Crossland, J., and Shaw, G. G., 1975, The action of spermidine and spermine on the central nervous system, *Neuropharmacology* **14**:571.

Banik, N. L., Davidson, A. N., Palo, J., and Savolainen, H., 1975, Biochemical studies on myelin isolated from the brains of patients with Down's syndrome, *Brain* **98**:213.

Bartholini, G., Bates, H. M., Burkard, W. P., and Pletscher, A., 1967, Increase of cerebral catecholamines caused by 3,4-dihydroxyphenylalanine after inhibition of peripheral decarboxylase, *Nature* **215**:852.

Battistini, L., Grynbaun, A., and Lajtha, A., 1969, Distribution and uptake of amino acids in various regions of the cat brain *in vitro, J. Neurochem.* **16**:1459.

Bauman, N. A., Jacque, C. M., Pollet, S. A., and Harpin, M. L., 1968, Fatty acid and lipid composition of the brain of a myelin deficient mutant the 'quaking' mouse, *Eur. J. Biochem.* **4**:340.

Bauman, N. A., Harpin, M. L., and Bourre, J. M., 1970, Long chain fatty acid formation: Key step in myelination studied in mutant mice, *Nature* **227**:960.

Bauman, N. A., Bourre, J. M., Jacque, C., and Pollet, S., 1972, Genetic disorders of myelination, in: *Lipids, Malnutrition and the Developing Brain,* CIBA Foundation Symposia, pp. 91–105, Elsevier, Amsterdam.

Baumblatt, M. J., and Winston, F., 1970, Pyridoxine and the pill, *Lancet* **1**:832.

Baysal, A., Johnson, B. A., and Linkswiler, H., 1966, Vitamin B_6 depletion in man: Blood vitamin B_6, plasma pyridoxal-phosphate, serum cholesterol, serum transaminases and urinary vitamin B_6 and 4-pyridoxic acid, *J. Nutr.* **89**:19.

Bejsovec, M., Knlenda, Z., and Ponca, E., 1967, Familial intrauterine convulsions in pyridoxine dependency, *Arch. Dis. Child.* **42**:201.

Bergamini, L., Mutani, R., Delsedime, M., and Durelli, L., 1974, First clinical experience on the antiepileptic action of taurine, *Eur. Neurol.* **11**:261.

Berson, E. L., Hayes, K. C., Rabin, A. R., Schmidt, S. Y., and Watson, G., 1976, Retinal degeneration in cats fed casein. 2. Supplementation with methionine, cysteine or taurine, *Invest. Ophthalmol.* **15**:52.

Bessey, D. A., Adam, D. J. D., and Hansen, A. E., 1957, Intake of vitamin B_6 and infantile convulsions: A first approximation of requirements of pyridoxine in infants, *Pediatrics* **20**:33.

Bhuvaneswaran, C., and Dakshinamurti, K., 1972, Oxidative phosphorylation by pyridoxine deficient rat brain mitochondria, *J. Neurochem.* **19**:149.

Bloom, F. E., 1972, Amino acids and polypeptides in neuronal function, *Neurosci. Res. Program Bull.* **10**:121.

Boone, S. C., and Wakil, S. J., 1970, *In vitro* synthesis of lignoceric and nervonic acids in mammalian liver and brain, *Biochemistry* **9**:1470.

Borg, J., Balear, J. V., and Mandel, P., 1977, High affinity uptake of taurine by neuronal and glial cells, *Brain Res.* **118**:514.

Boyd, A. E. III, Lebovitz, H. E., and Pfeiffer, J. B., 1970, Stimulation of human-growth-hormone secretion by L-DOPA, *N. Engl. J. Med.* **283**:1425.

Brady, R. O., and Koval, G. J., 1958, The enzymatic synthesis of sphingosine, *J. Biol. Chem.* **233**:26.

Braun, P., and Snell, E., 1968, Biosynthesis of sphingolipid bases. II. Keto-intermediates in synthesis of sphingosine and dihydrosphingosine by cell-free extracts of *Hansenula ciferri, J. Biol. Chem.* **243**:3775.

Brin, M., 1971, Abnormal tryptophan metabolism in pregnancy and with the oral contraceptive pill I. Specific effects of an oral estrogenic contraceptive steroid on the tryptophan oxygenase and two aminotransferase activities in livers of ovariectomized-adrenalectomized rats, *Am. J. Clin. Nutr.* **24**:699.

Brophy, M. H., and Siiteri, P. K., 1975, Pyridoxal phosphate and hypertensive disorders of pregnancy, *Am. J. Obstet. Gynecol.* **121**:1075.

Brophy, P. J., and Vance, D. E., 1975, Elongation of fatty acids by microsomal fractions from the brain of the developing rat, *Biochem. J.* **152**:495.

Brown, R. R., Rose, D. P., Price, J. M., and Wolf, H., 1968, Tryptophan metabolism as affected by anovulatory agents, *Ann. N.Y. Acad. Sci.* **166**:44.

Cake, M. H., Goidl, J. A., Parchman, G., and Litwack, G., 1976, Involvement of a low molecular weight component(s) in the mechanism of action of the glucocorticoid receptor, *Biochem. Biophys. Res. Commun.* **71**:45.

Cake, M. H., DiSorbo, D. M., and Litwack, G., 1978, Effect of pyridoxal phosphate on the DNA binding site of activated hepatic glucocorticoid receptor, *J. Biol. Chem.* **253**:4886.

Canham, J. G., Baker, E. M., Harding, R. S., Sauberlich, H. E., and Plaugh, I. C., 1968, Dietary protein: Its relationship to vitamin B_6 requirements and function, *Ann. N.Y. Acad. Sci.* **166**:16.

Cantrill, R. C., and Carey, E. M., 1975, Changes in the activities of *de novo* fatty acid synthesis and palmitoyl-CoA synthetase in relation to myelination in rabbit brain, *Biochim. Biophys. Acta* **380**:165.

Carney, M. W. P., 1967, Serum folate values in 423 psychiatric patients, *Br. Med. J.* **4**:512.

Casola, L., and DiMatteo, G., 1972, Studies on the dansylation reaction by the use of [^{14}C-]dansyl chloride: Application to the analysis of free amino acids in the rat optic nerve, *Anal. Biochem.* **49**:416.

Chabner, B., and Livingston, D., 1970, A simple enzymic assay for pyridoxal phosphate, *Anal. Biochem.* **34**:413.

Chase, T. N., and Walters, J. R., 1976, Pharmacologic approaches to the manipulation of GABA-mediated synaptic function in man, in: *GABA in Nervous System Function* (E. Roberts, T. N. Chase, and D. B. Tower, eds.), pp. 497–513, Raven Press, New York.

Chauhan, M. S., and Dakshinamurti, K., 1977, The *in vitro* elongation of fatty acyl coenzyme A by rat brain sub-cellular fractions, in *Proceedings, VI International Meeting International Society for Neurochemistry,* p. 495, Copenhagen.

Chauhan, M. S., and Dakshinamurti, K., 1979a, Fluorometric assay of pyridoxal, in *Methods in Enzymology,* Vol. 62 (D. B. McCormick and L. D. Wright, eds.), pp. 405–407, Academic Press, New York.

Chauhan, M. S., and Dakshinamurti, K., 1979b, Flurormetric assay of pyridoxal and pyridoxal phosphate, *Anal. Biochem.* **96**:426.

Chauhan, M. S., and Dakshinamurti, K., 1979c, The elongation of fatty acids by microsomes and mitrochondria from normal and pyridoxine-deficient rat brain, *Exp. Brain Res.* **36**:265.

Chen, H. J., and Meites, J., 1975, Effects of biogenic amines and TRH on release of prolactin and TSH in the rat, *Endocrinology* **96**:10.

Clausen, J., 1969, The effect of vitamin A deficiency on myelination in the central nervous system of the rat, *Eur. J. Biochem.* **7**:575.

Cleary, R. E., Lumeng, L., and Li, T.-K., 1975, Maternal and fetal plasma levels of pyridoxal phosphate at term: Adequacy of vitamin B_6 supplementation during pregnancy, *Am. J. Obstet. Gynecol.* **121**:25.

Cohen, A. A., 1968, Pyridoxine in the prevention and treatment of convulsions and neurotoxicity due to cycloserine, *Ann. N.Y. Acad. Sci.* **166**:346.

Contractor, S. F., and Shane, B., 1970, Blood and urine levels of vitamin B_6 in the mother and fetus before and after loading of the mother with vitamin B_6, *Am. J. Obstet. Gynecol.* **107**:635.

Coursin, D. B., 1954, Convulsive seizures in infants with pyridoxine-deficient diet, *J. Am. Med. Assoc.* **154**:406.

Coyer, J. R., and Nicholson, D. P., 1976, Isoniazid-induced convulsions: Part I—Clinical, *South. Med. J.* **69**:294.

Craig, C. R., and Hartman, E. R., 1973, Concentration of amino acids in the brain of cobalt epileptic rat, *Epilepsia* **14**:409.

Curtis, D. R., and Johnson, G. A. R., 1970, Amino acid transmitters, in: *Handbook Neurochemistry,* Vol. 4 (A. Lajtha, ed.), pp. 115–134, Plenum Press, New York.

Curtis, D. R., and Tebécis, A. K., 1972, Bicuculline and thalamic inhibition, *Exp. Brain Res.* **16**:210.

Curtis, D. R., and Watkins, J. C., 1960, The excitation and depression of spinal neurons by structurally related amino acids, *J. Neurochem.* **6**:117.

Curtis, D. R., Hösli, L., and Johnston, G. A. R., 1968, A pharmacological study of the depression of spinal neurons by glycine and related amino acids, *Exp. Brain Res.* **6**:1.

Curtis, D. R., Duggan, A. W., Felix, D., and Johnston, G. A. R., 1971a, Bicuculline, an antagonist of GABA and synaptic inhibition in the spinal cord of the cat, *Brain Res.* **32**:69.

Curtis, D. R., Duggan, A. W., Felix, D., Johnston, G. A. R., and McLennan, H., 1971b, Antagonism between bicuculline and GABA in the cat brain, *Brain Res.* **33**:57.

Dakshinamurti, K., 1977. B vitamins and nervous system function, in: *Nutrition and the Brain*, Vol. 1 (R. J. Wurtman and J. J. Wurtman, eds.), pp. 251-318, Raven Press, New York.

Dakshinamurti, K., and Stephens, M. C., 1969, Pyridoxine deficiency in the neonatal rat, *J. Neurochem.* **16**:1515.

Dakshinamurti, K., and Stephens, M. C., 1971, Myelin lipids in pyridoxine deficiency, in: *Proceedings: III International Meeting, International Society for Neurochemistry*, p. 347, Budapest.

Dakshinamurti, K., Stephens, M. C., and Mokashi, S., 1973, Cerebral fatty acids in pyridoxine deficient young rats, in: *Proceedings: IV International Meeting. International Society for Neurochemistry*, p. 423, Tokyo.

Dakshinamurti, K., Le Blancq, W. D., Herchl, R., and Havlicek, V., 1976, Nonparallel changes in brain monoamines of pyridoxine-deficient growing rats, *Exp. Brain Res.* **26**:355.

DeBelleroche, J. S., and Bradford, H. F., 1973, Amino acids in synaptic vesicles from mammalian cerebral cortex: A reappraisal, *J. Neurochem.* **21**:441.

Deckardt, K., Pujol, J.-F., Belin, M.-F., Seiler, N., and Jonvet, M., 1978, Increase of ornithine decarboxylase activity elicited by reserpine in the peripheral and central monoaminergic systems of the rat, *Neurochem. Res.* **3**:745.

Delitala, G., Masala, A., Alagna, S., and Devilla, L., 1976, Effect of pyridoxine on human hypophyseal trophic hormone release: A possible stimulation of hypothalamic dopaminergic pathway, *J. Clin. Endocrinol. Metab.* **42**:603.

Delitala, G., Rovasio, P., and Lotti, G., 1977, Suppression of thyrotropin (TSH) and prolactin (PRL) release by pyridoxine in chronic primary hypothyroidism, *J. Clin. Endocrinol. Metab.* **45**:1019.

DiSorbo, D. M., Phelps, D. S., Ohl, V. S., and Litwack, G., 1980, Pyridoxine deficiency influences the behaviour of the glucocorticoid-receptor complex, *J. Biol. Chem.* **255**:3866.

Dixon, H. G., Browne, J. C. M., and Davey, D. A., 1963, Choriodecidual and myometrial bloodflow, *Lancet* **2**:369.

Dobbelstein, H. W. F., Korner, W., Mempel, H., Grosse, W., and Edel, H. H., 1974, Vitamin B_6 deficiency in uremia and its implications for the depression of immune response, *Kidney Int.* **5**:233.

Dobbing, J., 1974, The later growth of the brain and its vulnerability, *Pediatrics* **53**:2.

Driskell, J. A., Geders, J. M., and Urban, M. C., 1976, Vitamin B_6 status of young men, women and women using oral contraceptives, *J. Lab Clin. Med.* **87**:813.

Durko, I., Vladovska-Yukhnovska, Y., and Ivanov, Ch. P., 1973, A new fluorometric method for the determination of vitamin B_6 in blood, *Clin. Chim. Acta* **40**:407.

Ebadi, M., and Costa, E., 1972, *Role of Vitamin B_6 in Neurobiology*, Raven Press, New York.

Ebadi, M., and Govitrapong, P., 1979, Biogenic amine-mediated alteration of pyridoxal phosphate formation in rat brain, *J. Neurochem.* **32**:845.

Ebadi, M. S., Russel, R. L., and McCoy, E. E., 1968, The inverse relationship between the activity of pyridoxal kinase and the level of biogenic amines in rabbit brain, *J. Neurochem.* **15**:659.

Ebadi, M. S., McCoy, E. E., and Kugel, R. B., 1970, Interrelationships between pyridoxal phosphate and pyridoxal kinase in rabbit brain, *J. Neurochem.* **17**:941.

Enna, S. J., Kuhar, M. J., and Synder, S. H., 1975, Regional distribution of post synaptic receptor binding for γ-amino butyric acid (GABA) in monkey brain, *Brain Res.* **93**:168.

Fishman, M. A., Madyastha, P., and Prensky, A. L., 1971, The effect of undernutrition on the development of myelin in the rat central nervous system, *Lipids* **6**:458.

Foukas, M. D., 1973, An antilactogenic effect of pyridoxine, *J. Obstet. Gynecol. Br. Commonw.* **80**:718.

Gaull, G. E., Rassin, D. K., Raiha, N. C. R., and Heinonen, K., 1977, Milk protein quantity and quality in low birth weight infants. 3. Effects on sulfur amino acids in plasma and urine, *J. Pediatr.* **90**:348.

Gaynor, R., and Dempsey, W. B., 1972, Vitamin B_6 enzymes in normal and preeclamptic human placentae, *Clin. Chim. Acta* **37**:411.

Gerstl, B., Malamud, N., Eng, L. F., and Hayman, A. B., 1967, Lipid alterations in human brain in phenylketonuria, *Neurology* **17**:51.

Gilbert, J. B., Ku, Y., Rogers, L. L., and Williams, R. J., 1960, The increase in urinary taurine after intraperitoneal administration of amino acids to the mouse, *J. Biol. Chem.* **235**:1055.

Goldberg, I., Schecter, I., and Bloch, K., 1973, Fatty acyl coenzyme A elongation in brain of normal and quaking mice, *Science* **182**:497.

Green, A. R., Joseph, M. H., and Gurzon, G., 1970, Oral contraceptives, depression, and amino acid metabolism, *Lancet* **1**:1288.

Greentree, L. B., 1979, Dangers of vitamin B_6 in nursing mothers, *N. Engl. J. Med.* **300**:141.

Gries, C. L., and Scott, M. L., 1972, The pathology of pyridoxine deficiency in chicks, *J. Nutr.* **102**:1259.

Haas, H. L., and Hösli, L., 1973, The depression of brain stem neurons by taurine and its interaction with strychnine and bicuculline, *Brain Res.* **52**:399.

Haber, B., Kuriyama, K., and Roberts, E., 1970, An anion stimulated L-glutamic acid decarboxylase in non-neural tissues, *Biochem. Pharmacol.* **19**:1119.

Halliday, C. A., and Shaw, C. G., 1978, Clearance of the polyamines from the perfused cerebroventricular system of the rabbit, *J. Neurochem.* **30**:807.

Hamfelt, A., and Tuvemo, T., 1972, Pyridoxal phosphate and folic acid concentration in blood and erythrocyte aspartate aminotransferase activity during pregnancy, *Clin. Chim. Acta* **41**:287.

Harik, S. I., and Snyder, S. H., 1974, Putrescine: Regional distribution in the nervous system of the rat and the cat, *Brain Res.* **66**:328.

Harris, J. W., and Horrigan, D. L., 1964, Pyridoxine-responsive anemia-prototype and variations on the theme, *Vitam. Horm.* **22**:721.

Harris, R. S., Wool, J. G., and Lorraine, J. A., 1964, International symposium on vitamin B_6 in honour of Professor Paul György, *Vitam. Horm.* **22**:361.

Haskell, B. E., and Snell, E. E., 1972, An improved apotryptophanase assay for pyridoxal phosphate, *Anal. Biochem.* **45**:567.

Hayes, K. D., Carey, R. E., and Schmidt, S. V., 1975, Retinal degeneration associated with taurine deficiency in the cat, *Science* **188**:949.

Hedrick, J. L., 1972, The role of pyridoxal-5′-phosphate in the structure and function of glycogen phosphorylase, *Adv. Biochem. Psychopharmacol.* **4**:23.

Heller, S., Salkeld, R. M., and Körner, W. F., 1973, Vitamin B_6 status in pregnancy, *Am. J. Clin. Nutr.* **26**:1339.

Herschkowitz, N., Vassella, F., and Bischoff, A., 1971, Myelin differences in the central and peripheral nervous system in the 'jimpy'mouse, *J. Neurochem.* **18**:1361.

Hökfelt, T., Jonsson, G., and Ljungdahl, A., 1970, Regional uptake and subcellular localization of (^{3}H)-γ-aminobutyric acid in rat brain slices, *Life Sci.* **9**:203.

Hope, D. B., 1957, The persistence of taurine in the brains of pyridoxine deficient rats, *J. Neurochem.* **1**:364.

Hornykiewicz, O., Lloyd, K. G., and Davidson, L., 1976, The GABA system, function of the basal ganglia, and Parkinson's disease, in: *GABA in Nervous System Function* (E. Roberts, T. N. Chase, and D. B. Tower, eds.), pp. 479–485, Raven Press, New York.

Hunt, A. D., Jr., Strokes, J., Jr., McCrory, W. W., and Stroud, H. H., 1954, Pyridoxine dependency: Report of a case of intractable convulsions in an infant controlled by pyridoxine, *Pediatrics* **13**:140.

Ingoglia, N. A., Sturman, J. A., and Eisner, R. A., 1977, Axonal transport of putrescine, spermidine and spermine in normal and regenerating goldfish optic nerves, *Brain Res.* **130**:433.

Izumi, K., Donaldson, J., Minnich, J. L., and Barbeau, A., 1973, Ouabain induced seizures in rats. Suppressive effects of taurine and γ-aminobutyric acid, *Can. J. Physiol. Pharmacol.* **51**:885.

Izumi, K., Igisu, H., and Fukuda, T., 1974, Suppression of seizures by taurine—specific or nonspecific, *Brain Res.* **76**:171.

Jasper, H. H., Khan, R. T., and Elliott, K. A. C., 1965, Amino acids released from the cerebral cortex in relation to its state of activation, *Science* **147**:1448.

Jouvet, M., 1972, The role of monoamines and acetylcholine-containing neurons in the regulation of the sleep-waking cycle, *Ergeb. Physiol.* **64**:166.

Kaczmarek, L. K., and Davison, A. N., 1972, Uptake and release of taurine from rat brain slices, *J. Neurochem.* **19**:2355.

Kamberi, I. A., Mical, R. S., and Porter, J. C., 1970, Effect of anterior pituitary perfusion and intraventricular injection of catecholamines and indoleamines on LH release, *Endocrinology* **87**:1.

Kamberi, I. A., Mical, R. S., and Porter, J. C., 1971a, Effect of anterior pituitary perfusion and intraventricular injection of catecholamines on FSH release, *Endocrinology* **88**:1003.

Kamberi, I. A., Mical, R. S., and Porter, J. C., 1971b, Effects of melatonin and serotonin on the release of FSH and prolactin, *Endocrinology* **88**:1288.

Karlin, R., and Dumont, M., 1963, Contribution à l'étude du taux de vitamine B_6 pendant l'accouchement, dans le sang total de la mère et dans le sang total du cordon, *Gynec. Obstet.* **62**:281.

Karlin, R., Croizat, P., Revol, L., Pommatau, E., Viala, J.-J., and Dumont, M., 1968, Recherches sur des carences en vitamine B_6 pendant la gestation et dans divers états pathologiques a l'aide d'une épreuve de surcharge en pyridoxine, *Pathol. Biol.* **16**:917.

Kelly, J. S., and Renaud, L. P., 1971, Post-synaptic inhibition in the cuneate blocked by GABA antagonist, *Nature* [*New Biol.*] **232**:25.

Kelly, J. S., Gottesfeld, Z., and Schon, F., 1973, Reduction in GADI activity from the dorsal lateral region of the deafferented rat spinal cord, *Brain Res.* **62**:581.

Kelsal, M. A., 1969, Vitamin B_6 in metabolism of the nervous system, *Ann. N.Y. Acad. Sci.* **166**:1.

Kelsay, J., Baysal, A., and Linkswiler, H., 1968, Effect of vitamin B_6 depletion on the pyridoxal, pyridoxamine and pyridoxine content of the blood and urine of men, *J. Nutr.* **94**:490.

Kiianma, K., and Fuxe, K., 1977, The effects of 5,7-dihydroxytryptamine-induced lesions of the ascending 5-hydroxytryptamine pathways on the sleep-wakefulness cycle, *Brain Res.* **131**:287.

Kishi, H., Kishi, T., Williams, R. H., and Folkers, K., 1975, Human deficiencies of vitamin B_6. I. Studies on parameters of the assay of glutamic oxaloacetic transaminase by the CAS principle, *Res. Commun. Chem. Pathol. Pharmacol.* **12**:557.

Kleinberg, D. L., Noel, G. L., and Frantz, A. G., 1977, Galactorrhea: 235 cases including 48 with pituitary tumors, *N. Engl. J. Med.* **296**:589.

Klieger, J. A., Altshuler, C. H., Krakow, G., and Hollister, C., 1969, Abnormal pyridoxine metabolism in toxemia of pregnancy, *Ann. N.Y. Acad. Sci.* **168**:288.

Kordon, C., Blake, C. A., Terkel, J., and Sawyer, C. H., 1973/74, Participation of serotonin-containing neurons in the suckling-induced rise in plasma prolactin levels in lactating rats, *Neuroendocrinology* **13**:213.

Kravitz, E. A., and Potter, D. D., 1965, A further study of the distribution of γ-aminobutyric acid between excitatory and inhibitory axons of the lobster, *J. Neurochem.* **12**:323.

Krishnaswamy, K., 1974, Isonicotinic acid hydrazide and pyridoxine deficiency, *Int. J. Vitam. Nutr. Res.* **44**:457.

Krnjević, K., 1974, Chemical nature of synaptic transmission in vertebrates, *Physiol. Rev.* **54**:418.

Krnjević, K., and Phillis, J. W., 1963, Ionotophoretic studies of neurones in the mammalian cerebral cortex, *J. Physiol.* (*London*) **165**:274.

Krnjević, K., and Puil, E., 1976, Electrophysiological studies on actions of taurine, in: *Taurine* (R. Huxtable and A. Barbeau, eds.), pp. 179–190, Raven Press, New York.

Krnjević, K., and Schwartz, S., 1967, The action of γ-aminobutyric acid on cortical neurones, *Exp. Brain Res.* **3**:320.

Krulich, L., 1979, Central neurotransmitters and the secretion of prolactin, GH, LH and TSH, *Annu. Rev. Physiol.* **41**:603.

Kuriyama, K., and Yoneda, Y., 1978, Morphine induced alterations of γ-aminobutyric and taurine contents and L-glutamate decarboxylase activity in rat spinal cord and thalamus: Possible correlates with analgesic action of morphine, *Brain Res.* **148**:163.

Kurtz, D. J., and Kanfer, J. N., 1973, Composition of myelin lipids and synthesis of 3-keto dihydrosphingosine in the vitamin B_6-deficient developing rat, *J. Neurochem.* **20**:963.

Lancranjan, I., Wirz-Justice, A., Pühringer, W., and Del Pozo, E., 1977, Effect of 1,5-hydroxytryptophan infusion on growth hormone and prolactin secretion in man, *J. Clin. Endrocrinol. Metab.* **45**:588.

Lande, N. I., 1979, More on dangers of vitamin B_6 in nursing mothers, *N. Engl. J. Med.* **300**:926.

Leklem, J. E., Brown, R. R., Rose, D. P., and Linkswiler, H., 1975, Vitamin B_6 requirements of women using oral contraceptives, *Am. J. Clin. Nutr.* **28**:535.

Levi, C., Kandera, J., and Lajtha, A., 1967, Control of cerebral metabolite levels. I. Amino acid uptake and levels in various species, *Arch. Biochem. Biophys.* **119**:303.

Levy, L., 1969, Mechanism of drug-induced vitamin B_6 deficiency, *Ann. N.Y. Acad. Sci.* **168**:184.

Li, T. K., and Lumeng, L., 1974, Regulation of hepatic pyridoxal phosphate content: A role of alkaline phosphatase, *Fed. Proc.* **33**:1546.

Lloyd, K. G., and Hornykiewicz, O., 1973, L-Glutamic acid decarboxylase in Parkinson's disease: Effect of L-dopa therapy, *Nature* **243**:521.

Loo, Y. H., and Badger, L., 1969, Spectrofluormetric assay of vitamin B_6 analogues in brain tissue, *J. Neurochem.* **18**:801.

Lott, I. T., Coulombe, T., DiPaolo, R. V., Richardson, E. P., Jr., and Levy, H. L., 1978, Vitamin B_6-dependent seizures: Pathology and chemical findings in brain, *Neurology* **28**:47.

Lubby, A. L., Brin, M., Gordon, M., Davis, P., Murphy, M., and Spiegel, H., 1971, Vitamin B_6 metabolism in users of oral contraceptive agents. I. Abnormal urinary xanthurenic acid excretion and its correction by pyridoxine, *Am. J. Clin. Nutr.* **24**:684.

Lumeng, J., Cleary, R. I., and Li, T. K., 1974, Effect of oral contraceptives on the plasma concentration of pyridoxal phosphate, *Am. J. Clin. Nutr.* **27**:326.

Lust, N. D., Kupperburg, H. J., Passonneau, J. V., and Penry, J. K., 1975, Brain cyclic nucleotides and γ-aminobutyric acid: Effect of anticonvulsant agents, *Trans. Am. Soc. Neurochem.* **6**:170.

Makris, A., and Gershoff, S. N., 1973, Growth hormone levels in vitamin B_6-deficient rats, *Horm. Metab. Res.* **5**:457.

Malone, M. J., Bosman, N. P., Szoke, M., and Davis, D., 1975, Myelination of brain in experimental hypothyroidism. An electron-microscopic and biochemical study of purified myelin isolates, *J. Neurol. Sci.* **26**:1.

Mandel, P., and Pasantes-Morales, H., 1978, Taurine in the nervous system, *Rev. Neurosci.* **3**:158.

Maniero, G., Toffano, G., Vecchia, P., and Orlando, P., 1973, Intervention of brain cortex phospholipids in pyridoxal phosphate-dependent reactions, *J. Neurochem.* **20**:1401.

Mao, C. C., Guidotti, A., and Costa, E., 1975, Evidence for an involvement of GABA in the mediation of the cerebellar c GMP decrease and the anticonvulsant action of diazepam, *Naunyn Schmiedbergs Arch. Pharmacol.* **289**:369.

Marie, J., Hennequet, A., Lyon, G., Debris, P., and Le Balle, J. C., 1961, La pyridoxino-dépendance, maladie metabolique s'exprimant par des crisis convulsives pyridoxino-sensibles, *Rev. Neurol. (Paris)* **105**:406.

Martin, W. G., Truex, C. R., Tarka, S., Gorby, W., and Hill, L., 1974, The synthesis of taurine from sulfate. VI. Vitamin B_6 deficiency and taurine synthesis in the rat, *Proc. Soc. Exp. Biol. Med.* **147**:835.

Matsuda, T., Wu, J.-Y., and Roberts, E., 1973, Immunochemical studies on glutamic acid decarboxylase (EC 4.1.1.15) from mouse brain, *J. Neurochem.* **21**:159.
Maudsley, D. V., 1979, Regulation of polyamine biosynthesis, *Biochem. Pharmacol.* **28**:153.
McCann, V. J., and Davis, R. E., 1978, Serum pyridoxal phosphate concentrations in patients with diabetic neuropathy, *Aust. N.Z. J. Med.* **8**:259.
McGeer, P. L., McGeer, E. G., and Fibiger, H. C., 1973, Choline acetylase and glutamic acid decarboxylase in Huntington's chorea, *Neurology (Minneap.)* **23**:912.
Meinardi, H., 1971, Clinical trials of anti-epileptic drugs, *Psychiatr. Neurol. Neurochir.* **74**:141.
Mendelson, W. B., Reichman, J., and Othmer, E., 1975, Serotonin inhibition and sleep, *Biol. Psychiatry* **10**:459.
Miller, L. P., and Walters, J. R., 1979, Effects of depolarization on cofactor regulation of glutamic acid decarboxylase in substantia nigra synaptosomes, *J. Neurochem.* **33**:533.
Milogram, E., and Atger, M., 1975, Receptor translocation inhibitor and apparent saturability of the nuclear acceptor, *J. Steroid Biochem.* **6**:487.
Miyasaki, K., Matsumoto, J., Murao, S., Nakamura, K., Yokoyama, S., Hayano, M., and Nakamura, H., 1978, Infantile convulsion suspected of pyridoxine responsive seizures, *Acta Pathol. Jpn.* **28**:741.
Modigh, K., 1974, Functional aspects of 5-hydroxytryptamine turnover in the central nervous system, *Acta Physiol. Scand.* [*Suppl.*] **403**:1.
Molony, C. J., and Parmelee, A. H., 1954, Convulsions in young infants as a result of pyridoxine deficiency, *J.A.M.A.* **154**:405.
Moore, D. M., and Kirksey, A., 1978, The effect of a dietary deficiency of vitamin B_6 on the specific activity of 2′,3′-cyclic nucleotide 3′-phosphohydrolase of neonatal rat brain, *Brain Res.* **146**:200.
Myers, R. D., 1975, Impairment of thermoregulation, food and water intakes in the rat after hypothalamic injection of 5,6-dihydroxytryptamine, *Brain Res.* **94**:491.
Myslivecek, J., 1970, Electrophysiology of the developing brain, in *Developmental Neurobiology* (W. A. Hinwich, ed.), pp. 475–528, Charles C. Thomas, Springfield, Ill.
Nakhasi, H. L., Toews, A. D., and Horrocks, L. A., 1975, Effects of a postnatal protein deficiency on the content and composition of myelin from brain of weanling rats, *Brain Res.* **83**:176.
Neary, J. T., Meneely, R. L., Grever, M. R., and Diven, W. F., 1972, The interactions between biogenic amines and pyridoxal, pyridoxal phosphate and pyridoxal kinase, *Arch. Biochem. Biophys.* **151**:42.
Novelli, G. D., Schmetz, F., and Kaplan, N. O., 1954, Enzymatic degradation and resynthesis of coenzyme A, *J. Biol. Chem.* **206**:533.
O'Brien, J. S., 1965, Stability of the myelin membrane, *Science* **147**:1099.
Orr, H. T., Cohen, A. T., and Lowry, O. H., 1976, The distribution of taurine in the vertebrate retina, *J. Neurochem.* **26**:609.
Pajunen, A. E. I., Hietala, O. A., Virransalo, E.-L., and Piha, R. S., 1978, Ornithine decarboxylase and adenosylmethionine decarboxylase in mouse brain—effect of electrical stimulation, *J. Neurochem.* **30**:281.
Parchman, L. G., and Litwack, G., 1977, Resolution of activated and unactivated forms of glucocorticoid receptor from rat liver, *Arch. Biochem. Biophys.* **183**:374.
Pasantes-Morales, H., Bonaventure, N., Wioland, N., and Mandel, P., 1973a, Effect of intravitreal injections of taurine and GABA on chicken ERG, *Int. J. Neurosci.* **5**:235.
Pasantes-Morales, H., Urban, P. F., Klethi, J., and Mandel, P., 1973b, Light stimulates release of ^{35}S-taurine from chicken retina, *Brain Res.* **51**:375.
Pasantes-Morales, H., Klethi, J., Urban, P. F., and Mandel, P., 1974, The effect of electrical stimulation, light and amino acids on the efflux of ^{35}S taurine from the retina of domestic fowl, *Exp. Brain Res.* **19**:131.
Pasantes-Morales, H., Mapes, C., Tapia, R., and Mandel, P., 1976, Properties of soluble and

particulate cysteine sulfinate decarboxylase of adult and developing rat brain, *Brain Res.* **107**:575.
Pegg, A. E., 1977, Role of pyridoxal phosphate in mammalian polyamine biosynthesis: Lack of requirement for mammalian 5-adenosylmethionine decarboxylase activity, *Biochem. J.* **166**:81.
Pegg, A. E., and Williams-Ashman, H. G., 1968, Biosynthesis of putrescine in the prostrate gland of the rat, *Biochem. J.* **108**:533.
Phillis, J. W., 1978, Overview of neurochemical and neurophysiological actions of taurine, in *Taurine and Neurological Disorders* (A. Barbeau and R. J. Huxtable, eds.), pp. 289–303, Raven Press, New York.
Puskar, T., and Tryfiates, G. P., 1974, Induction of tyrosine transaminase activity by hydrocortisone in vitamin B_6-deficient rats, *J. Nutr.* **104**:1407.
Rando, R. R., 1974, β,γ-Unsaturated amino acids as irreversible enzyme inhibitors, *Nature* **250**:586.
Rassin, D. K., and Sturman, J. A., 1975, Cysteine sulfinic acid decarboxylase in rat brain: Effect of vitamin B_6-deficiency on soluble and particulate components, *Life Sci.* **16**:875.
Roberts, E., and Eidelberg, E., 1960, Metabolic and neurophysiological roles of γ-aminobutyric acid, *Int. Rev. Neurobiol.* **2**:279.
Roberts, E., Wein, J., and Simonsen, D. G., 1964, γ-Aminobutyric acid, vitamin B_6 and neuronal function—a speculative synthesis, *Vitam. Horm.* **22**:503.
Robins, E., Robins, J. M., Croninger, A. B., Moses, S. G., Spencer, S. J., and Hudgens, R. W., 1967, The low level of 5-hydroxytryptophan decarboxylase in human brain, *Biochem. Med.* **1**:240.
Rose, D. P., and Braidman, I. P., 1971, Excretion of tryptophan metabolites as affected by pregnancy, contraceptive steroids, and steroid hormones, *Am. J. Clin. Nutr.* **24**:673.
Rosman, N. P., Malone, M. J., Helfenstein, M., and Kraft, E., 1972, The effect of thyroid deficiency on myelination of brain, *Neurology* **22**:99.
Russel, D. H., and Meier, H., 1975, Alterations in the accumulation patterns of polyamines in brains of myelin-deficient mice, *J. Neurobiol.* **6**:267.
Russel, D. H., Fgeller, E., Marton, L. J., and LeGendre, S. M., 1974, Distribution of putrescine, spermidine and spermine in rhesus monkey brain: Decrease in spermidine and spermine concentrations in motor cortex after electrical stimulation, *J. Neurobiol.* **5**:349.
Sakurada, T., Imai, M., Tadano, T., and Kisara, K., 1976, Effect of bilateral olfactory bulb ablations on the polyamine levels in rat brain, *Jpn. J. Pharmacol.* **26**:509.
Salceda, R., and Pasantes-Morales, H., 1975, A calcium coupled release of taurine from retina, *Brain Res.* **96**:206.
Sanberg, P. R., Staines, W., and McGeer, E. G., 1979, Chronic taurine effects on various neurochemical indices in control and kainic acid—lesioned neostriatum, *Brain Res.* **161**:367.
Saraswathi, S., and Bachhawat, B. K., 1963, Phosphatases from human brain—purification and properties of pyridoxal phosphate phosphatase, *J. Neurochem.* **10**:127.
Schmidt, S. Y., Berson, E. L., and Hayes, K. C., 1976, Retinal degeneration in cats fed casein. I. Taurine deficiency, *Invest. Ophthalmol.* **15**:47.
Schulte, F. J., Hinze, G., and Schrempf, G., 1971, Maternal toxemia, fetal malnutrition and bioelectric brain activity of the newborn, *Neuropaediatrie* **2**:439.
Scriver, C. R., 1964, Comment on vitamin B_6 deficiency and dependency syndromes, *Yearbook of Pediatrics* (S. Gellis, ed.), pp. 46–48, Year Book, Chicago.
Scriver, C. R., and Whelan, D. T., 1969, Glutamic acid decarboxylase (GAD) in mammalian tissue outside the central nervous system and its possible relevance to hereditary vitamin B_6 dependency with seizures, *Ann. N.Y. Acad. Sci.* **166**:83.
Seiler, N., and Schmidt-Glenwinkel, 1975, Regional distribution of putrescine, spermidine and spermine in relation to the distribution of RNA and DNA in the rat nervous system, *J. Neurochem.* **24**:791.

Seiler, N., Bink, G., and Grove, J., 1979, Regulatory interrelations between GABA and polyamines. I. Brain GABA levels and polyamine metabolism, *Neurochem. Res.* **4**:425.

Shaw, G. G., 1977, Evidence against the view that the central action of polyamines are indirectly mediated, *Biochem. Pharmacol.* **26**:1450.

Shaw, G. G., 1979, The polyamines in the central nervous system, *Biochem. Pharmacol.* **28**:1.

Shoemaker, W. J., and Wurtman, R. J., 1971, Prenatal undernutrition: Accumulation of catecholamines in rat brain, *Science* **171**:1017.

Simler, S., Ciesielski, L., Maitre, M., Randrianarisoa, H., and Mandel, P., 1973, Effect of sodium *n*-dipropylacetate on audiogenic seizures and brain γ-aminobutyric acid level, *Biochem. Pharmacol.* **22**:1701.

Simon, D., and Penry, J. K., 1975, Sodium di-*N*-propylacetate (DPA) in the treatment of epilepsy: A review, *Epilepsia* **16**:549.

Simons, S. S., Jr., Martinez, H. M., Garcia, R. L., Baxter, J. D., and Tomkins, G. M., 1976, Interaction of glucocorticoid receptor. Steroid complexes with acceptor sites, *J. Biol. Chem.* **251**:334.

Smith, D. B., and Gallagher, B. B., 1970, The effect of penicillamine on seizure threshold: The role of pyridoxine, *Arch. Neurol.* **23**:59.

Snyderman, S. E., Holt, L. E., Jr., Carretero, R., and Jacob, K., 1953, Pyridoxine deficiency in a human infant, *Am. J. Clin. Nutr.* **1**:200.

Sourkes, T. L., 1972, Influence of specific nutrients on catecholamine synthesis and metabolism, *Pharmacol. Rev.* **25**:349.

Spector, R., 1977, Vitamin homeostasis in the central nervous system, *N. Engl. Med.* **296**:1393.

Spector, R., 1978a, Vitamin B_6 transport in the central nervous system: *In vivo* studies, *J. Neurochem.* **30**:881.

Spector, R., 1978b, Vitamin B_6 transport in the central nervous system: *In vitro* studies, *J. Neurochem.* **30**:889.

Spector, R., 1979, Development of the vitamin transport systems in choroid plexus and brain, *J. Neurochem.* **33**:1317.

Spector, R., Cancilla, P., and Damasio, A., 1979, Is idiopathic dementia a regional vitamin deficiency state? *Med. Hypotheses* **5**:763.

Stephens, M. C., and Dakshinamurti, K., 1975, Brain lipids in pyridoxine-deficient young rats, *Neurobiology* **5**:262.

Stephens, M. C., and Dakshinamurti, K., 1976, Galactolipid fatty acids in brain of pyridoxine-deficient young rats, *Exp. Brain Res.* **25**:465.

Stephens, M. C., Havlicek, V., and Dakshinamurti, K., 1971, Pyridoxine deficiency and development of the central nervous system, *J. Neurochem.* **18**:2407.

Stewart, C. N., Coursin, D. B., and Bhagavan, H. N., 1973, Cortical-evoked responses in pyridoxine-deficient rats, *J. Nutr.* **103**:462.

Stone, W. J., Warnock, L. G., and Wagner, C., 1975, Vitamin B_6 deficiency in anemia, *Am. J. Clin. Nutr.* **28**:950.

Storvick, C. A., and Peters, J. M., 1964, Methods for the determination of vitamin B_6 in biological materials, *Vitam. Horm.* **22**:833.

Sturman, J. A., 1973, Taurine pool sizes in the rat: Effects of vitamin B_6 deficiency and high taurine diet, *J. Nutr.* **103**:1566.

Sturman, J. A., and Gaull, G. E., 1975, Taurine in the brain and liver of the developing human and monkey, *J. Neurochem.* **25**:831.

Sturman, J. A., Cohen, P. A., and Gaull, G. E., 1969, Effects of deficiency of vitamin B_6 on transsulfuration, *Biochem. Med.* **3**:244.

Sturman, J. A., Rassin, D. K., and Gaull, G. E., 1978, Taurine in the development of the central nervous system, in: *Taurine and Neurological Disorders* (A. Barbeau and R. J. Huxtable, eds.), pp. 49–71, Raven Press, New York.

Sze, P. Y., 1970, Possible repression of L-glutamic acid decarboxylase by gamma-aminobutyric acid in developing brain, *Brain Res.* **19**:322.

Sze, P. Y., 1979, L-Glutamate decarboxylase, In: *GABA—Biochemistry and CNS Function* (P. Mandel and F. V. DeFeudis, eds.), pp. 59–78, Plenum Press, New York.

Tabor, C. W., and Tabor, H., 1976, 1,4-Diaminobutane (putrescine), spermidine and spermine, *Annu. Rev. Biochem.* **45**:285.

Thiele, V. F., and Brin, M., 1968, Availability of vitamin B_6 vitamers fed orally to Long-Evans rats as determined by tissue transaminase activity and vitamin B_6 assay, *J. Nutr.* **94**:237.

Toepfer, E. W., and Lehmann, J., 1961, Procedure for chromatographic separation and microbiological assay of pyridoxine, pyridoxal and pyridoxamine in food extracts, *J. Assoc. Off. Anal. Chem.* **44**:426.

Tomono, I., Abe, M., and Matsuda, M., 1973, Effect of penicillamine on pyridoxal enzymes, *J. Biochem.* (*Tokyo*) **74**:587.

Trapp, B. D., and Bernsohn, J., 1978, Essential fatty acid deficiency and CNS myelin, *J. Neurol. Sci.* **37**:249.

Tunnicliff, G., Wimer, R. E., and Roberts, E., 1972, Pyridoxine dietary levels and open field activity in inbred mice, *Brain Res.* **42**:234.

Tzeng, S., and Ho, I. K., 1977, Effects of acute and continuous phenobarbital administration on the γ-aminobutyric acid system, *Biochem. Pharmacol.* **26**:699.

Urban, D. F., Dreyfus, H., and Mandel, P., 1976, Influence of various amino acids on the bioelectrical response to light stimulation of a superfused frog retina, *Life Sci.* **18**:473.

Van Gelder, N. M., 1972, Antagonism by taurine of cobalt induced epilepsy in cat and mouse, *Brain Res.* **47**:157.

Van Gelder, N. M., 1978, Taurine, the compartmentalized metabolism of glutamic acid, and the epilepsies, *Can. J. Physiol. Pharmacol.* **56**:362.

Van Gelder, N. M., Sherwin, A. L., and Rasmussen, T., 1972, Amino acid content of epileptogenic human brain: Focal versus surrounding regions, *Brain Res.* **40**:385.

Volpe, T. J., and Kishimoto, Y., 1972, Fatty acid synthetase of brain: Development, influence of nutritional and hormonal factors and comparison with liver enzyme, *J. Neurochem.* **19**:737.

Walshe, J. M., 1956, Penacillamine, a new oral therapy for Wilson's disease, *Am. J. Med.* **21**:487.

Weiner, R. I., and Ganong, W. F., 1978, Role of brain monoamines and histamine in regulation of anterior pituitary secretion, *Physiol. Rev.* **58**:905.

Wiggins, R. C., Miller, S. L., Benjamins, J. A., Krigman, M. B., and Morell, P., 1976, Myelin synthesis during postnatal nutritional deprivation and subsequent rehabilitation, *Brain Res.* **107**:257.

Williamson, B., and Coniglio, J. G., 1971, The effect of pyridoxine deficiency and of caloric restriction on lipids in the developing brain, *J. Neurochem.* **18**:267.

Wood, J. G., 1975, The role of γ-aminobutyric acid in the mechanism of seizures, *Prog. Neurobiol.* **5**:77.

Wood, J. G., and Peeskar, S. J., 1974, Development of an expression which relates the excitable state of the brain to the level of GAD activity and GABA content with particular reference to the action of hydrazine and its derivatives, *J. Neurochem.* **23**:703.

Wood, J. G., McLaughlin, B. J., and Vaughn, J. E., 1976, Immunocytochemical localization of GAD in electronmicroscopic preparation of rodent CNS, in: *GABA in Nervous System Function* (E. Roberts, T. N. Chase, and D. B. Tower, eds.), pp. 133–148, Raven Press, New York.

Wyatt, R. J., 1972, The serotonin-catecholamine-dream bicycle: A clinical study, *Biol. Psychiatry* **5**:33.

Wynn, V., Adams, P. W., Folkard, J., and Seed, M., 1975, Tryptophan, depression and steroidal contraception, *J. Steroid Biochem.* **6**:965.

Yoshida, T., Tada, K., and Arakawa, T., 1971, Vitamin B_6 dependency of glutamic acid decarboxylase in the kidney from a patient with vitamin B_6 dependent convulsions, *Tohoku J. Exp. Med.* **104**:195.

Chapter 7

Carnitine Biosynthesis Nutritional Implications

Harry P. Broquist and Peggy R. Borum

1. Introduction

Much interest in carnitine metabolism and function has been shown in recent years with the recognition of its catalytic role in the intramitochondrial transport of fatty acids. Figure 1 shows a series of events requisite for the activation and transport of extramitochondrial long-chain fatty acids such as palmitic acid to the site of β-oxidation in the mitochondrial matrix [cf. Bremer (1977) for discussion and relevant references]. Palmitic acid released from adipose tissue or derived from the diet is activated by outer membrane ATP-dependent palmitoyl-CoA synthetase to form palmitoyl-CoA. Such long-chain fatty acyl-CoA esters have only a limited ability to cross the mitochondrial membrane, but their entry is facilitated by "outer" carnitine palmitoyl transferase which catalyzes a transesterification reaction in which the palmitoyl moiety from CoA is transferred to carnitine, forming palmitoylcarnitine. The latter ester is then thought to cross the inner mitochondrial membrane through the action of a translocase. A second transesterification reaction now takes place wherein "inner" palmitoyltransferase, located in the inner mitochondrial membrane, regenerates palmitoyl-CoA for subsequent β-oxidation and releases carnitine for a repetition of its catalytic role in overall fatty acid transport.

It is thus apparent from a consideration of the events of Fig. 1 that carnitine is mandatory for the initiation of events of fatty acid oxidation in the mitochon-

Harry P. Broquist and Peggy R. Borum • Division of Nutrition, Department of Biochemistry, Vanderbilt University, Nashville, Tennessee 37232.

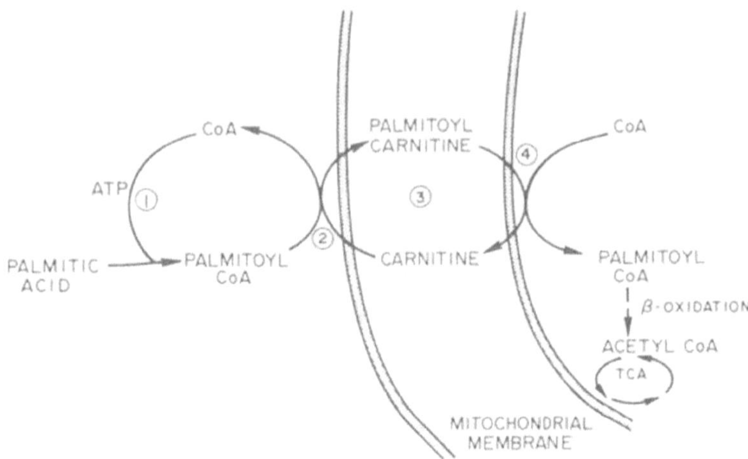

Fig. 1. Carnitine and intramitochondrial transport of long-chain fatty acids. See Bremer (1977) for discussion and relevant references.

dria from where energy demands of critical tissues such as the heart and skeletal muscle ultimately derive. Medium- and short-chain carnitine acyltransferases also exist, although their function in intermediary metabolism, particularly that of carnitine acetyltransferase, is not as well understood as in the case of the long-chain carnitine acyltransferases. Also, it should be recognized that additional functions of carnitine may well exist.

The requirement of higher animals for carnitine must be met either by the diet, particularly animal protein foods, and/or from *de novo* synthesis. Hence, any consideration of carnitine as a nutrient must take into account the relative contribution of these two sources. Mitchell (1978a) has pointed out that information concerning the carnitine content of foods is presently scarce and unsatisfactory in many ways. She assembled available analytical data for about 50 foodstuffs, illustrating that, in general, carnitine is low in foods of plant origin and high in animal foods. For example, the edible portion of beef tenderloin, beef shoulder, and beef rump were reported to contain 59.8, 67.40, and 61.60 mg carnitine per 100 g. In contrast, no carnitine was detected in a 100-g portion of a vegetable protein mixture (soyameal, 50 : rice, 30 : pinto beans, 20). Moreover, it is just those foodstuffs that are deficient in carnitine that are also limiting in its amino acid precursors, lysine and methionine. Thus, as is well known, such important cereal grains as wheat, corn, and rice are limiting in lysine, whereas the legumes are limiting in methionine. Such considerations have prompted a number of investigators in the field to consider the effect of consumption of cereal grain diets in animals and in man in calling forth carnitine deficiency as indicated by reduced levels of carnitine in physiological fluids and body tissues and impairment in lipid metabolism. Examples of such studies will be considered herein.

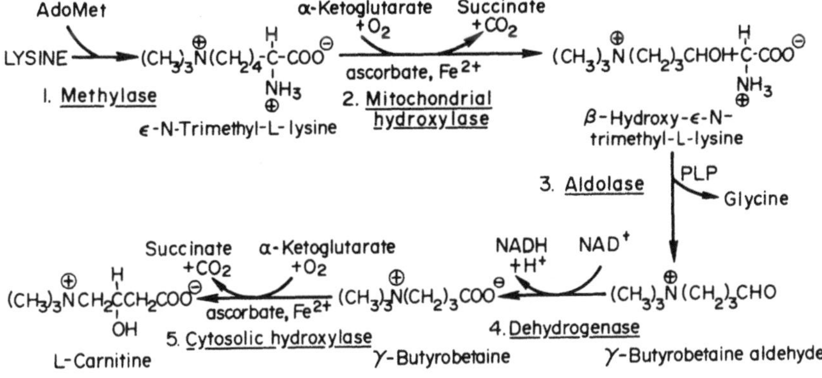

Fig. 2. The biogenesis of carnitine from lysine.

The overall pathway of carnitine biosynthesis from lysine and methionine now appears to be established as is shown in Fig. 2. The evidence for this pathway comes principally from isotopic and enzymatic studies in fungi and the rat and will be briefly discussed. Reviews dealing with aspects of carnitine biosynthesis, physiology, and nutriture have appeared (Mitchell, 1978a–c; Frenkel and McGarry, 1980; Broquist and Borum, 1977). This article will principally focus on the nutritional implications arising from present knowledge of carnitine biosynthesis.

2. Carnitine Biosynthesis and Enzymology

In *Neurospora crassa,* a homogeneous enzyme protein, *S*-adenosylmethionine : ε-*N*-L-lysine methyltransferase, has been isolated and shown to carry out the stepwise ε-*N*-methylation of lysine, reactions 1, 2, and 3, respectively, yielding ε-*N*-trimethyllysine, a committed intermediate of carnitine biogenesis (Borum and Broquist, 1977a). *S*-adenosylmethionine (AdoMet) is the methyl donor in these reactions, and the biosynthetic role of the essential amino acids, lysine and methionine, in the formation of trimethyllysine and ultimately carnitine thus becomes apparent.

$$\text{lysine} \xrightarrow[\text{AdoMet}]{} \epsilon\text{-}N\text{-monomethyllysine} \quad (1)$$

$$\epsilon\text{-}N\text{-monomethyllysine} \xrightarrow[\text{AdoMet}]{} \epsilon\text{-}N\text{-dimethyllysine} \quad (2)$$

$$\epsilon\text{-}N\text{-dimethyllysine} \xrightarrow[\text{AdoMet}]{} \epsilon\text{-}N\text{-trimethyllysine} \quad (3)$$

It should also be noted that the biosynthesis of the key carnitine intermediate, trimethyllysine, is energy dependent (reaction 4) as 3 mol of ATP are required in the synthesis of 3 mol of AdoMet needed in steps 1, 2, and 3 above.

$$\text{Methionine} + \text{ATP} \longrightarrow \text{AdoMet} + \text{Pyrophosphate} \quad (4)$$

Trimethyllysine present in tissues can arise from the hydrolysis of certain select proteins containing trimethyllysine in peptide linkage. Protein methylation is one of several posttranslational modification reactions of polypeptides to have been extensively studied in recent years (Paik and Kim, 1975). Trimethyllysine has been found in such diverse proteins as histone, myosin, actin, and cytochrome C. The enzyme responsible for methylating protein lysine residues has been designated protein methylase III, and this enzyme is found widely distributed throughout various rat tissues.

The relevance of protein methylation to carnitine biosynthesis comes from the work of LeBadie *et al.* (1976) who chemically modified asialofetuin to contain [CH_3-^{14}C]trimethyllysine. When this protein was administered to rats, it was rapidly degraded by proteolytic enzymes present in hepatic lysosomes to yield free trimethyllysine. Moreover, carnitine was found in 35% yield in the rat carcass only 3 hr after administration of methyl ^{14}C-labeled fetuin. This experiment demonstrates then that an important source of the carnitine precursor trimethyllysine in the animal is via protein catabolism. Whether this is the exclusive source of trimethyllysine destined for carnitine synthesis in mammalian systems is not known, but it may be an academic question if normal body pools of trimethyllysine resulting from protein degradation are adequate for the synthesis of carnitine which is needed only in catalytic quantities in the cell.

Trimethyllysine is then hydroxylated in a complex manner to be discussed to yield β-hydroxy-ϵ-N-trimethyllysine. This α-amino-β-hydroxy acid may be viewed as an analogue of serine, and, indeed, it was found to be cleaved by a PLP-requiring aldolase, serine transhydroxymethylase, to give γ-N-trimethylaminobutyraldehyde and glycine (Hulse *et al.*, 1978). The aldehyde is then oxidized by an apparently specific NAD-requiring dehydrogenase to γ-butyrobetaine (Hulse and Henderson, 1979), long recognized as the immediate precursor of carntine (Fig. 2). Direct evidence for the participation of β-hydroxytrimethyllysine and γ-N-trimethylaminobutyraldehyde in carnitine biosynthesis was provided by Kaufman and Broquist (1977) who synthesized these compounds with tritium in the N-methyl groups and showed that they were efficiently utilized for carnitine synthesis in *Neurospora crassa* and, further, that these compounds at high levels (nonradioactive) effectively blocked the conversion of ϵ-N-[methyl]-^3H]trimethyllysine to carnitine.

The complex oxygenase system required for the hydroxylation of γ-butyrobetaine to form carnitine (Fig. 2) has been well described in both rat liver (Lindstedt and Lindstedt, 1970) and a microbial system (Lindstedt *et al.*, 1970). The dioxygenase is present in the soluble portion of the cell and requires α-ketoglutarate, Fe^{2+}, molecular oxygen, and a reducing agent among which ascorbate gives maximal activity. α-Ketoglutarate and oxygen are required as substrates and react with γ-butyrobetaine, likely forming a peroxide intermediate which decomposes, forming carnitine, succinate, and CO_2 (Lindstedt and Lindstedt, 1970).

This hydroxylase is apparently absent in rat muscle and rat kidney as judged by the ability of such tissue slices to form γ-butyrobetaine but not carnitine from trimethyllysine (Haigler and Broquist, 1974; Cox and Hoppel, 1974). A surprising recent discovery (Hulse *et al.*, 1978) is that the enzyme hydroxylating trimethyllysine to form β-hydroxytrimethyllysine (Fig. 2) is also a dioxygenase having precisely the same cofactor requirements as γ-butyrobetaine hydroxylase but, in contrast, is present both in rat liver mitochondria (Hulse *et al.*, 1978) and rat kidney mitochondria (Sachan, 1978). The reason for such disparate cell compartmentations for these two similar hydroxylases is not clear at this time.

Recent work of Carter and Frenkel (1979) adds significance to the role of the kidney in the carnitine economy of the rat. They showed that following the administration of [methyl-^{3}H]trimethyllysine to the rat, it was very rapidly utilized by the kidney, in contrast to the liver, for γ-butyrobetaine synthesis. Subsequent transport of γ-butyrobetaine to the liver would yield carnitine. The kidney is viewed then as salvaging trimethyllysine released from appropriate tissue proteins to channel such trimethyllysine into the carnitine biosynthetic pathway.

Significant species differences exist with respect to capacity of the liver and kidney for total synthesis of carnitine from trimethyllysine. Thus, Englard and Carnicero (1978) found that crude extracts from kidneys of cat, hamster, and rabbit had levels of γ-butyrobetaine hydroxylase activity equal to or exceeding that in liver, in contrast to the dog, guinea pig, mouse, and rat kidney, which exhibited little or no capacity to hydroxylate γ-butyrobetaine. Of particular interest was the finding that γ-butyrobetaine hydroxylase activity in crude extracts of kidney of Rhesus monkeys and human autopsy material was significantly higher than in respective extracts of liver (Englard, 1979). In other studies, Rebouche (1980) monitored enzymes 2–5 (Fig. 2) of carnitine biosynthesis including γ-butyrobetaine hydroxylase in a number of human autopsy tissues and found that all were present in both liver and kidney. An interesting finding was that γ-butyrobetaine hydroxylase activity in humans apparently depends on the age of the subject; for example, in three infants the activity was approximately 12% of the normal adult mean. This latter finding may have relevance in consideration of the carnitine nutriture of infants as will be discussed below. The ability of the kidney to synthesize carnitine from trimethyllysine in adult man is of interest and may indicate an efficient mechanism by the kidney to salvage excreted trimethyllysine by tubular reabsorption with subsequent metabolism to form carnitine.

From a nutritional viewpoint, the foregoing information may be summarized by saying that in mammalian systems, studies of carnitine biosynthesis have shown (1) that two essential amino acids, lysine and methionine, contribute all the carbon and nitrogen atoms of carnitine; (2) that protein synthesis is required to form protein-bound trimethyllysine which becomes available for carnitine biosynthesis following proteolytic release by lysosomal enzymes; (3) that five enzymes, namely, a methylase, a mitochondrial hydroxylase, an aldolase, a dehydrogenase, and cytosolic hydroxylase are necessary for catalysis of the biosynthetic transformation of Fig. 2; and (4) that three B-complex vitamins, ascorbate, niacin, and vitamin B_6, plus a metal ion, reduced iron, are accessory factors in latter steps of carnitine biosynthesis. It is apparent then that a nutritional potpourri is required for carnitine biosynthesis in the tissues, and it accordingly follows that malnutrition in a broad sense might be expected to jeopardize in diverse ways the carnitine status of an individual.

3. Protein Malnutrition and Carnitine Status in the Rat

In view of the lysine : carnitine, precursor : product relationship in the rat, the effect of a lysine deficiency on carnitine nutriture in the rat has been extensively studied. A series of studies are summarized in Table I in which rats were maintained for many weeks on diets in which the sole protein source was wheat gluten, wheat flour, or rice. Such incomplete proteins are limiting in lysine and contain negligible carnitine, and hence carnitine synthesis is demanded by the rat under conditions in which lysine is limiting. Even though about 0.1% of the lysine requirement of the rat may be consigned for carnitine synthesis (Tanphaichitr and Broquist, 1973b), one might reasonably expect that the consumption of a diet marginal in lysine would result in a proportional decrease in the amount of lysine available for carnitine synthesis. At time of sacrifice, the rats in all four experiments (Table I) were severely lysine deficient as judged by poor growth and other criteria detailed in the original papers. However, inspection of the carnitine findings indicates, in general, only a marginal carnitine deficiency. For example, in experiments A, C, and D, the skeletal muscle and heart muscle contained about 70% of the carnitine of the lysine-supplemented rats. Only the epididymis, a tissue that is extraordinarily high in carnitine, was markedly reduced (44%) in carnitine content in the lysine-deficient rats (experiment B). Somewhat surprisingly, the liver of lysine-deficient rats contains more carnitine than those of lysine-supplemented rats. Possibly, this reflects the absence of a carrier protein required for carnitine transport to the tissues.

Does the mild carnitine deficiency induced by feeding the lysine-deficient diets of Table I cause a significant impairment in lipid metabolism? In experiments C and D (Table I), liver lipids and liver triglycerides were also examined under varying dietary conditions as shown in Table II. Consumption of either the wheat flour diet or the rice diet caused a marked elevation in liver lipids in each instance which was significantly lowered in those groups receiving either carnitine or its precursor, lysine. It is interesting that although carnitine liver levels are elevated in these instances (experiments C and D, Table I), a fatty liver results under such dietary situations which can be partially relieved by administering additional carnitine (Table II).

Table I. Carnitine Status in Lysine-Deficient Rats

Experiment	Description of experiment	Final weight (Gp. I as % Gp. II)	Carnitine status				Reference
			Tissue examined	nmole carnitine/g[a]		Gp. I as % Gp. II	
				Gp. I	Gp. II		
A	Group I: male weanling rats fed 20% wheat gluten diet 90 days, sacrificed, and relevant tissues examined for carnitine content. Group II: as Group I but diet supplemented with 1% lysine.	26	Skeletal muscle Heart muscle Liver	580 679 259	827 963 198	70 71 131	Tanphaichitr and Broquist (1973a)
B	Group I: male weanling rats fed 20% wheat gluten diet for 78 days, sacrificed, and relevant tissues examined for carnitine content. Group II: as Group I but diet supplemented with 0.8% lysine.	28	Skeletal muscle Heart muscle Liver Epididymis Plasma	592 675 102 4,840 44	679 812 79 10,965 52	87 83 129 44 86	Borum and Broquist (1977b)
C	Group I: male weanling rats fed 56% wheat flour diet 70 days, sacrificed, and relevant tissues examined for carnitine content. Group II: as Group I but diet supplemented with 0.2% lysine and 0.2% threonine. Group II rats pair-fed to Group I rats.	69	Skeletal muscle Heart muscle Liver Plasma	1,000 1,500 630 149	1,490 1,900 590 114	67 79 106 131	Khan and Bamji (1979)
D	Group I: male weanling rats fed 72% rice diet 35 days, then sacrificed, and relevant tissues examined for carnitine content. Group II: as Group I but diet supplemented with 0.45% lysine and 0.36% threonine.	28	Skeletal muscle Kidney Liver	447 174 247	607 232 213	74 75 115	Tamphaichitr *et al.* (1976)

[a] As per gram wet weight of tissue, or as per milliliter plasma.

Table II. Cereal Grain Diets and Liver Lipid Metabolism

Experiment	Diet	Liver lipids (mg/g)	%[a]	Liver triglycerides (mg/g)	%[a]
C (Table I)	56% wheat flour	109		37	
	56% wheat flour + 0.2% carnitine	85	78	18	49
	56% wheat flour + 0.2% lysine + 0.2% threonine	67	61	16	43
D (Table I)	72% rice	207		125	
	72% rice + 0.2% carnitine	157	76	102	82
	72% rice + 0.45% lysine + 0.36% threonine	70	34	31	25

[a]Relative to basal group.

In a further refinement of experiment D (Tables I and II), Khan and Bamji (1979) showed that palmitic acid oxidation by heart homogenates was lowest in rats consuming the unsupplemented wheat flour diet, 8.3 μmole/min per g protein. In rats receiving the carnitine supplement, for example, the oxidation rate significantly increased to 53.0 μmole/min per g protein. From these data and those of Table II, it seems clear that a significant impairment in lipid metabolism does indeed result in rats that are only marginally deficient in carnitine.

It is now known that the first enzyme of lysine catabolism, lysine-ketoglutarate reductase (reaction 5), is under dietary control.

$$\text{Lysine} + \alpha\text{-Ketoglutarate} \xrightarrow{\text{NADPH,H}^+ \;\to\; \text{NAD}^+} \text{Saccharopine} + H_2O \quad (5)$$

Thus, Chu and Hegsted (1976) found that feeding a lysine-free diet or a 10% wheat gluten diet to rats significantly decreased the activity of the enzyme. These findings may well explain the data of Table I that the feeding of diets marginal in lysine results in only a mild carnitine deficiency. For if lysine-ketoglutarate reductase is turned off under these stringent dietary conditions, more lysine should be available for carnitine synthesis. This situation, coupled with an efficient mechanism in the kidney for salvaging trimethyllysine from tissue catabolism for carnitine synthesis (Carter and Frenkel, 1979), would both operate in favor of maintaining tissue levels of carnitine.

The role of methionine in carnitine biogenesis is to provide the *N*-methyl groups. One might expect that diets marginal in this amino acid or deficient in one or more of the factors of one-carbon metabolism involved in *de novo* synthesis of the *S*-methyl group of methionine would have a deleterious effect on carnitine synthesis. Khairallah and Wolf (1965) found that rats fed low-protein diets limiting in methionine grew poorly and developed fatty livers. In contrast, rats on this diet, when supplemented with 0.2% carnitine, gave a substantial growth response with a marked reduction in fatty liver. These workers concluded that carnitine spares methionine when low-protein diets (methioine limiting) are fed and pointed out the possible significance of their findings to protein malnutrition in man. These intriguing studies need confirmation and extensions, as little research emphasis has been given to methionine–carnitine nutritional relationships.

4. Carnitine Nutriture in Man

4.1. In Protein Malnutrition

Several papers have now appeared from the Middle East, India, and Thailand that show, in general, lowered blood carnitine levels in patients suffering from severe protein malnutrition. Mikhail and Monsour (1976) studied a series of patients with schistosomal infection with associated signs of anemia and protein malnutrition (low serum albumin). Such patients had abnormally low serum carnitine levels as compared to healthy controls; these levels markedly improved following nutritional repletion (hospital diet plus milk and meat). Before hospitalization, the diet of these patients was deficient in animal protein and consisted mainly of cereals and grain. The study indicates a relationship between the nutritional status of patients with schistosomiasis and serum carnitine levels. In this regard, it is interesting to note that advanced cases of schistosomiasis are usually accompanied by disturbances in lipid metabolism, e.g., high levels of free fatty acids in the serum.

Kahn and Bamji (1977) examined plasma carnitine and serum albumin levels in 13 children with kwashiorkor, 12 children suffering from marasmus, and ten other children judged to be undernourished from body weight considerations. All the children were between 1 and 5 years of age. The results of the study are shown in Table III and illustrate significantly lowered carnitine levels compared with controls in all of the malnourished children. Furthermore, there is good correlation between carnitine and serum albumin levels, further emphasizing a relation between carnitine level and protein nutritional status. Seventeen of these children were reexamined after treatment (4 weeks) with a high-protein diet (4 g protein, mainly skim milk) and 200 kcal/kg body weight per day. Such dietary repletion produced marked improvment in carnitine as well as albumin levels. Khan and Bamji (1977) point out that the functional significance of low plasma carnitine levels in their protein-malnourished children remain to be established; but in this regard, the accumulation of fat in the liver in protein calorie malnutrition is particularly intriguing. It should be recognized, of course, that in

Table III. Plasma Carnitine and Albumin Levels of Control and Malnourished Children[a]

	Carnitine (μmol/100 ml)	Albumin (g/100 ml)
Control	9.0 ± 0.6 (8)	3.5 ± 0.1 (8)
Undernourished	6.4 ± 0.9 (10)*	2.7 ± 0.2 (5)
Marasmus	3.7 ± 0.5 (12)**	2.7 ± 0.2 (8)***
Kwashiorkor	2.6 ± 0.5 (13)**	1.7 ± 0.1 (9)**

[a]Data are from Khan and Bamji (1977). Figures are represented as means ± S.E. Values in parentheses indicate the number of subjects. Significance, p, compared to control: *$p < 0.1$; **$p < 0.01$; ***$p < 0.001$.

both of the aforementioned studies the carnitine deficiency encountered could well have arisen from multiple deficiencies, i.e., lysine and methionine needed as substrates for carnitine synthesis, but also diverse cofactors concerned in carnitine biogenesis as well. For example, the patients with schistosomiasis studied by Mikhail and Monsour (1976) were not only protein malnourished but were anemic; hence, iron, needed at two different loci of carnitine biosynthesis (enzymes 2 and 5, Fig. 2) could well have been limiting.

Tanphaichitr *et al.* (1980) have made an interesting survey of the carnitine nutrition of Thai adults living either in Bangkok, the capitol of Thailand, or in Ubol, a rural province in northeast Thailand. Since rice is the staple food of the Thais and is limiting in lysine and carnitine, one might anticipate degrees of carnitine deficiency in the population depending on the extent of animal or fish protein sources in the primarily vegetable diet. The mean plasma carnitine level of the Bangkok adults as compared to Ubol adults was 56.6 vs. 50.3 nmole/liter, and urinary carnitine levels were 161 vs. 127 nmole/liter, respectively. It was found that sex affected plasma carnitine levels in the Bangkok adults and the urinary carnitine excretion in both groups. Thus, if only the data for the males are considered, the carnitine levels in plasma and in urine of the Ubols were 81% and 75%, respectively, of the Bangkok adults. The nutritional status of the Ubol adults was inadequate as shown by significantly lower levels of urinary creatinine excretion, serum albumin, and hematocrit as compared to these nutritional parameters in the Bangkok adults. Such findings were also borne out by dietary assessment. Although the extent of carnitine deficiency in the Ubol population does not appear to be severe, the physiological consequences could be significant. But the situation could be much more serious, of course, in the infant population where the carnitine requirement could be increased in a period where rapid growth and concommitant increased energy are demanded.

4.2. In Cirrhosis

The three foregoing accounts from the Middle East and Asia indicate a degree of carnitine deficiency associated with protein calorie malnutrition. Recently, we have heard much of undernutrition existing in U.S. hospital populations (Butterworth and Blackburn, 1975). For example, recent surveys report protein calorie undernutrition in about one-third of patients on the medical and surgical services in two hospitals (Bistrian *et al.,* 1974; Bollet and Owens, 1973). Such considerations prompted Rudman *et al.* (1977, 1980) to monitor fasting serum total carnitine in 16 normal controls and in 255 hospital patients. Only in the patients with cirrhosis was there a notable hypocarnitinemia (<55 μM). Of this group, 20 of 60 cirrhotics had hypocarnitinemia and hypocarnitinuria. Such patients exhibited severe hepatocellular disease and protein calorie starvation as indicated by creatinine/height index, mid-arm muscle circumference, and triceps skin-fold thickness measurements. Significantly, the dietary intake of lysine, methionine, and carnitine of the hypocarnitinemic cirrhotics was only 30% of the intake of controls (normal individuals). But when the latter group and the normal controls were given ample lysine and methionine (orally and i.v.) without exogenous carnitine, the controls maintained normal serum carnitine levels, but the cirrhotics remained hypocarnitinemic. In other instances, patients with hypocarnitinemic cirrhosis died during the course of the study, thus permitting a postmortem examination of tissue levels of carnitine and triglycerides. Such studies revealed reduced levels of carnitine in the hypocarnitinemics and elevated triglycerides in the liver and muscle when compared with similar analysis in postmortem examinations of patients dying from cardiovascular causes. From this study, it was concluded that carnitine depletion is common in patients hospitalized for advanced cirrhosis and that it may result from marginal intakes of dietary carnitine and its precursors, lysine and methionine, together with a loss of capacity by the liver to synthesize carnitine from lysine and methionine.

4.3. In Renal Disease

In a study of patients with renal insufficiency, 17 of 26 cases had plasma carnitine concentrations that exceeded the normal limit (Chen and Lincoln, 1977). Several laboratories have shown that hemodialysis for renal failure dramatically decreases the plasma carnitine concentration (Bohmer *et al.*, 1978; Bartel *et al.*, 1977; Battistella *et al.*, 1978; Bizzi *et al.*, 1978). The loss of carnitine in the dialysate greatly exceeds the normal loss in urine (Bohmer *et al.*, 1978). Prolonged longitudinal plasma carnitine measurements have subdivided patients into two groups, one in which there is a chronic plasma carnitine deficiency and one in which there is a return to normal or higher than normal plasma carnitine concentrations (Battistella *et al.*, 1978). Even for the patients whose plasma carnitine will return to normal after dialysis, approximately 6 hr are needed for the recovery. However if the patients are given 3 g of D,L-carnitine orally at the end of the dialysis, the plasma carnitine concentrations return to normal within 2 hr (Bizzi *et al.*, 1978). If L-carnitine is added to the dialysate at a final concentration of 65 nmole/ml the decrease in plasma carnitine during hemodialysis is completely prevented (Bizzi *et al.*, 1979).

In eight of nine patients, muscle carnitine concentration after hemodialysis was only 10% of the concentration in controls (Bohmer *et al.*, 1978). The authors of this report feel that the carnitine concentration of the cardiac muscle of these patients is probably also low and that cardiac carnitine deficiency induced by hemodialysis may help to explain the clinical syndrome of cardiomyopathy and cardiac failure which has been observed in some patients treated for a long time with intermittant hemodialysis.

There is a high incidence of hypertriglyceridemia in uremic patients. Six patients were studied (Bougneres *et al.*, 1979) who were hemodialyzed three times a week and had displayed a type IV hypertriglyceridemia for 6–24 months. D,L-Carnitine was given, 2.4 g daily, with no other hypolipemic drugs. In all patients, triglycerides fell to normal levels in 14 days and remained normal as long as the carnitine was given. No adverse effects were observed. However, in another study (Bazzato *et al.*, 1979) three of 15 patients receiving an extremely high dose of carnitine (2 g of D,L-carnitine injected intravenously for 45 days) developed myasthenialike symptoms.

These studies indicate that if the body's natural reabsorption of carnitine is not operating properly or if the normal reabsorption of carnitine is surpassed by hemodialysis, nutritional therapy of the patient with carnitine may be advised.

4.4. In Infants

Fatty acid metabolism is of major importance during fetal and perinatal development. Although fatty acids are not a major source of energy for the fetus, they are a central structural component of the developing cells and are required for deposition of triglycerides in adipose tissue. At 7 months of gestation, fat content in the human is 3.5% of body weight. At birth, adipose tissue of a full-term human newborn comprises approximately 16% of the body weight. Therefore, during the last trimester of human pregnancy, there are great requirements for lipogenesis and considerable accumulation of lipids in adipose tissue (Warshaw, 1979). The mechanism of transport of fatty acids across the placenta from mother to fetus is not known; however, it has been suggested that fatty acids could be transported at least in part as acylcarnitines (Karp *et al.*, 1971). Blood levels of carnitine are lower in pregnant than in nonpregnant women. Although the human newborn has much lower tissue carnitine concentrations than the adult, carnitine can be detected in most tissues at birth. It has been suggested that the placenta may play a role in carnitine transport from the mother to the fetus and that fetal blood and amniotic fluid carnitine levels may reflect retention of carnitine by fetal tissues. Carnitine levels are found to be higher in cord blood than in maternal blood and usually are higher in the umbilical artery than vein (Hahn *et al.*, 1977).

The accumulation of carnitine in tissues from birth to adulthood has been studied in great detail in the rat. Tissue carnitine concentration varies with both age and sex (Borum, 1978). Further investigations have shown that during development after weaning the plasma carnitine levels are normally regulated, at least in part, by androgens and estrogens (Borum, 1980). In the rat, the primary source of carnitine during the suckling period is probably milk, with a carnitine concentration in milk highest during the first 2 to 3 days of suckling (Robles-Valdes *et al.*, 1976).

The total carnitine content of human breast milk increases during the first week post-partum. After 1 month of lactation, the carnitine values of milk decrease to those of samples obtained during the first 3 days post-partum (Schmidt-Sommerfeld *et al.*, 1978). Liquid formulas and special diets whose main protein source is soy protein isolate, casein, or egg white protein have very low to undetectable concentrations of carnitine. Soy-based infant formulas contain no detectable carnitine (Borum *et al.*, 1979). If the human infant normally obtains carnitine from the mother's milk, one would expect infants fed a soy-based protein formula to be somewhat carnitine deficient. Indeed, the levels of plasma carnitine are lower in infants receiving soy-protein-based formula than in infants receiving human breast milk or cow's milk (Novak *et al.*, 1979). High levels of hepatic carnitine have been shown to be required for ketogenesis to occur (McGarry *et al.*, 1975). Thus, one would expect an infant who is fed a soy-protein-based formula to have impaired production of ketone bodies compared to an infant receiving dietary carnitine. The plasma concentration of β-hydroxybutyrate has been shown to be significantly lower in infants maintained on a soy-based formula (Wieser *et al.*, 1978).

It has been suggested that the carnitine in breast milk may be better absorbed than carnitine in commercial formulas (Curry and Warshaw, 1978). One group of newborns was fed human breast milk, and another group was fed commercial formulas containing approximately the same concentration of carnitine. At 42 hr of age, breast-fed newborns had higher plasma carnitine levels than newborns fed commercial formulas. Correspondingly, by 42 hr of age the serum keton body concentrations were higher in breast-fed as compared with formula-fed infants.

Although fatty acid oxidation is not a major source of energy for the fetus, fatty acids become the preferred substrate for metabolically active tissues such as the heart and kidney during the first few days of life. The development of fatty acids and ketone body oxidation spares glucose oxidation and contributes to glucose homeostasis. Thus, the development of fatty acid oxidation is of great importance to the overall energy economy. Experiments have not been performed that would determine if reduced oral intake of carnitine will adversely affect energy production or whether carnitine administration will be of benefit to the newborn. However, the exogenous supply of carnitine to the premature infant may have a significant influence on the ability to stimulate optimal fatty acid oxidation. Solutions commonly used for intravenous feedings in the newborn infant contain no carnitine. However, when Intralipid® is used as a major source of calories in the newborn, fatty acid oxidation must be operating in order for the newborn infant to obtain adequate energy. Studies have shown that infants maintained on intravenous feedings have significantly lower total, free, and acylcarnitine levels than when they are fed orally with human milk or with commercial formulas containing carnitine (Schiff *et al.*, 1979). Further studies are needed to determine the possible therapeutic benefits of carnitine supplementation in the infant receiving intravenous fluids or a carnitine-free diet.

4.5. In Muscle Weakness and Associated Lipid Myopathies

Seven years ago, A. G. Engel and co-workers (Engel and Angelini, 1973) described a new syndrome characterized by progressive muscle weakness and lipid infiltration in the skeletal muscle, particularly in Type I fibers. A skeletal muscle biopsy of the patient had an extremely low concentration of carnitine accompanied by an impaired ability to oxidize long-chain fatty acids, and both could be restored to normal with the addition of exogenous carnitine. Since the metabolic lesion resulting in most myopathies is not well documented, this syndrome was considered interesting but probably rare. However, during the past 7 years, at least 28 patients suffering from the syndrome have been documented and described in the literature (See Table IV for patient number and references). The disease is a very serious one as shown by the fact that it proved fatal for 10 of the 28 patients described. In addition, patients suffering from other myopathies such as Duchenne dystrophy and Becker dystrophy have muscle carnitine concentrations that are lower than normal (Borum *et al.*, 1977).

A cursory examination of Table IV illustrates that the syndrome of human carnitine deficiency is a family of syndromes. The common features of the 28 patients described in Table IV are that (1) all patients suffered from progressive muscle weakness, (2) histological examination of skeletal muscle biopsy of all patients showed lipid accumulation, usually in the Type I fibers, and (3) biochemical analysis of the skeletal muscle biopsy of all patients showed an abnormally low concentration of carnitine in the muscle. Beyond these three common characteristics, the descriptions of the patients become quite variable. Skeletal muscle carnitine deficiency is equally distributed between the sexes (15 female and 13 male), and the age of onset of symptoms ranges from birth to 48 years of age. Of the patients studied, 13 patients were between the ages of 0–12 years, seven patients between the ages of 13–24 years, six patients between the ages of 25–50 years, and two patients between the ages of 51–75 years. Thus, although most of patients are under 25 years of age, all age groups are affected.

Although the reported muscle carnitine concentrations for all the patients are low, the carnitine concentration in other tissues may be low or may be normal.

The carnitine concentration was determined in either a biopsy or autopsy liver sample in nine of the patients. Normal carnitine concentrations were found in three of the patients' livers, and low carnitine concentrations were found in six of the patients' livers. Plasma carnitine concentration was determined in 22 of the patients. Exactly 50% of the patients had at least a normal concentration of plasma carnitine, and 50% had low concentrations of plasma carnitine.

The variable concentration of carnitine seen in tissues other than skeletal muscle leads to a variety of hypotheses concerning the metabolic lesions that may result in a low skeletal muscle carnitine. Patients (#3, 12, 13, 18) with low carnitine concentration in plasma, liver, and skeletal muscle are said to have a systemic carnitine deficiency which is most likely caused by a defect in the biosynthesis of carnitine. At autopsy, low concentrations of carnitine were found in the hearts of patients #12 and 13, which is consistent with the defect in carnitine biosynthesis. However, the human kidney has also been shown to synthesize carnitine, and at autopsy the kidneys of both patients 12 and 13 had normal carnitine concentration. Patients such as #26 and #28 with low skeletal muscle carnitine but normal liver and plasma carnitine concentrations may have normal carnitine biosynthesis but a defective transport system for the uptake of carnitine by muscle. At autopsy, low concentrations of carnitine were also found in the heart of patient #26, indicating that the carnitine transport system was defective in both skeletal muscle and cardiac muscle. Patient #23 had low liver and low skeletal muscle carnitine concentration but normal plasma carnitine concentration, which could be the result of a defect in carnitine biosynthetic capability. Patient #1 had a normal liver carnitine concentration but low plasma and low muscle carnitine concentrations, which could be the result of an inability of the liver to release the carnitine into the blood after it is synthesized. All of the above types of patients are represented in the fatal cases. It is interesting to note that acidosis, coma, hypoglycemia, respiratory distress, and cardiac arrest are repeatedly listed as cause of death in the fatal cases. Thus, carnitine deficiency can lead to weakness of the muscles needed for respiration, heart failure, and complications in lipid and carbohydrate metabolism that are severe enough to result in death.

Carnitine palmitoyl transferase has been measured in eight (#1, 7, 9, 10, 23, 25, 26, 28) of the patients listed in Table IV, and in all cases, the activity was as high as or higher than normal. Activity of γ-butyrobetaine hydroxylase was measured in patients #1 and 3 are found to be normal. The conditions of two patients (#7 and 19) deteriorated rapidly when they became pregnant, whereas a third patient, #25, had temporary remission of symptoms during two pregnancies. Two patients (#2 and 23) also suffered from insulin-requiring diabetes mellitus. In addition to muscle weakness, several patients suffered from abnormal liver metabolism. Eight patients (#3, 4, 7, 12, 14, 15, 16, and 21) had an enlarged liver at some stage of the disease. Seven patients had documented bouts of hypoglycemia (#3, 11, 12, 18, 21, 22, 27), and six additional patients (#4,

Table IV. Effect of Corticosteroids and Carnitine Administration in a Series of Patients with Muscle Weakness and Associated Lipid Myopathies

			Age (years) at		Carnitine in muscle[b]		Carnitine in liver[b]		Carnitine in plasma (nmol/ml)		
Patient	Ref[a]	Sex	Onset	Study	Control	Patient	Control	Patient	Control	Patient	Treatment and outcome[c]
1	(a,b,c)	F	19	24	13.92 ± 0.92	2.25	6.8 ± 0.7	7.05	51.6 ± 2.5	17.7	Pred + Carn—improve
2	(d)	F	38	61	7.96 - 22.86	1.92			23–70	49	
3	(e)	M	3.5	11	13.92 ± 0.92	0.50	6.8 ± 0.7	0.73	51.6 ± 2.5	7.9	Carn—improve
4	(f,g)	M	5.5	11	11.6–15.7[d]	2.61[d]					Carn—improve
5	(h)	M	1.5	8	2640 ± 610[e]	240[e]			46 ± 13	32	Pred—improve
6	(i,j)	F	7	10	15.55 ± 1.45	0.81			23–55	32	Carn—improve
7	(k)	F	Child	20	13.92 ± 0.92	0.49	6.8 ± 0.7	0.99[f]			Fatal
8	(l)	M	18	20	3750 ± 400[e]	235[e]			54.9 ± 2.4	46.6	Pred + Propran—improve
9	(m)	F	48	51	1640-3430[e]	680[e]			23–70	64.7	Pred—improve
10	(n)	M	34	36	9–18	2.75			25–62	14.8	Pred—no improve; Carn—improve
11	(o)	F	16	28	16.0 ± 2.6	4.0					Fatal
12	(o)	M	3.5	8	16.0 ± 2.6	3.9	8.3 ± 2.3	4.6[f]	57.9 ± 12.0	24.7	Carn—fatal
13	(o)	F	16	19	16.0 ± 2.6	1.5	8.3 ± 2.3	5.1[f]	57.9 ± 12.0	14.8	Fatal
14	(p)	M	22 mo	5	14.57 ± 0.93	0.94					Fatal
15	(q)	M	Birth	2–2.5	2010 ± 470[e]	410[e]			45–55	47.2	Carn—fatal
16	(r)	F	Birth	2	22.2 ± 5.0	16.1					Carn—fatal

17	(s)	F	23	25	6.0 ± 0.18	0.13			26.5–34.5	18.1, 19.8	Pred + Estro—no improve
18	(t)	F	11 mo	5	7.96–22.86	4.22	3.3–10.4	0.40	27.9–67.2	8.66	
19	(u)	F	16	16	10–23	1.53			41.8 ± 5.75	19.8	Carn—improve
20	(v)	M	35	35	9–18	2.75			25–70	14.8	Steroids—no improve; Carn—improve
21	(v)	M	3	12	9–18	2.50			25–70	13.0	Fatal
22	(w)	M	12	14	654–830[e]	303[e,f]			45–58	45.6	Fatal
23	(x)	F	Birth	3	1600[e]	290[e]	522[e]	57[e,f]	53 ± 21	79.2	
24	(y)	M	6 wk	13 wk	1340 ± 330[e]	520[e]					Fatal
25	(z)	F	22	33	17.23 ± 4.19	2.36			58.3 ± 11.8	60, 31.1	Carn—no improve; Pred—improve
26	(aa)	F	Birth	15	2640 ± 460[e]	440[e]	800 ± 100[e]	960[e,f]	37 ± 2	46	Fatal
27	(bb)	F	5	5	3070 ± 790[e]	270[e]			46 ± 6.9	9	Carn—improve
28	(cc)	M	Teens	38	20.2	3.0	9.96	11.5	42.8	65.8	

[a]References to table: (a) Engel and Angelini (1973), (b) Engel and Siekert (1972), (c) Engel *et al.* (1974), (d) Markesbery *et al.* (1974), (e) Karpati *et al.* (1975), (f) Smyth *et al.* (1975), (g) Hosking *et al.* (1977), (h) Vandyke *et al.* (1975), (i) Angelini *et al.* (1976), (j) Angelini (1975), (k) Boudin *et al.* (1976), (l) Isaacs *et al.* (1976), (m) Whitaker *et al.* (1977), (n) Scarlato *et al.* (1977), (o) Cornelio *et al.* (1977), (p) Engel *et al.* (1977), (q) Hart *et al.* (1978), (r) Di Donato *et al.* (1978), (s) Bradley *et al.* (1978), (t) Glasgow *et al.* (1978), (u) Angelini *et al.* (1978), (v) Scarlato *et al.* (1978), (w) Ware *et al.* (1978), (x) Koski *et al.* (1978), (y) Esiri *et al.* (1979), (z) Willner *et al.* (1979), (aa) Scholte *et al.* (1979), (bb) Morand *et al.* (1979), (cc) Di Donato *et al.* (1979).

[b]All values are expressed as nmol/mg noncollagen protein unless otherwise indicated.

[c]Abbreviations used: Carn, carnitine; Pred, prednisone; Propran, propranolol; Estro, estradiol; Steroids, corticosteroids.

[d]Expressed as nmol/g dry weight.

[e]Expressed as nmol/g wet weight.

[f]Tissues obtained at autopsy.

7, 13, 14, 26, 28) had symptoms such as repeated vomiting, lethargy, metabolic acidosis, etc. which would have been caused by hypoglycemia, but blood glucose levels were not measured.

Carnitine deficiency appears to be hereditary. Consanguinity was present in the family of three of the patients (#17, 25, 26). The muscle carnitine concentration in the parents of patient #5 was lower than normal although not as low as that of the patient. The muscle biopsy of the mother of patient #6 had a low concentration of carnitine and numerous lipid droplets. The muscle biopsy of the father of patient #15 had a carnitine concentration somewhat lower than normal. The only sister of patient #12 had died at the age of 5 years, and autopsy examination showed fatty infiltration of liver, heart, and skeletal muscle. The 7-year-old sister of patient #26 had lipid infiltration and low carnitine concentration in her muscle biopsy as well as a low plasma carnitine concentration.

Since polymyositis was the tentative diagnosis in some of the patients, corticosteroid treatment was begun. Two patients (#5 and 9) improved with prednisone treatment alone, one patient (#8) improved with a combination of prednisone and propranolol treatment, one patient (#1) improved with a combination of prednisone and carnitine treatment, and one patient (#17) did not improve with a combination of predinsone and estradiol treatment. One patient (#25) did not improve with carnitine treatment alone but did improve with prednisone treatment alone. Two patients (#10 and 20) did not improve with prednisone treatment alone but did improve with carnitine treatment alone. In two cases that proved fatal (#12 and 15), carnitine treatment gave no improvement. But in six cases (#3, 4, 6, 16, 19, 27), oral carnitine treatment alone gave marked improvement. The molecular mechanism by which any of these treatments induces clinical improvement is not understood. Carnitine treatment usually increases the plasma carnitine concentration to normal if it is low, but, although the skeletal muscles are clinically dramatically stronger, the carnitine concentration of the muscles is not returned to normal. However, histologically, the muscle of patients #3, 6, and 9 appeared more normal after treatment than before treatment. The liver biopsy of patient #9 also looked more normal after treatment. All signs of heart failure disappeared in patient #27 after carnitine treatment. In the 12 cases where oral carnitine treatment has been tried nine cases showed dramatic improvement, and only three cases showed no improvement.

5. Concluding Remarks

As interest in carnitine metabolism and function increases, the number of published reports concerning carnitine concentration in normal and abnormal physiological conditions is also rapidly increasing. We suggest that unless two very important guidelines discussed below are strictly adhered to by workers in the field, the present confusion in the literature concerning the meaning and interpretation of "carnitine concentration" will also rapidly increase (see also discussion by Mitchell, 1978a on this point).

The two most commonly used methods for carnitine determination [spectrophotometric assay (Marquis and Fritz, 1964), radioisotopic assay (Cederblad and Lindstedt, 1972)] measure only free carnitine. Thus, acid precipation and alkaline hydrolysis performed prior to the actual measurement of carnitine determines if the measurement is for free carnitine, short-chain carnitine, long-chain carnitine, or total carnitine. All carnitine determinations should be reported so that it is very clear what fraction of the cell's carnitine was measured. Thus, we suggest as a useful guideline that the term "carnitine concentration" be used only when total carnitine is measured, and that the terms free carnitine, short-chain carnitine, and long-chain carnitine be used when only a particular fraction of the cell's cernitine is measured.

Many investigators are seemingly studying two different metabolic processes at the same time. Measurement and comparison of the ratios of the different carnitine fractions of the cell is important in studying fatty acid metabolism. For example, a decrease in free carnitine with a concomitant increase in long-chain carnitine is a reflection of cell metabolism switching to fatty acid oxidation as a source of energy but should not be termed carnitine deficiency, since there is no net change in the overall carnitine concentration in this instance. Hence, the second suggested guideline is that the term "carnitine deficiency" be used only when the total number of carnitine molecules per test sample is significantly lower than normal, i.e., total carnitine must be determined and shown to be decreased from normal before the term "carnitine deficiency" is used. There has been outstanding progress in recent years in an understanding of both carnitine metabolism and the use of nutritional therapy to prevent clinical problems resulting from carnitine deficiency. It would be a needless tragedy for progress to be slowed by imprecise communication between laboratories.

It is evident from considering the examples of "carnitine deficiency" discussed herein that such deficiency may arise variously as follows: (1) *a nutritional deficiency* wherein carnitine, its amino acid precursors, or the cofactors concerned in carnitine biosynthesis are limiting in the diet in varying degrees; (2) *a functional carnitine deficiency* wherein a protein concerned, for example, in carnitine function, transport, or uptake by the tissues is rendered inoperative by genetic damage; and (3) genetic and/or nutritional factors mediating to alter the level or activity of the carnitine biosynthetic enzymes (Fig. 2), thus limiting the production of carnitine *in vivo*. In all these instances, the net result is an impairment in fatty acid oxidation and ultimate energy release. In this regard, numerous examples were cited in this review wherein a carnitine deficiency invoked via mechanisms such as (1), (2), or (3) above did point to aberrations in lipid metabolism. Much progress has been made in identifying the nutritional factors concerned in carnitine biosynthesis (Fig. 2) as well as in understanding the role of carnitine in the intramitochondrial transport of fatty acids (Fig. 1). But little is known at present about mechanisms of carnitine transport and uptake by the tissues, processes that obviously relate importantly to carnitine function. Clearly, this is an area for research in the future.

This chapter also illustrates that it serves no useful purpose to adhere to strict definitions of "essential" vs. "nonessential" nutrients in considering carnitine nutriture in man. In this regard, the view of Harper (1974) may be relevant: in a delightful commentary, "Nonessential Amino Acids," he chides the nutrition establishment with the statement that the term "nonessential" with respect to the amino acids is a misnomer from nutritional, physiological, and biochemical points of view. Thus, in the present instance, even though higher animals including man can synthesize carnitine under suitable circumstances, varying degrees of carnitine deficiency are nevertheless being reported in man from all over the world. Such deficiency arises via any number of diverse ways, e.g., mechanisms (1), (2), and (3). The physiological outcome, an ultimate assault on energy production, is serious regardless of the mechanism involving the deficiency of the "nonessential" nutrient, carnitine.

References

Angelini, C., 1975, Carnitine deficiency, *Lancet* **2**:554.

Angelini, C., Lucke, S., and Cantarutti, F., 1976, Carnitine deficiency of skeletal muscle: Report of a treated case, *Neurology* **26**:633.

Angelini, C., Govoni, E., Bragaglia, M. M., and Vergani, L., 1978, Carnitine deficiency: Acute postpartum crisis, *Ann. Neurol.* **4**:558.

Bartel, L., Hussey, J., Ewart, R., and Shrago, E., 1977, Serum carnitine levels during hemodialysis, *Clin. Res.* **25**:627A.

Battistella, P. A., Angelini, C., Vergani, L., Bertoli, M., and Lorenzi, S., 1978, Carnitine deficiency induced during haemodialysis, *Lancet* **1**:939.

Bazzato, G., Mezzina, C., Ciman, M., and Guarnieri, G., 1979, Myasthenia-like syndrome associated with carnitine in patients on long-term haemodialysis, *Lancet* **1**:1041.

Bistrian, B. R., Blackburn, G. L., Hallowell, E., and Heddle, R., 1974, Protein status of general surgical patients, *J. Am. Med. Assoc.* **230**:858.

Bizzi, A., Mingardi, G., Codegoni, A. M., Mecca, G., and Garattini, S., 1978, Accelerated recovery of post-dialysis plasma carnitine by oral carnitine, *Biomedicine* **29**:183.

Bizzi, A., Cini, M., Garattini, S., Mingardi, G., Licini, L., and Mecca, G., 1979, L-Carnitine addition to haemodialysis fluid prevents plasma-carnitine deficiency during dialysis, *Lancet* **1**:882.

Bohmer, T., Bergrem, H., and Eiklid, K., 1978, Carnitine deficiency induced during intermittent haemodialysis for renal failure, *Lancet* **1**:126.

Bollet, A. J., and Owens, S., 1973, Evaluation of nutritional status of selected hospitalized patients, *Am. J. Clin. Nutr.* **26**:931.

Borum, P. R., 1978, Variation in tissue carnitine concentrations with age and sex in the rat, *Biochem. J.* **176**:563.

Borum, P. R., 1980, Regulation of the carnitine concentration in plasma, in: *O'Hara Biochemical Research Symposia, Biosynthesis, Metabolism, and Functions of Carnitine, Dallas, 1979* (R. A. Frenkel and J. D. McGarry, eds.), pp. 115–126, Academic Press, New York.

Borum, P. R., and Broquist, H. P., 1977a, Purification of *S*-adenosylmethionine : ε-*N*-L-lysine methyltransferase, *J. Biol. Chem.* **252**:5651.

Borum, P. R., and Broquist, H. P., 1977b, Lysine deficiency and carnitine in male and female rats, *J. Nutr.* **107**:1209.

Borum, P. R., Broquist, H. P., and Roelofs, R. I., 1977, Muscle carnitine levels in neuromuscular disease, *J. Neurol. Sci.* **34**:279.

Borum, P. R., York, C. M., and Broquist, H. P., 1979, Carnitine content of liquid formulas and special diets, *Am. J. Clin. Nutr.* **32**:2272.

Boudin, G., Mikol, J., Guillard, A., and Engel, A. G., 1976, Fatal systemic carnitine deficiency with lipid storage in skeletal muscle, heart, liver and kidney, *J. Neurol. Sci.* **30**:313.

Bougneres, P. F., Lacour, B., Di Giulio, S., and Assan, R., 1979, Hypolipaemic effect of carnitine in uraemic patients, *Lancet* **1**:1401.

Bradley, W. G., Tomlinson, B. E., and Hardy, M., 1978, Further studies of mitochondrial and lipid storage myopathies, *J. Neurol. Sci.* **35**:201.

Bremer, J., 1977, Carnitine and its role in fatty acid metabolism, *Trends Biochem. Sci.* **2**:207.

Broquist, H. P., and Borum, P. R., 1977, Some aspects of carnitine nutriture, *Compr. Ther.* **3**:66.

Butterworth, C. E., and Blackburn, G. L., 1975, Hospital malnutrition, *Nutr. Today* **10**:8.

Carter, A. L., and Frenkel, R., 1979, The role of the kidney in the biosynthesis of carnitine in the rat, *J. Biol. Chem.* **254**:10670.

Cederblad, G., and Lindstedt, S., 1972, A method for the determination of carnitine in the picomole range, *Clin. Chim. Acta* **37**:235.

Chen, S., and Lincoln, S. D., 1977, Increased serum carnitine concentration in renal insufficiency, *Clin. Chem.* **23**:278.

Chu, S. W., and Hegsted, D. M., 1976, Adaptive response of lysine and threonine degrading enzymes in adult rats, *J. Nutr.* **106**:1089.

Cornelio, F., Di Donato, S., Peluchetti, D., Bizzi, A., Bertagnolio, B., D'Angelo, A., and Wiesmann, U., 1977, Fatal cases of lipid storage myopathy with carnitine deficiency, *J. Neurol. Neurosurg. Psychiatry* **40**:170.

Cox, R. A., and Hoppel, C. L., 1974, Carnitine and trimethylaminobutyrate synthesis in rat tissues, *Biochem. J.* **142**:699.

Curry, E., and Warshaw, J. B., 1978, Higher serum carnitine levels and ketogenesis in breast-fed as compared to formula-fed infants, *Pediatr. Res.* **12**:504.

De Donato, S., Cornelio, F., Balestrini, M. R., Bertagnolio, B., and Peluchetti, D., 1978, Mitochondria-lipid-glycogen myopathy, hyperlactacidemia, and carnitine deficiency, *Neurology* **28**:1110.

Di Donato, S., Cornelio, F., Storchi, G., and Rimoldi, M., 1979, Hepatic ketogenesis and muscle carnitine deficiency, *Neurology* **29**:780.

Engel, A. G., and Angelini, C., 1973, Carnitine deficiency of human skeletal muscle with associated lipid storage myopathy: A new syndrome, *Science* **179**:899.

Engel, A. G., and Siekert, R. G., 1972, Lipid storage myopathy responsive to prednisone, *Arch. Neurol.* **27**:174.

Engel, A. G., Angelini, C., and Nelson, R. A., 1974, Identification of carnitine deficiency as a cause of human lipid storage myopathy, in: *Exploratory Concepts in Muscular Dystrophy II* (A. T. Milhorat, ed.), pp. 601–617, Excerpta Medica, Amsterdam.

Engel, A. G., Banker, B. Q., and Eiken, R. M., 1977, Carnitine deficiency: Clinical, morphological, and biochemical observations in a fatal case, *J. Neurol. Neurosurg. Psychiatry* **40**:313.

Englard, S., 1979, Hydroxylation of γ-butyrobetaine to carnitine in human and monkey tissues, *FEBS Lett.* **102**:297.

Englard, S., and Carnicero, H. H., 1978, γ-Butyrobetaine hydroxylation to carnitine in mammalian kidney, *Arch. Biochem. Biophys.* **190**:361.

Esiri, M. M., Bower, B. D., and Ross, B. D., 1979, Fatal lipid storage myopathy in an infant, case report and autopsy findings, *J. Neurol. Sci.* **41**:93.

Frenkel, R. A., and McGarry, J. D. (eds.), 1980, *O'Hara Biochemical Research Symposia, Biosynthesis, Metabolism, and Functions of Carnitine, Dallas, 1979*, Academic Press, New York.

Glasgow, A. M., Eng. G., and Engel, A. G., 1978, Systemic carnitine deficiency: A cause of recurrent Reyes syndrome, *Pediatr. Res.* **12**:1129.

Hahn, P., Skala, J. P., Seccomke, D. W., Frohlich, J., Penn-Walker, D., Novak, M., Hynie, I., and Towell, M. E., 1977, Carnitine content of blood and amniotic fluid, *Pediatr. Res.* **11**:878.

Haigler, H. T., and Broquist, H. P., 1974, Carnitine synthesis in rat tissue slices, *Biochem. Biophys. Res. Commun.* **56**:676.

Harper, A. E., 1974, Editorial: Nonessential Amino Acids, *Nutr. Rev.* **104**:965.

Hart, Z. H., Chang, C. H., Di Mauro, S., Farooki, Q., and Ayyar, R., 1978, Muscle carnitine deficiency and fatal cardiomyopathy, *Neurology (Minneap.)* **28**:147.

Hosking, G. P., Cavanagh, N. P. C., Smyth, D. P. L., and Wilson, J., 1977, Oral treatment of carnitine myopathy, *Lancet* **1**:853.

Hulse, J. D., and Henderson, L. M., 1979, Isolation and characterization of an aldehyde dehydrogenase exhibiting preference for 4-*N*-trimethyl aminobutyraldehyde as substrate, *Fed. Proc.* **38**:2359.

Hulse, J. D., Ellis, S. R., and Henderson, L. M., 1978, Carnitine biosynthesis, β-hydroxylation of trimethyllysine by an α-ketoglutarate-dependent mitochondrial dioxygenase, *J. Biol. Chem.* **253**:1654.

Isaacs, H., Heffron, J. J. A., Badenhorst, M., and Pickering, A., 1976, Weakness associated with the pathological presence of lipid in skeletal muscle: A detailed study of a patient with carnitine deficiency, *J. Neurol. Neurosurg. Psychiatry* **39**:1114.

Karp, W., Sprecher, H., and Robertson, A., 1971, Carnitine palmityltransferase activity in the human placenta, *Biol. Neonate* **18**:341.

Karpati, G., Carpenter, S., Engel, A. G., Watters, G., Allen, J., Rothman, S., Klassen, G., and Mamer, O. A., 1975, The syndrome of systemic carnitine deficiency, *Neurology (Minneap.)* **25**:16.

Kaufman, R. A., and Broquist, H. P., 1977, Biosynthesis of carnitine in *Neurospora crassa, J. Biol. Chem.* **252**:7437.

Khairallah, E. A., and Wolf, G., 1965, Growth-promoting and lipotropic effect of carnitine in rats fed diets limited in protein and methionine, *J. Nutr.* **87**:469.

Khan, L., and Bamji, M. S., 1977, Plasma carnitine levels in children with protein-calorie malnutrition before and after rehabilitation, *Clin. Chim. Acta.* **75**:163.

Khan, L., and Bamji, M. S., 1979, Tissue carnitine deficiency due to dietary lysine deficiency: Triglyceride accumulation and concomitant impairment in fatty acid oxidation, *J. Nutr.* **109**:24.

Koski, C., Gumbinas, M., Ozand, P., Bejar, R., and McLaughlin, J., 1978, Muscle carnitine deficiency in a patient with persistent neonatal diabetes and hypokalemia, in: *IV International Congress on Neuromuscular Disease Abstract*, p. 476.

LaBadie, J. H., Dunn, W. A., and Aronson, N. N., Jr., 1976, Hepatic synthesis of carnitine from protein bound trimethyllysine. Lysosomal digestion of methyl-lysine labelled asialo-fetuin, *Biochem. J.* **160**:85.

Lindstedt, G., and Lindstedt, S., 1970, Cofactor requirements of γ-butyrobetaine hydroxylase from rat liver, *J. Biol. Chem.* **245**:4178.

Lindstedt, G., Lindstedt, S., and Tofft, M., 1970, γ-Butyrobetaine hydroxylase from *Pseudomonas* sp AK 1, *Biochemistry* **9**:4336.

Markesbery, W. R., McQuillen, M. P., Procopis, P. G., Harrison, A. R., and Engel, A. G., 1974, Muscle carnitine deficiency, association with lipid myopathy, vacuolar neuropathy and vacuolated leukocytes, *Arch. Neurol.* **31**:320.

Marquis, N. R., and Fritz, I. B., 1964, Enzymological determination of free carnitine concentrations in rat tissues, *J. Lipid Res.* **5**:184.

McGarry, J. D., Robles-Valdes, C., and Foster, D. W., 1975, Role of carnitine in hepatic ketogenesis, *Proc. Natl. Acad. Sci. U.S.A.* **72**:4385.

Mikhail, M. M., and Mansour, M. M., 1976, The relationship between serum carnitine levels and the nutritional status of patients with schistosomiasis, *Clin. Chim. Acta* **71**:207.

Mitchell, M. E., 1978a, Carnitine metabolism in human subjects, I. Normal metabolism, *Am. J. Clin. Nutr.* **31**:293.

Mitchell, M. E., 1978b, Carnitine metabolism in human subjects, II. Values of carnitine in biological fluids and tissues of "normal" subjects, *Am. J. Clin. Nutr.* **31**:481.

Mitchell, M. E., 1978c, Carnitine metabolism in human subjects, III. Metabolism in disease, *Am. J. Clin. Nutr.* **31**:645.

Morand, P., Despert, F., Carrier, H. N., Saudubray, B. M., Fardeau, M., Romieux, B., Fauchier, C., and Combe, P., 1979, Myopathie lipidique avec cardiomyopathie severe par deficit generalise en carnitine, *Arch. Mal. Coeur* **72**:536.

Novak, M., Wieser, P. B., Buch, M., and Hahn, P., 1979, Acetylcarnitine and free carnitine in body fluids before and after birth, *Pediatr. Res.* **13**:10.

Paik, W. K., and Kim, S., 1975, Protein methylation: Chemical, enzymological, and biological significance, *Adv. Enzymol.* **42**:227.

Rebouche, C. J., 1980, Comparative aspects of carnitine biosynthesis in microorganisms and mammals with attention to carnitine biosynthesis in man, in: *O'Hara Biochemical Research Symposia, Biosynthesis, Metabolism, and Functions of Carnitine, Dallas, 1979,* (R. A. Frenkel and J. D. McGarry, eds.), p. 57, Academic Press, New York.

Robles-Valdes, C., McGarry, J. D., and Foster, D. W., 1976, Maternal-fetal carnitine relationships and neonatal ketosis in the rat, *J. Biol. Chem.* **251**:6007.

Rudman, D., Sewell, C. W., and Ansley, J. D., 1977, Deficiency of carnitine in cachectic cirrhotic patients, *J. Clin. Invest.* **60**:716.

Rudman, D., Ansley, J. D., and Sewell, C. V., 1980, Carnitine deficiency in cirrhosis, in: *O'Hara Biochemical Research Symposia, Biosynthesis, Metabolism, and Functions of Carnitine, Dallas, 1979* (R. A. Frenkel and J. D. McGarry, eds.), p. 307, Academic Press, New York.

Sachan, D., 1978, Carnitine biosynthesis: Hydroxylation of 6-*N*-trimethyllysine, *Fed. Proc.* **37**:2462.

Scarlato, G., Albizzati, M. G., Bassi, S., Cerri, C., and Frattola, L., 1977, A case of lipid storage myopathy with carnitine deficiency, *Eur. Neurol.* **16**:222.

Scarlato, G., Pellegrini, G., Cerri, C., Meola, G., and Veicsteinas, A., 1978, The syndrome of carnitine deficiency: Morphological and metabolic correlations in two cases, *Can. J. Neurol. Sci.* **5**:205.

Schiff, D., Chaw, G., Seccomke, D., and Hahn, P., 1979, Plasma carnitine levels during intravenous feeding of the neonate, *J. Pediatr.* **95**:1043.

Schmidt-Sommerfeld, E., Novak, M., Penn, D., Wieser, P. B., Buch, M., and Hahn, P., 1978, Carnitine and development of newborn adipose tissue, *Pediatr. Res.* **12**:660.

Scholte, H. R., Meijer, A. E. F. H., Van Wijngaarden, G. K., and Leenders, K. L., 1979, Familial carnitine deficiency—a fatal case and subclinical state in a sister, *J. Neurol. Sci.* **42**:87.

Smyth, D. P. L., Lake, B. D., MacDermot, J., and Wilson, J., 1975, Inborn error of carnitine metabolism, *Lancet* **1**:1198.

Tanphaichitr, V., and Broquist, H. P., 1973a, Lysine deficiency in the rat: Concomitant impairment in carnitine biosynthesis, *J. Nutr.* **103**:80.

Tanphaichitr, V., and Broquist, H. P., 1973b, Role of lysine and ϵ-*N*-trimethyllysine in carnitine biosynthesis, II. Studies in the rat, *J. Biol. Chem.* **248**:2176.

Tanphaichitr, V., Zaklama, M. S., and Broquist, H. P., 1976, Dietary lysine and carnitine: Relation to growth and fatty livers in rats, *J. Nutr.* **106**:111.

Tanphaichitr, V., Lerdvuthisopon, N., Dhanamitta, S., and Broquist, H. P., 1980, Carnitine status in Thai adults, *Am. J. Clin. Nutr.* **33**:876.

Vandyke, D. H., Griggs, R. C., Markesbery, W., and DiMauro, S., 1975, Hereditary carnitine deficiency of muscle, *Neurology (Minneap.)* **25**:154.
Ware, A. J., Burton, W. C., McGarry, J. D., Marks, J. F., and Weinberg, A. G., 1978, Systemic carnitine deficiency—report of a fatal case with multisystemic manifestations, *J. Ped.* **93**:959.
Warshaw, J. B., 1979, Fatty acid metabolism during development, *Semin. Perinatol.* **3**:131.
Whitaker, J. N., DiMauro, S., Solomon, S. S., Sakesin, S., Duckworth, W. C., and Mendell, J. R., 1977, Corticosteroid-responsive skeletal muscle disease associated with partial carnitine deficiency, *Am. J. Med.* **63**:805.
Wieser, P. B., Buch, M., and Novak, M., 1978, Effect of carnitine on ketone body production in human newborns, *Pediatr. Res.* **12**:224.
Willner, J., DiMauro, S., Eastwood, A., Hays, A., Roshi, F., and Lovelace, R., 1979, Muscle carnitine deficiency—genetic heterogeneity, *J. Neurol. Sci.* **41**:235.

Chapter 8

Insect Nutrition
A Comparative Perspective

W. G. Friend and R. H. Dadd

1. Introduction

Developments in the field of insect nutrition have closely paralleled those of vertebrate nutrition over the last few decades, and similar techniques have been applied in both fields. The information emerging from these studies indicates great similarities in nutritional requirements among insects and sufficient similarities between insects and mammals that data obtained with insects may have wide applicability in certain areas of vertebrate nutrition. Insects, because of their small size and rapid life cycles, might provide the animal of choice in many areas of general nutrition. Few students of vertebrate nutrition, however, learn much about insects, and consequently, the integration of information on insect nutrition into the general field of nutrition is often hampered. In this chapter, we describe the special structural and physiological features of insects that affect their feeding and nutrition, briefly summarize important historical developments, and selectively review nutritional literature of relevance to vertebrate nutrition.

Detailed reports of the nutrient requirements of insects are available in many general reviews (Dadd, 1970a, 1973; David, 1967; House, 1961, 1962, 1965a, 1974; Lipke and Fraenkel, 1956; Rodriguez, 1966), and recently, the results

---

W. G. Friend • Department of Zoology, University of Toronto, Toronto, Ontario M5S 1A1, Canada R. H. Dadd • Division of Entomology and Parasitology, University of California, Berkeley, California 94720. This work was supported by grants from the National Sciences and Engineering Council of Canada and the California State Mosquito Control Program.

from research on the 78 species yielding the most clear-cut results have been summarized in tabular form by Dadd (1977a). More detailed treatment is available in reviews dealing with phytophagous insects (Chippendale and Beck, 1968; Friend, 1958; Vanderzant, 1966), plant-sucking insects (Auclair, 1963, 1969), plant-insect interactions (Harborne, 1978), parasitic insects (House, 1958), nutritional pathology (House, 1963), nutrition and resistance to biochemicals (Gordon, 1961), insect biochemistry (Rockstein, 1978), water balance (Arlian and Veselica, 1979), Diptera (Friend, 1968), grasshoppers (Dadd, 1963), mosquitoes (Clements, 1963), bees (Haydak, 1970), and silkworms (Ito, 1967; Yokoyama, 1963). Information is available on the digestibility and utilization of natural and synthetic foods (Gordon, 1968; Waldbauer, 1968), quantitative nutrition (Gordon, 1972), the effects of diet acceptability on nutritional studies (Beck and Chippendale, 1968; Dadd, 1968; Davis, 1968; Vanderzant, 1969a), and the effects of symbiotic microorganisms on nutritional requirements (Brooks, 1964; Ehrhardt, 1968b; Henry, 1962). Insect digestion and the absorption of nutrients have been reviewed (Dadd, 1970b; Gilmour, 1963; House, 1965b; Treherne, 1967), as has the role played by phagostimulants (Beck, 1965; Fraenkel, 1969; Schoonhoven, 1968; Thorsteinson, 1960; Wood *et al.*, 1970). To help in diet formulation, there are a bibliography of artificial diets used between 1900 and 1970 (Singh, 1972), two books that give the composition of all of the artificial diets used for insects up to the beginning of 1976 (House *et al.*, 1971; Singh, 1977), and a paper that describes interactions among dietary components during formation (Mittler, 1972).

2. Special Features of Insects Affecting Their Nutrition and Its Study

Insects and vertebrates evolved along completely different pathways and consequently differ in many ways. Peculiarities of insect anatomy, development and growth patterns, behaviors (particularly feeding behaviors), and physiology and metabolism have obvious effects on their nutrition. Consequently, the development of the field of insect nutrition and the concerns of many of the researchers have differed from those of vertebrate nutrition.

2.1. Tracheal Breathing and Its Effects on Size and Water Loss

Insects are smaller than vertebrates. The upper limit to their size is probably set by the way they breathe. Air is carried directly to the tissues by diffusion through a system of tubes, the trachea and tracheoles. These tubes, which are formed by invaginations of the exoskeleton, can constitute up to 50% of the total body volume. Insect hemolymph plays no role in oxygen transport. As an animal increases in size, its demand for oxygen increases very rapidly, somewhere between the square and the cube of the original value. The oxygen supplied by a tracheal system can increase only in direct proportion to the increase in size, depending as it does on the cross-sectional area of the tracheal tubes which cannot increase as fast as the insect's mass, and the fact that in a larger animal diffusion distances are increased and, consequently, oxygen supply is hampered. Although the largest and most active insects have evolved mechanisms of pumping the abdomen, ventilating the larger tracheal trunks, and consequently providing more oxygen than would be available by simple diffusion, there are still limits to the amount of oxygen that can be made available. Carbon dioxide, which diffuses much more easily than oxygen, is excreted by the tracheal system, which necessitates the system being exposed to the atmosphere at least some of the time. The internal surface area of the tracheal system is extremely large and, like all respiratory surfaces, must be kept moist. This presents for insects a serious problem of water loss, compounded by the fact that small animals have a disproportionately high external surface-area-to-weight ratio. Insects limit water loss by having an extremely impervious exoskeleton with an outer waxy layer and by having the tracheal system closed off from the external atmosphere much of the time by means of valves that open only when necessary. Nonetheless, water requirements are seriously affected by tracheal breathing. Certain insects require only metabolic water formed by the oxidation of dietary carbohydrates or fats; others have special features that allow them to take water from unsaturated ambient air (see Section 4.7).

2.2. Behavior

Limitations in size also limit the amount of nervous tissue an animal can possess, and since neurons of all animals are approximately the same size, insects have fewer nerve cells than vertebrates. The nervous system of insects lacks the large association areas that allow vertebrates to learn and to modify their behavior. Much of an insect's behavior is programmed into the system genetically. An insect cannot think; it reacts to specific stimuli in stereotyped and highly nonmodifiable ways. The behavioral reactions are modified by internal physiological variables (such as the state of hunger), by external variables (such as the physical state of the food), and, in some cases, by prior conditioning that might be considered a very crude kind of learning.

Many specific stimuli (most of them chemical) can affect the feeding behavior. As one would expect, the feeding behavior of phytophagous insects is quite different from that of hematophagous insects, both of which have been extensively studied. Little is known about the dietary requirements of predatory insects, although some of them can be reared on synthetic diets (Hagen and Tasson, 1966; Smith, 1965; Vanderzant, 1969b). We also know little about the nutrition of honeybees despite the great efforts spent in studying this insect (Haydak, 1970).

2.2.1. Food Selection in Phytophagous Insects

Early workers recognized that many insects vary widely in the range of plants they would accept as food: some were polyphagous, with extremely broad food tolerances; others were stenophagous, eating only a narrow range of plants; still others were oligophagous, accepting only one group of plants; and some, like the silkworm, were monophagous, eating only one plant species. It was generally believed that these restrictions in host range were the result of specific plants containing both specific stimuli and special nutritional factors required by the insect (Uvarov, 1928). The belief that the nutritional requirements of insects might differ from those of vertebrates in many significant ways was supported by the early findings that blowfly larvae required cholesterol as an essential nutrient (Hobson, 1935).

As the fields of biochemistry and nutrition developed in the 1930s and 1940s (see Fraenkel, 1959a; Chauvin, 1956; Trager, 1947, 1953; Wigglesworth, 1972), it became apparent that insects did not differ greatly among themselves in their qualitative nutrient requirements and that they showed fundamental similarities to vertebrates in their nutritional requirements. Although there were shown to be special requirements for sterol, choline, carnitine, and some other water-soluble growth factors, the close links between plant-feeding insects and their hosts obviously did not depend on the plants containing any exotic nutrient. The similarities in nutritional requirements led Fraenkel (1959b) to propose that food habits and host plant specificity are established by the presence of secondary plant chemicals specifically affecting the insect's feeding behavior by acting as attractants, repellants, and/or phagostimulants. He further postulated that the nutritional requirements of all phytophagous insects were identical and that any green plant would serve as a satisfactory source of nutrients if the insect could be induced to eat enough of it. This theory caused a shift in emphasis away from determinations of nutrient requirements to broader considerations of insect dietetics, including the effects on behavior of the secondary plant chemicals.

Modifications of Fraenkel's original theory became necessary when it was shown that, once the block to feeding was overcome, certain nonhost plant tissues supported much better growth and survival than did others (Waldbauer, 1962, 1964; House, 1961, 1969; Bongers, 1970). Work with grasshoppers, a broadly stenophagous insect, has shown that certain nutrients (sugars, ascorbic acid, thiamine, betaine, various amino acids, oxaloacetic and citric acids, and potassium hydrogen phosphate) stimulate feeding (Thorsteinson, 1960). Thorsteinson suggests that feeding is primarily stimulated by universally distributed plant substances including nutrients and water, and host plant ranges for insects that are oligophagous are determined by the presence or absence of factors that inhibit feeding (Thorsteinson, 1960). Dadd (1963) has added to the Thorsteinson theory by noting that many species of grasshoppers do not require taxonomically specific phagostimulants and that starvation causes acceptance of plants normally refused.

Currently, there is a considerable broadening of opinion as to how an enormous number of plant biochemicals regulate insect feeding. Besides embracing both the original concept of attractants and phagostimulants (both nutrient and token) and Thorsteinson's (1960) emphasis on repellants and phagoinhibitants, we now must take account of antifeedants (which diminish feeding via internal physiological routes rather than via external sensory perception) (Nunakata, 1970) and possibly other antimetabolic mechanisms such as enzyme inhibitors. The plant biochemicals involved in these relationships are termed allelochemics (see Harborne, 1978; Wood *et al.*, 1970).

2.2.2. Feeding in Hematophagous Insects

The habit of blood feeding has evolved independently in a wide range of insects. Differences in feeding mechanisms, sensory receptors, and host characteristics all affect feeding, but there are many features in common. Feeding results as the culmination of a series of stimulus–response events that include detecting the host, alighting on it, probing, piercing, or penetrating, locating blood, taking blood into part or parts of the gut, and ceasing feeding. Factors affecting feeding may determine individual events in the sequence or may modify parts or all of it. These factors have been reviewed (Friend and Smith, 1977; Galun, 1977), and there is evidence that experimental attempts to isolate subsets of the sequence often distort the feeding process (Friend, 1978, 1981).

Attempts to determine the nutritional requirements of hematophagous insects have been severely hampered by a lack of suitable chemically defined diets. As a food, blood is high in protein, water, and sodium, and relatively poor in B vitamins. Insects that are blood feeders throughout their life cycle depend on their ever-present symbiotes for vitamins (Lake and Friend, 1968) and use sodium to excrete water via the Malpighian tubules (phytophagous insects use potassium) (Maddrell, 1972). Normal blood feeding occurs only after piercing, and artificial diets are best presented covered by a suitable membrane through which feeding can take place (Friend, 1978, 1981; Friend and Smith, 1977).

The phagostimulants for all the hematophagous insects tested to date (six spp. of mosquito, two spp. of tsetse fly, two spp. of blood-sucking bug, the rat flea, and a tick) are ATP, ADP, or AMP (see Friend, 1978; Galun, 1977). The bug *Rhodnius prolixus* responds to the widest range of nucleotide phagostimulants. In descending order of potency, the compounds tested rank A(Tetra)P $>$ ATP $>>$ deoxyATP $>$ CTP $=$ ADP $=$ GTP $\geq$ CDP $\geq$ ITP $>$ cAMP $=$ UTP $\geq$ deoxyADP $\geq$ IDP $\geq$ GDP $>>$ AMP. Potencies range from 3.2 μM for A(Tetra)P and 3.8 μM for ATP to 0.63 mM for AMP; these values are doses that elicit 50% gorging in test populations (Smith and Friend, 1976). The rat flea, *Xenopsylla cheopis*, has the most restricted range tested to date; it only responds to ATP (Galun, 1966).

There is great interest in mass rearing of tsetse flies in order to apply the sterile male technique for control. Pig blood has been shown to be an adequate food, but the blood from other vertebrates and birds does not support complete development (Mews *et al.*, 1977).

2.3. Insect Exoskeleton

Unlike vertebrates, insects lack bones. The skeleton, which is external to the body, consists of chitin, an aminopolysaccharide of high molecular weight (see Chippendale, 1978). A matrix of arthropodin, a special protein, infiltrates the chitin. The degree of hardness of the exoskeleton depends on the amount of "tanning," or cross linking of the arthropodin molecules. Tyrosine is involved in this process. The amount of tyrosine required exceeds the amount of free tyrosine that could be held in solution in the blood. A much more soluble dipeptide, β-alanyl-L-tyrosine, is used as the tyrosine carrier in the fleshfly *Sarcophaga bullata* (Bodnaryk and Levenbook, 1969).

Protein deficiency affects the skeletal structure in insects. The honeybee, *Apis mellifera,* depletes exoskeleton proteins to maintain nitrogen balance; this results in a brittle integument and, finally, general paralysis (Butler, 1943). Imbalance or absence of certain dietary amino acids can affect cuticle formation. Excess tryptophan produced deformed heads and tarsi in *Drosophila melanogaster* (Hinton *et al.*, 1951), and lack of cystine caused abnormal pupal formation in *D. melanogaster* and *Phormia sericata* and considerable mortality during moulting in the cockroach *Blattella germanica* and during adult emergence in the mosquito *Aedes aegypti* (Lafon, 1938; Michelbacher *et al.*, 1932; House, 1949; Golberg and DeMeillon, 1948b). These cystine effects were probably caused by dietary imbalances possibly involving limitations of methionine and/or sulfate in the diet. No studies done after 1950 support the view that cystine is an essential amino acid (see Section 4.1).

Insect nutritional requirements are also affected by the lack of bone. The D vitamins, which function in vertebrates as antirachitic factors influencing calcium absorption and bone formation, are not required, and there is a low requirement for calcium. Calciferol has been tested in many insects as a substitute for sterol, always with negative results (Dadd, 1973).

2.4. Metamorphosis

An exoskeleton is mechanically advantageous in small animals because a tubular casing, such as occurs in an insect leg, provides much greater rigidity than a bone of equal weight. However, this rigid outer coating means that an insect must moult, shedding the old exoskeleton and producing a new larger one, in order to grow. Most insect species moult several times before attaining a mature size. At each moult, there is a temporary cessation of feeding, and many morphological changes take place.

There are two major types of metamorphosis. In incomplete metamorphosis, the immature forms, or nymphs, resemble the adults and feed in the same way. Representatives of this group used in insect nutrition studies include Orthoptera (grasshoppers and crickets), Blattoidea (cockroaches), Isoptera (termites); Hemiptera ("true bugs," e.g., the milkweed bug, *Oncopeltus fasciatus*, and the blood-feeder *Rhodnius prolixus*); and Homoptera (aphids and scale insects). Complete metamorphosis involves very drastic changes in form, feeding habits, and nutritional requirements. The immature form, be it beetle grub, fly maggot, or moth caterpillar, is specialized as an eating and growing machine. This specialization allows growth rates that far exceed any shown by vertebrates. Fly maggots can increase their biomass a thousandfold in 4 or 5 days (House, 1974). During the pupal stage, in which no feeding takes place, the larval tissues are drastically altered, and adult organs are formed. The adult stage is specialized for dispersal and reproduction. Some adults (mayflies and some moths) do not eat; many adult Diptera require only water and sugar; and as a general rule, the nutritional requirements of the adults that do feed differ markedly from those of the larvae (House, 1974). Nutritional studies on insects that undergo complete metamorphosis are usually restricted to larval requirements. Information is available on species of Coleoptera (beetles), Lepidoptera (butterflies and moths), Diptera (true flies), Siphonaptera (fleas), and Hymenoptera (ants, bees, and wasps).

2.5. Special Metabolic Features

Insects are uricotelic, synthesizing uric acid in the same way as birds (Bursell, 1967; Barrett and Friend, 1970). Glycine is used heavily in the formation of uric acid either directly as glycine in carbon atoms 4 and 5 or indirectly as formate in carbons 2 and 8 (Barrett and Friend, 1970). This may partially explain a requirement for glycine in many insects (See Section 4.1). Uric acid, with its low solubility and consequently low toxicity, is an extremely efficient molecule to use as an end product of nitrogen metabolism in animals where water conservation is a problem. The ability to produce "dry urine" is a major evolutionary adaptation for terrestrial life. Considering this, the similarities of uric acid metabolism found in such diverse groups as insects, birds, reptiles, and land snails become less surprising.

Another special feature of insect metabolism is that phosphoarginine (not phosphocreatine) is the energy-storing phosphagen in insect muscle. Analogues of arginine, homoarginine, and canavanine along with ornithine were investigated as possible metabolic antagonists in the boll weevil, *Anthonomus grandis*. Canavanine acted as a potentially reversible growth inhibitor, but homoarginine augmented the effects of arginine in reversing the canavanine effect (Vanderzant and Chremos, 1971). Canavanine does not inhibit reproduction in the fly *Pseudosarcophaga* (*Agria*) *affinis* (*housei*) (Hegdekar, 1970).

Insect hemolymph contains the disaccharide trehalose (α-D-glucopyranosylα-D-glucopyranoside) which acts as a storage form of glucose and is an important reserve carbohydrate in insects. The concentration of trehalose in the hemolymph ranges from 23 to 175 mM (8–60 mg/ml), depending on the species, developmental stage, and sex (Wyatt, 1967). The adaptive advantages provided by these high concentrations seem to be (1) few nonspecific glucosidations because of the masking of the aldehyde group, (2) lower osmotic effects than from equivalent glucose concentrations, and (3) a promotion of glucose absorption by facilitated diffusion (see Chippendale, 1978).

3. Techniques

Three main types of approach have provided most of our knowledge of the nutritional requirements of insects. The most direct technique is the classic deletion method, whereby one measures the effects of eliminating one specific component from a chemically defined (holidic) diet on which the insect can develop under sterile (axenic) conditions. A variation on the deletion method is the determination of nutrient utilizability by substituting various analogues within a class of essential nutrients. The third approach is the determination of the metabolic capabilities of the insect by supplying various precursor materials and later testing for the endogenous production of physiologically essential substances. Often isotopes are used to indicate metabolic conversions.

The limitations of each of these methods have been fully reviewed recently (see Dadd, 1977a) and consequently will be only briefly summarized here. The deletion method requires a holidic diet allowing good growth, development, and reproduction under axenic conditions. There are many problems associated with the development of such diets. Each nutrient must be present at a level that satisfies the requirements for structural materials and allows the efficient metabolism of other nutrients in the diet. Diet balance has been discussed by Gordon (1959), Sang (1959), and Friend (1968). Gordon (1959) states, "A deficiency of any one essential nutrient lowers the rate of utilization of many other nutrients (and so, in effect, lowers the nutritional requirements for them). The balance of essential nutrients is the dominant quantitative factor in any diet; it is likely that an organism must destroy surplus essential nutrients until they are restored to optimum balance with the most deficient essential nutrient." Because of this interdependence of nutrients, determinations of the optimal or minimal requirements for any one nutrient must take into account the other nutrients in the diet. This makes comparisons of the minimal or optimal levels of nutrients required by different species almost meaningless. The problem is further complicated by some larvae that, allowed to feed *ad libitum,* may adjust their total food intake to compensate for dietary imbalances (Sang, 1962; House, 1965c); also the great plasticity of the metabolic activities of insects can affect nutrient requirements. Sang (1959) states that he could probably select a strain of the blowfly *Phormia regina* that would have the same vitamin requirements as the fruit fly *Drosophila melanogaster,* although these animals differ markedly in the food that they normally eat.

Utilizability studies, which attempt to determine the molecular structures necessary to support metabolism in any particular class of nutrient, have primarily been applied to the sterols and carbohydrates. Because it is difficult to exclude some utilization of nonrequired dietary components, detailed metabolic studies are sometimes necessary to fully establish the nutritional role played by various analogues of required nutrients.

Metabolic studies have relied heavily on the use of appropriate radioactive precursors which are fed or injected; after a suitable period to allow for metabolic cycling, the distribution of the label is determined. Heavily labeled end products are usually considered nonessential, and unlabeled molecules are usually considered to be beyond the metabolic capabilities of the organism and consequently must be supplied from an extrinsic source. Isotope studies have confirmed deletion studies and provide the only information on qualitative requirements when the deletion technique cannot be applied because of the lack of suitable diets or axenic rearing conditions.

The difficulties in nutritional research techniques (see Dadd, 1977a) apply to studies of vertebrate nutrition as well as insect nutrition, but a major difference between the two fields exists. Most work on vertebrates is done by groups of people working in many laboratories on relatively few species; consequently, confirmatory information is readily available. In contrast, many of the insect species have been studied in only one laboratory, often by a single individual, so confirmatory checks are often lacking.

An attitudinal factor also differentiates the fields of insect and vertebrate nutrition. Many of the nutritional studies done on insects are motivated by a desire to "know the enemy" so that one can control or destroy pests. This leads to development of food supplies that are satisfactory for vertebrates yet not phagostimulatory to insects or able to support their normal metabolism or development. This is a far cry from the beef growers' goal of maximum weight gain in minimum time with minimum food cost.

4. Nutritional Requirements

4.1. Amino Acids

Amino acid requirements are best determined by the classical deletion method using axenic insects and chemically defined diets in which the protein component has been replaced by a suitable mixture of L-amino acids. The composition of the amino acid mixture is often based on the relative concentrations of the amino acids found in casein or in the insect's natural food (Dadd, 1977a).

Recently Dadd (1978) demonstrated that a mixture in which all the amino acids (except for tyrosine which is rather insoluble) were present in equal proportions by weight would support good growth of larvae of the mosquito *Culex pipiens*. Dadd's mixture also contained glutamine and asparagine, fortuitous inclusions since asparagine was shown to be essential. Some insects such as the stored products beetle *Tenebrio molitor* (Davis, 1971; Leclercq and Lopez-Francos, 1966, 1967), and the flea *Xenopsylla cheopis* (Pausch and Fraenkel, 1966) do not grow well when amino acid mixtures are substituted for proteins.

Radioisotope methods have been used extensively to determine the essential amino acids for many insects lacking satisfactory synthetic diets. A suitable precursor, such as [U-^{14}C]glucose, is fed or injected; after an interval to allow for metabolic cycling, the presence or absence of the label on various amino acids is determined. Heavily labeled amino acids are considered nonessential, and unlabeled amino acids are usually considered essential. Dadd (1977a) notes that, in addition to the general problems associated with radioisotope methods discussed earlier, amino acids present special problems. Tryptophan is destroyed during the hydrolysis of the extracted protein. Tyrosine can usually be synthesized from phenylalanine which will pick up little or no isotope because it is essential; consequently tyrosine synthesized in this way would contain no label. Because of the uncertainties associated with the isotope method, Dadd (1977a) uses the terms synthesized and nonsynthesized instead of nonessential and essential to describe the amino acid results obtained by this method.

All aposymbiotic insects to which the deletion technique has been applied have been shown to require the same ten amino acids that are required for growth by the rat: arginine, histidine, isoleucine, leucine, lysine, methionine, phenylalanine, threonine, tryptophan, and valine. Of the 24 species listed by Dadd (1977a) and reported subsequently, the deletion technique has shown that nine species (five Diptera, three Coleoptera, and one Lepidoptera) require amino acids other than the "rat ten." In addition, the mosquito *Culex pipiens* requires asparagine (Dadd, 1978), and it is of interest that pregnant and weanling rats may also require this amide in their food (Newburg *et al.*, 1975; Newburg and Fillios, 1979). Proline is essential for some strains of the blowfly *Phormia regina* (Cheldelin and Newberg, 1959; Rock *et al.*, 1975) in spite of the fact that proline synthesis by this insect has been demonstrated by the use of isotopes (Kasting and McGinnis, 1958, 1960). The screwworm fly *Cochliomyia hominivorax* (Gingrich, 1964) and another fly *Agria housei* (House, 1954) also require proline. Lack of proline affected larval growth of the mosquitoes *Aedes aegypti* (Lea and Delong, 1958; Singh and Brown, 1957) and *Culex pipiens* (Dadd, 1978) to the extent that it is considered essential. The silkworm *Bombyx mori* is the only nondipteran shown to require proline thus far. Proline biosynthesis does occur in this species but too slowly to support normal growth (Arai and Ito, 1967; Inokuchi *et al.*, 1967, 1969).

Glycine may be essential for certain insects as it is for chicks. Certain workers (Lea and Delong, 1958) consider it essential for the mosquito *A. aegypti,* whereas others (Singh and Brown, 1957) detected no effect when it was omitted from their amino acid mixture. Omission of glycine markedly retarded development of larvae of another mosquito, *Culex pipiens,* but the animals survived to become adults (Dadd, 1978). It is apparently required by aposymbiotic *Stegobium paniceum* beetles (Pant *et al.,* 1960). There are other reports that indicate that deletion of glycine has detrimental effects on insects developing on chemically defined diets (see Dadd, 1973; House, 1974; Friend, 1968), but it has not been shown to be absolutely essential. Dadd (1978) concludes that the beneficial effects of glycine probably depend on its ability to adjust nutrient imbalances by acting as a precursor for nonessential amino acids. Glycine has been shown to ameliorate the toxic effects of certain D-amino acids when racemic mixtures were used (see Hinton *et al.,* 1951; Friend, 1968; House, 1974).

The fly *P. regina* (Cheldelin and Newberg, 1959) and the silkworm *B. mori* (Ito and Arai, 1966) require either aspartic or glutamic acids in conjunction with the ten essential amino acids for full growth. The aphid *Myzus persicae* requires large supplements of glutamic acid, alanine, or serine in addition to the essential amino acids, plus cysteine for good growth (Dadd and Krieger, 1968). Addition of aspartic or glutamic acid improved the growth rates of the flour beetle *Tribolium confusum* (Naylor, 1963) and the red-banded leafroller *Argyrotaenia velutinana* (Rock and King, 1967a,b).

There are many other reports that additions of nonessential amino acids, particularly aspartic and glutamic acids, to mixtures of the ten essential ones resulted in improved growth and development (see Dadd, 1973, 1977a; Friend, 1968; House, 1974). The beneficial effects probably result from these nonessential amino acids supplying some of the necessary amino groups for the formation of other amino acids required for tissue formation but absent from the amino acid mixture. This process would involve transamination reactions that require glutamic and aspartic acids; consequently, the improvement of growth rates observed when these acids are supplied is easily understood.

To date, only the mosquitos *C. pipiens* and *Culiseta incidens* have been shown to require asparagine. Asparagine is not required for *A. aegypti,* the other mosquito whose nutrition has been studied intensively (Dadd, 1978; Dadd *et al.,* 1980). Dadd (1978) notes that certain types of leukemic mammalian cells lack asparagine synthetase, the enzyme that converts aspartic acid to asparagine, and consequently, the amine must be supplied as it must for certain tissue-cultured cell lines of the fruit fly *Drosophila melanogaster.* Lack of this enzyme in some or all of the tissues of *C. pipiens* or *C. incidens* would result in the demonstrated requirement for preformed asparagine.

Certain aphids containing intracellular symbiotes can use inorganic sulfate to replace or spare certain amino acids. In *Myzus persicae,* it can replace cysteine (Dadd and Krieger, 1968); in *Neomyzus circumflexis,* it can completely replace both methionine and cysteine (Ehrhardt, 1969); and it can spare but not replace both methionine and cysteine in *Acyrthosiphon pisum* (Markkula and Laurema, 1967; Retnakaran and Beck, 1968). Inorganic sulfate neither spares nor replaces methionine or cysteine in *Aphis gossypii* (Turner, 1971) or in the red-banded leafroller *Argyrotaenia velutinana* (Sharma *et al.*, 1972). The degree of nutritional independence that the presence of symbiotes confers is also emphasized by the observation that *M. persicae* with symbiotes requires only histidine, isoleucine, and methionine for development (Dadd and Krieger, 1968), whereas aposymbiotic *M. persicae* requires the usual ten essential amino acids (Mittler, 1971). The complications introduced by symbiotes constitute an area of nutrition that is as confusing as it is specialized, and it is beyond the scope of this review.

4.2. B Vitamins and Other Water-Soluble Growth Factors

Insects' requirements for B vitamins and other water-soluble growth factors such as choline, carnitine, ascorbic acid, inositol, and nucleic acids vary depending on the insect species and whether or not the animals were aposymbiotic. There is ample evidence to show that the vitamin requirements of insects that normally have specific symbiotic microorganisms (e.g., aphids, blood-sucking hemipterans, cockroaches, certain stored products beetles, and grasshoppers) increase when these symbiotes are eliminated. Determinations of vitamin requirements are also affected by initial reserves in the egg or newly hatched larvae in which the effects of vitamin deficiency do not appear until the later stages of development. In some cases, vitamin requirements can only be shown if the species is reared for more than one generation on the deficient diet. Interpretation of some of the early work is complicated by the possibility that the casein or yeast fractions used might have been contaminated with low levels of vitamins. Even the highly refined diets developed by House (1954a,b) for the fly *Agria housei* (*Pseudosarcophaga affinis*), on which no requirement for pyridoxine could be demonstrated, were later shown to contain traces of this vitamin sufficient to satisfy an extremely low requirement (Barlow, 1962). Perhaps the claim that the screwworm *Cochliomyia hominivorax* can develop without pyridoxine (Gingrich, 1964) has a similar explanation. Dadd (1977a) notes that the requirements for folic acid, biotin, and B_{12} are minute and could easily be masked by their presence as contaminants.

In general, aposymbiotic insects tested on chemically defined diets require choline and, with the exception of B_{12}, all of the B vitamins (thiamine, riboflavin, nicotinamide, pyridoxine, pantothenate, folic acid, and biotin) (see Dadd, 1973, 1977a; House, 1974). Many species, particularly plant feeders, require inositol at much higher levels than the B vitamins. The effects of inositol deficiency appear late in development and sometimes are not manifested until at least one generation of deprivation. This requirement is not restricted to any particular insect order; it has been shown necessary for plant-feeding Orthopterans (Dadd, 1963), Coleoptera (Galford, 1972; Vanderzant, 1963; Vanderzant and Richardson,

1964), Homoptera (Dadd *et al.*, 1967; Ehrhardt, 1968a), and for several Lepidoptera (Horie *et al.*, 1966; Kasting and McGinnis, 1967; Vanderzant, 1968); however, other leaf-feeding lepidopterans have been shown to have no requirement for inositol (Ouye and Vanderzant, 1964; Rock *et al.*, 1964). The stored products beetle *Tribolium castaneum* requires it (Applebaum and Lubin, 1967), but a closely related species *T. confusum* does not. No Dipterans studied to date have been shown to need inositol (Friend, 1968), but these studies did not include any true plant feeders and were conducted for only one generation.

Like inositol, the lipogenic growth factor choline is required in amounts that greatly exceed the requirements for B vitamins; this suggests that in addition to whatever catalytic role it may play, choline is probably also forming important structural elements as components of phospholipids, phosphatidyl cholines, and phosphatidyl inositols (Fast, 1970). All insects studied critically have demonstrated a requirement for choline. Several compounds with quaternary ammonium or ethanolamine configurations such as 2-, 2-methyl-, and 2-dimethylaminoethanols; mono-, di-, and triethylcholines; sulfocholine; β-monomethyl- and α,α-dimethylcholines; homocholine; carnitine and betaine aldehyde (but not betaine) have been shown to spare the choline requirement in *Drosophila melanogaster* (Geer and Vovis, 1965; Geer *et al.*, 1968); but only the various choline esters functioned as well as choline. The metabolic implications of the choline requirement in insects have been discussed by Dadd (1973).

Carnitine, a compound closely resembling choline, is required as well as choline by beetles of the family Tenebrionidae. Lack of carnitine produces symptoms of fatty degeneration and eventually death, conditions also produced when the choline analogue γ-butyrobetaine, which is metabolically antagonistic to carnitine, is included in diets containing carnitine (Naton, 1967). Insects that do not require carnitine are not so affected by butyrobetaine. Factors influencing the production of carnitine deficiency symptoms in *T. molitor* included the type of casein used, the strain of the species, and whether zinc and potassium were limited in the diet (Fraenkel, 1958).

Gilbert (1967) postulated that carnitine functions in the activation of intermediates in fatty acid oxidation and in the biosynthesis of phospholipids. Carnitine effects in species other than the Tenebrionidae have been shown. The beetle *Oryzaephilus surinamensis* requires it for optimum growth and pupation (Davis, 1964). It is also required for maximum larval growth and adult activity in *D. melanogaster;* the larvae can synthesize a suboptimal quantity of this molecule (Geer *et al.*, 1971).

In recent years, many species of insects that eat fresh plant tissues have been shown to require ascorbic acid, including various Orthopterans (Dadd, 1963; Nayar, 1964), Lepidopterans (Chippendale and Beck, 1964; Chippendale *et al.*, 1965; Ito and Arai, 1965; Levinson and Navon, 1969; Reddy and Chippendale, 1972; Rock, 1967; Vanderzant and Richardson, 1963; Vinson, 1967), Coleopterans (Vanderzant *et al.*, 1962; Wardojo, 1969), and Homopterans (Dadd *et al.*,

1967; Ehrhart, 1968a). Other phytophagous species do not require vitamin C (Levinson and Gothilf, 1965; Rock, 1967; Vanderzant and Richardson, 1964), and no insect that does not normally eat fresh plant food has been shown to require it. The requirement for an exogenous source of ascorbic acid apparently arose in insects, as it did in the vertebrates, as a result of some species that normally eat food with a high ascorbic acid content losing the ability to synthesize this molecule. In vertebrates that require ascorbic acid (primates, guinea pigs, fruit-eating bats, and certain frugivorous birds), the ultimate enzymatic step in the conversion of hexose sugar via D-glucuronolactone, L-gulonolactone, and 2-ketogulonolactone to L-ascorbic acid is lost. These lactones were ineffective substitutes for ascorbic acid in the silkworm *Bombyx mori* (Ito and Arai, 1965), the Egyption cotton leafworm *Prodenia litura* (Levinson and Navon, 1969), and the aphid *Myzus persicae* (Mittler *et al.,* 1970), suggesting that these insects also have a terminal block in biosynthesis. Oxidized (L-dehydro-) ascorbic acid, which is antiscorbutic in mammals, is only half as effective for growth of *B. mori* and *M. persicae*. D-araboascorbic acid is as effective as ascorbic acid in *B. mori* and *M. persicae,* but, as in mammals, it is not utilized by *P. litura*.

The development of chemically defined diets for phytophagous insects became possible about two decades ago after it was realized that ascorbic acid was required by many species and that special precautions had to be taken to prevent it from being destroyed by oxidation (Dadd *et al.*, 1967; Ito and Arai, 1965). Diets must be frequently changed or high concentrations of the vitamin supplied to allow for its oxidation (Chippendale and Beck, 1964).

Ribonucleic acid or its components have been shown to act as growth-promoting substances for almost all species of Diptera tested critically (Friend, 1968; Dadd, 1973, 1977a), the exception being *Phormia regina* (Brust and Fraenkel, 1955). The effects on larvae range from a requirement for optimal growth to an absolute requirement in order to complete development. Biosynthesis of purine and perhaps other nucleic acid components is apparently the limiting reaction for dipterous larvae. The following species grow slowly in the absence of nucleic acid but are able to complete their developments: *Musca domestica* (Brookes and Fraenkel, 1958), *Aedes aegypti* (Singh and Brown, 1957; Lea *et al.,* 1956; Akov, 1962), *Agria affinis* (= *Agria housei*) (House, 1954, 1964; House and Barlow, 1957), and various species and strains of *Drosophila* (Hinton *et al.,* 1951; Hinton, 1956, 1959; Sang, 1956; Royes and Robertson, 1964; Falk and Nash, 1972). The mosquito *Culex pipiens* (Dadd and Kleinjan, 1977), the screwworm *C. hominivorax* (Gingrich, 1964), and certain RNA-auxotrophic strains of *Drosophila* (Geer, 1964; Falk and Nash, 1972) fail to complete development without RNA in the diet. The dependance on RNA of various strains of *Drosophila* is influenced by dietary levels of proteins, certain amino acids, and folic acid, which are known to be involved in nucleic acid biosynthesis (Sang, 1959; Geer, 1963, 1964).

Components of RNA have been tested as replacements for this nucleic acid, but no common pattern of requirements has emerged. Mutant *Drosophila* show various degrees of requirements for adenosine and/or one of the pyrimidine nucleosides, cytidine or uridine, and a few mutants specifically require the purine nucleoside guanosine in addition to, or in place of, adenosine (Norby, 1971; Vyse and Sang, 1971; Falk and Nash, 1972, 1974). A requirement for thymine or thymine-containing compounds has not been demonstrated. Thymine is required by *C. pipiens,* and in this it differs from other Diptera tested this far. To substitute for RNA, *C. pipiens* requires the purine ribonucleotide adenylic acid; a pyrimidine ribonucleotide, either cytidylic or uridylic acid; and the pyrimidine deoxyribonucleoside thymidine (Dadd and Kleinjan, 1977; Dadd, 1979). *Agria affinis* (= *housei*) can utilize only nucleotides; any one or any combination of the four constituent ribonucleotides of either RNA or DNA replaced nucleic acid, but nucleoside bases or combinations thereof were unable to alleviate nucleic acid deficiency (House, 1964). Combinations of adenine, guanine, and cytosine or their corresponding nucleotides adenylic, guanylic, and cytidylic acids largely satisfied the nucleic acid requirement of *C. hominivorax;* uracil or uridylic acids were strongly inhibitory (Gingrich, 1964). The requirements for nucleic acid components in *M. domestica* were satisfied by adenine and guanine, although these tests were complicated by an interdependence between folic acid and nucleic acid requirements (Brookes and Fraenkel, 1958). *Musca domestica* females reared by conventional methods required dietary nucleic acid as adults to maintain egg production beyond the first ovarian cycle (Morrison and Davies, 1964).

RNA may benefit insects other than Diptera. Growth and survival of the beetle *Oryzaephilus surinamensis* were improved by dietary addition of RNA or a combination of guanine and cytosine (Davis, 1966), as was the growth of another beetle, *Tribolium castaneum* (Hogan, 1972).

4.3. Lipid Growth Factors

Perhaps the most noteworthy difference between the dietary requirements of insects, and indeed of arthropods in general (Dadd, 1977b), and those of vertebrates is the need for a simple sterol. Hobson's (1935) discovery of the blowfly sterol requirement marks the inauguration of precise nutritional study of insects, for it is now evident that until the necessity of dietary sterol was appreciated, no insect, other than those with symbiotes, could maintain good growth on chemically defined diets.

For a decade following Hobson's discovery, no further lipid requirements came to light, although several lipid growth factors already found necessary for larger animals (Vitamins A, D, E, and polyunsaturated fatty acids) were frequently tested on the increasing number of insect species for which good semi- or (nearly) fully synthetic diets became available. This lack of response to vertebrate lipid growth factors may have reflected the types of insect used. Many were dipterous larvae, which grow very well through one cycle of larval development with no lipid other than sterol. Others were mainly stored products beetles, moths, and cockroaches, reared on semisynthetic, casein-based dry diets. Often, their growth was known to depend on symbiotes or microorganisms adventitiously present even in dry diets which could have been providing lipid growth factors; and again, studies rarely extended beyond larval development. Some of these insects did grow better with natural oils or fats in the diet (Trager, 1953), but since the composition of such crude lipids was then imperfectly known, and since the lipid seems largely interchangeable with carbohydrate, such growth improvements tended to be ascribed to optimization of the balance of energy-producing nutrients. Whatever the case, during this period it was widely held that insects differed from vertebrates nutritionally in requiring no lipid nutrients other than a sterol.

This simple distinction faded with the discovery that certain flour moths, if reared as larvae on food lacking certain natural oils, suffered characteristic malformations when attempting to emerge from the pupal integument, even though they grew well up to pupation without oil (Fraenkel and Blewett, 1946b, 1947). Polyunsaturated fatty acids of the linoleic and linolenic families proved to be the essential lipid nutrients for the flour moth, as had been discovered for vertebrates.

The discovery of this insect fatty acid requirement would have emerged less readily had the deficiency not involved a characteristic morphological abnormality expressed at a very specific stage of development. Manifesting as it did only after the larval feeding and growing stages had been completed satisfactorily, it showed that other essential requirements might remain covert during the single larval growth cycle on a deficient diet that was then and still is the norm. Because insect eggs are often relatively large compared to the mature adult, it is probable that they carry maternally derived micronutrients sufficient for substantial or complete larval growth, especially so if the ultimate physiological requirement is primarily at a late developmental stage. Much information on insect lipid requirements has been obtained only in terms of dysfunction in the adult or subsequent generation. All insects studied so far require sterol, and a majority of those studied critically have been found to require unsaturated fatty acid. The determination of the particular sterols and fatty acids utilized by various insects and their metabolic interrelations is currently a very active area of insect nutrition. Many attempts to establish the requirement for other lipid factors have failed, perhaps because there are none, but possibly because of the difficulties of multigeneration studies.

4.3.1. Essential Fatty Acids

Polyunsaturated fatty acid requirements have been found for most of the approximately 50 species, from five orders, studied in this respect (Dadd, 1977a). Other species that needed dietary oil but were not tested with fatty acids as such are excluded from this count, as are the several aphid species reared for sequential generations on lipid-free diets, because of the possible lipogenic activities of their intracellular symbiotes.

Most Lepidoptera reveal fatty acid deficiency dramatically by failure of the pupal/adult ecdysis, as in the classic study of *Ephestia* flour moth (Fraenkel and Blewett, 1946b, 1947); development to the pharate adult is usually completed, but the adult fails to breach the pupal cuticle, emerging incompletely or with deformed legs; larval growth may be retarded but not always so. The critical physiological need for fatty acid at metamorphosis is confirmed by the complete alleviation of pupal/adult failure in some Lepidoptera when linolenic acid was provided only during late larval development (Rock *et al.*, 1965). As in Lepidoptera, deficiency manifests as pupal/adult failure in some Hymenoptera (Yasgan, 1972). Among Orthoptera, although hemimetabolous and lacking a pupal stage, grasshoppers also express fatty acid deficiency by the emergance of deformed adults at the final moult, sometimes preceded by markedly retarded nymphal growth (Dadd, 1963). However, another orthopteran, the cockroach *Blattella germanica*, showed no ill effect from deficiency imposed throughout larval development, but resulting females produced deformed oothecae or a second generation of weak, short-lived nymphs (Gordon, 1959). The Coleoptera shown to require essential fatty acid reveal the deficiency mainly by slow larval growth and decreased adult fecundity, with two generations required for this to become fully expressed in the boll weevil (Earle *et al.*, 1967).

These latter cases particularly emphasize the difficulties that may beset attempts to demonstrate essential fatty acid need. Most species that appear to have no such requirement are Diptera and Coleoptera studied before or about the time of the work with *Ephestia* that first demonstrated the requirement in an insect; the presumption of nonrequirement is thus based on good development to adulthood on diets without fat but with no information sought on adult or subsequent generational performance. Recent reexamination of the moth *Plodia interpunctella* (Morere, 1971a), the mosquito *Aedes aegypti* (Sneller and Dadd, 1981; Dadd, 1981), and the beetle *Tenebrio molitor* (Davis and Sosulski, 1973) now indicates that they may have a need for essential fatty acid. Multigenerational growth studies with the purer dietary ingredients now available may reveal the requirement to be general for insects.

Many of the foregoing studies with various insects indicated that either linoleic or linolenic acids adequately satisfied the requirement. This was so in the groundbreaking work with *Ephestia*, which also indicated that linolenic acid was more potent on a dosage basis than linoleic, a difference subsequently observed with another lepidopteran, *Pectinophthora gossypiella* (Vanderzant *et al.*, 1957). Especially in view of the questionable purity of fatty acids then available, the significance of such differences in potency was unclear, although suggestive of subtle functional differentiation within the overall polyunsaturated fatty acid requirement. Subsequent work with several Lepidoptera revealed that only linolenic acid could avert failure at pupal/adult ecdysis (e.g., Chippendale *et al.*, 1965; Turunen, 1974), though linoleic acid was sometimes independently necessary for an optimal larval growth rate; however, a recent study of the tea tortrix moth indicates that both linolenic and linoleic acids are necessary for normal adult emergence (Sivapalan and Gnanapragasam, 1979). A recent review (Dadd, 1981) notes that of the 18 species of Lepidoptera examined in this respect, three had no apparent requirement, nine could utilize either linolenic or linoleic (for three of which, however, linolenic was more potent), five could utilize only linolenic, and one required both fatty acids. As additional lepidopteran species are studied using the high-purity fatty acids now available, ever more are found to have a specific linolenic requirement, and Turunen (1974) surmises that this will prove general for the order.

Recently, the mosquito, *Culex pipiens* was shown to require arachidonic acid or certain structurally related long-chain polyunsaturates in the larval diet for teneral adults to fly and survive normally. Neither linoleic nor linolenic acids was able to satisfy the requirement, although these and other structurally related acids were considered semiactive. Flight-active fatty acids had in common a structure of three *cis* double bonds in divinyl methane rhythm terminating six carbons from the methyl end of the fatty acid carbon chain (the $\omega6$ position); semiactive fatty acids lacked the first of the group of three double bonds. In both cases, the presence of an additional double bond at the methyl end seemed immaterial, and so both active and semiactive fatty acids included both $\omega6$ and $\omega3$ members (Dadd and Kleinjan, 1979a; Dadd, 1980). Arachidonic acid appears to be an essential fatty acid for another five species of mosquito (Dadd *et al.*, 1980; Sneller and Dadd, 1981; Dadd, 1981); it may prove to be a general mosquito requirement.

These mosquito findings contrast with prior work on insect fatty acids in two important respects: as noted above, no Diptera have been found with any need for polyunsaturated fatty acids in the diet; and second, no other insects tested have been shown to require arachidonic acid. On the other hand, this essentiality is reminiscent of the centrality of arachidonic acid in the essential fatty acid requirement for metabolism and physiological function of vertebrates and thus offers some prospect of integrating the insect and vertebrate situations in some more general animal scheme. To assess current knowledge of insect fatty acids against the relatively well-understood vertebrate perspective, we offer the following brief synopsis compiled from recent reviews (Alfin-Slater and Aftergood, 1971; Guarnieri and Johnson, 1970; Holman, 1977; Lands *et al.*, 1977; Sprecher, 1977).

For warm-blooded vertebrates, the primary essential fatty acids are those of the linoleic series, which have *cis* double bonds in divinyl methane rhythm terminating on the sixth carbon from the methyl end, and hence termed $\omega6$ (or $n6$) fatty acids. The physiologically essential member is arachidonic acid ($\Delta5,8,11,14\text{-}C_{20:4}$, or 20:4$\omega6$) which, e.g., for rats, is also the most potent dietary fatty acid. Most mammals and birds can satisfy their requirement with the parent member of the ω6 series, linoleic acid ($\Delta 6,12\text{-}C_{18:2}$ or 18:2ω6), which they metabolize by chain elongation and further desaturations at the carboxyl end to the physiologically essential arachidonic acid, any of the intermediates, such as γ-linolenic or homo-γ-linolenic (18:3ω6 and 20:3ω6) also being fully adequate. An exception among mammals is the cat which, lacking appropriate desaturases, cannot derive arachidonic acid from linoleic acid and so must have long-chain polyunsaturates preformed in the food (Rivers *et al.*, 1975), a dietary situation analogous to that of mosquitos. Many, but not all, effects of fatty acid deficiency can be averted by dietary linolenic acid (18:3ω3), the parent member of the ω3 family whose double bond sequences terminate three carbons from the methyl end. Dietary linolenic acid is also elongated and further desaturated, giving e.g., 20:3ω3, 20:5ω3 and 22:6ω3, the last two of which, eicosapentaenoic and docosahexaenoic acids, are the characteristic long-chain fatty acids of many fish, for which class of vertebrates the linolenic rather than linoleic family of fatty acids is essential (Tinoco *et al.*, 1979). In passing, we note that the linolenic requirement of fish among vertebrates and of Lepidoptera among insects presents an interesting analogy.

In vertebrates, arachidonic acid or similar polyunsaturates are essential components of the phospholipids of cellular membranes, on the integrity of which normal membrane enzymatic and permeability functions depend. Also, arachidonic acids (and to a less important extent, homo-γ-linolenic and eicosapentaenoic acids) are precursors of prostaglandins, hormonelike entities found in most tissues and involved in the regulation of diverse functions. The gross abnormalities of vertebrate fatty acid deficiency such as poor growth, sterility, dermal lesions, and increased integumental water loss are thought to reflect these cellular biochemical abnormalities consequent on reduced phospholipid polyunsaturates. The essentiality of essential fatty acids arises basically from the fact that linoleic and linolenic acids, parent members of the ω6 and ω3 polyunsaturates, cannot be biosynthesized *de novo* or from other saturated or monoenoic fatty acids.

Many studies of fatty acid metabolism in diverse insects (Gilbert, 1967; Fast, 1970; Downer, 1978) show that, as in vertebrates, the ability to biosynthesize linoleic and linolenic acids is lacking, accounting for the essential fatty acid requirement. By analogy with the vertebrate situation, it is generally thought that the physiological need for unsaturated fatty acids involves questions of physical characteristics of phospholipids, although with little actual evidence beyond a predominance of linoleic and linolenic acids in phospholipids as compared to triglycerides. Are we then to assume that those species for which no fatty acid requirement could be demonstrated, sometimes taxonomically close to species with a requirement, have fundamentally different membrane or other phospholipid physiological functions? Possibly. But it seems more likely that in these cases, deprivation of dietary fatty acid was studied over too restricted a period of growth and development to exhaust reserves and allow symptoms of deficiency to appear.

With respect to subsequent metabolism of dietary linoleic and linolenic acids in insects, virtually nothing is known. The fact that the mosquito dietary requirement for arachidonic acid and related polyunsaturates resembles the vertebrate physiological requirement raises the question of whether linoleic and linolenic acids are required by the generality of insects as the actual physiological entities or as precursors of higher polyunsaturates and their metabolites, as is the vertebrate pattern. Fatty acid analyses of insect tissues, now available for scores of species (Fast, 1964, 1970), are notable for the almost complete absence of recorded fatty acids of longer chain length and greater unsaturation than linolenic acid. However, these negative findings were recently called in question on the basis of gas-chromatographic and bioassay evidence of low or trace levels of long-chain polyunsaturates in mosquitos and many other insects when these were specifically looked for with appropriately sensitive methods (Dadd, 1981; Stanley-Samuelson and Dadd, 1981; Stanley-Samuelson, 1980). Furthermore, prostaglandins were recently detected in crickets and the silkworm *Bombyx mori* (Destephano and Brady, 1977; Settya and Ramaiah, 1979, 1980), implying the presence, as a necessary precursor, of arachidonic acid. In the silkworm case, this particularly suggests the metabolism of arachidonic acid from linoleic acid, since silkworm food, mulberry leaves, would not provide dietary arachidonic acid.

It remains a stumbling block to the hypothesis of a general physiological need for higher polyunsaturates in insects that those few species tested with arachidonic or docosahexaenoic acid as substitutes for linoleic or linolenic acids gave negative results, in strong contrast to the corresponding position in vertebrates. It has been argued (Dadd and Kleinjan, 1979b; Dadd, 1981) that these negative findings, made two decades ago, perhaps using fatty acids of uncertain purity and with a high probability of oxidative loss during experimental runs of long duration, should not be taken as definitive unless rechecked using precautions against oxidation of test polyunsaturates. One of the species previously tested with arachidonic acid, the waxmoth *Galleria mellonella,* was recently reexamined using antioxidant-protected fatty acids offered in the diet only during late larval development (R. H. Dadd, unpublished data). The results confirmed the nonutilization of arachidonic acid and also of docosahexaenoic acid, not previously tested. Contrary to the previous study (Dadd, 1964), a difference was now found between the effects of linolenic and linoleic acids, linolenic being at least tenfold more potent. Most interesting, although none of four other $\omega6$ polyunsaturates was more effective than linoleic acid, two other $\omega3$ fatty acids, $C_{20:3}$ and $C_{22:3}$, could effectively substitute for linolenic acid, showing that the basic requirement was not for linolenic acid specifically but for a class of $\omega3$ fatty acids having less than the six double bonds of $C_{22:6}$. Although this leaves the question of a general insect physiological requirement for long-chain polyunsaturates murkier than ever, it does indicate that the problem is worth further investigation.

4.4. Sterol Nutrition and Metabolism

Many insects depend intimately on microorganisms to eke out the nutrients provided by otherwise inadequate food, and such relationships have been shown to include the provision in whole or in part of essential sterols (Ehrhardt, 1968c; Noda *et al.*, 1979). This aside, the requirements for dietary sterol found in all insects lacking steroidogenic symbiotes reflect the inability of arthropods to biosynthesize these chemicals and sets this phylum starkly apart, nutritionally, from the vertebrates, which biosynthesize whatever sterol they need from acetate. In spite of this, both phyla have analogous physiological needs for sterol in lipid biostructures and as steroid hormone precursors. Insects can synthesize many other isoprenoid chemicals (juvenile hormones, ubiquinones, defensive terpenoids, etc.) that share a common acetate-mevalonate-farnesyl phosphate biosynthetic pathway, so the biosynthetic branch route to sterols, as it is understood from vertebrate metabolism, must be broken, probably at multiple sites, from squalene cyclization onwards. Nevertheless, among sterols proper, interconversions are variously available to insects, and sterol nutrition has been largely concerned with documenting the range of dietary sterols that can be utilized by various species with full or partial efficiency and elucidating the metabolic steps in the derivation of those tissue sterols essential for physiological function from whatever sterols are available to particular insects in their food, be this a natural or synthetic diet.

The range of dietary sterols that satisfy the requirement has been documented for upwards of 40 species (Dadd, 1977a) and is discussed in several reviews on general insect nutrition cited in Section 1. Both nutritional and metabolic information available up to 1963 was surveyed exhaustively and critically in the important review of Clayton (1964), by which time it was established that the minimal structural features for utilizability in all insects are a closed planar ring system, a side chain at C-17 of 10–12 carbons, and a 3-β-hydroxyl function that can be variously esterified without loss of activity. Since then, work on sterol metabolism in particular has grown apace with the availability of increasingly refined analytical techniques, and numerous reviews now cover this burgeoning field (Robbins *et al.*, 1971; Thompson *et al.*, 1972; Svoboda *et al.*, 1975a, 1978). In attempting to impose some coherance on this currently very active and evolving field, we draw heavily on this series of reviews.

From the early growth studies, it appeared that all insects could satisfy their sterol requirement completely with only the typical animal sterol, cholesterol, in the diet, an eminently understandable situation when it was assumed that insects, like other animals, would have cholesterol as their main functional tissue sterol, though other sterols might be present adventitiously. However, if it were true that the main tissue sterol of insects should be cholesterol, what then of the situation of phytophagous insects ingesting various plant sterols but no cholesterol, thought at the time to be entirely absent from plants. Clearly, if they required cholesterol physiologically, they would need to dealkylate the ingested C_{28} and C_{29} plant sterols to C_{27} cholesterol. On the other hand, predaceous insects eating only animal food would ingest cholesterol directly and would need no dealkylation facility. Certain dermestid beetles, which feed on meat products, were early found to utilize cholesterol (and perhaps 7-dehydrocholesterol, also characteristic of animal tissues) but none of the common plant sterols, whereas most other insects studied in these early years and rather cavalierly grouped as phytophagous were able to develop well with a variety of phytosterols wholly replacing cholesterol. For several phytophagous insects, dealkylation to cholesterol was confirmed by an ingenious bioassay in which lipid extracts from phytophages were added to a basal lipid-free diet for *Dermestes granarius,* one of the beetles known to grow with cholesterol but not with plant sterols. *Dermestes* grew well on diets supplemented with most such extracts, showing that some of the particular plant sterols ingested by the phytophagous insects must have been metabolized to cholesterol, results confirmed by chromatography of the sterols found in phytophages when compared with sterols extracted from their respective food plants (Levinson, 1962). Thus arose the much-favored early belief that carnivorous insects utilized only animal sterols, whereas phytophagous insects had necessarily evolved the ability to dealkylate phytosterols to cholesterol and so had a wide dietary sterol versatility.

An important gloss on the foregoing hypothesis was introduced early to account for the fact that although *Dermestes* failed to grow with dietary phytosterols alone, if these sterols were provided in combination with a minute amount of cholesterol that alone was inadequate for growth, growth and complete development was then obtained (Clark and Bloch, 1959). Such observations using various insect species (Clayton, 1964) gave rise to the concept of "essential" and "sparing" sterols, based on the proposition that the overall sterol requirement could be broken down into a small metabolic component requiring cholesterol specifically and a major structural component for which the physiological requirement was less fastidious and so could be satisfied by a broader range of sterols, which were thus able to spare the greater part of the essential cholesterol requirement. The phenomenon of sterol sparing is well shown in a recent study with *Drosophila melanogaster,* which can use cholestanol, inadequate alone, to spare with complete efficiency the major part of the cholesterol requirement (Kircher and Grey, 1978). The sparing phenomenon is fundamentally important to the interpretation of insect sterol nutrition. The minute metabolic requirement was soon perceived as probably representing a need for a precursor for the ecdysone family of moulting hormones when it emerged that they were $\Delta 7$-polyhydroxyketosteroids, eventually shown to be derived in the insects via 7-dehydrocholesterol from cholesterol, whether this latter was obtained directly in the diet or by conversion from ingested phytosterols. The structural sterols, however, represent a much vaguer category. Evidence discussed in Clayton (1964), Dadd (1973), and Cooke and Sang (1970) suggested that this category must be considered at least bifunctional, with different restrictions on the ability of various sterols to subserve the two structural functions minimally postulated. Clayton noted there is likely to be further functional subdivision, and this might well entrain further degrees of fastidiousness with respect to internal sterol utilizability. Thus, the possibility arises that quite minor contamination by other sterols of the nominal sterols used in experimental studies might, via different restricted sparing actions at various levels in the heirarchy of structural as well as metabolic functions, greatly confound interpretation of results (Cooke and Sang, 1970).

Up to the time of Clayton's (1964) review, cholesterol had been found to completely satisfy the sterol requirement of all species studied, whether or not other sterols were utilized equally well or sometimes better. This pattern was broached with the discovery that *Drosophila pachea,* associated in nature with the senita cactus *Lophocereus schottii,* could not utilize cholesterol at all but required certain 7-dehydrosterols such as $\Delta7$-cholesten-3β-ol (lathosterol) or $\Delta5,7$-cholestadiene-3β-ol (7-dehydrocholesterol) related to the unusual sterol schottenol ($\Delta7$-stigmasten-3β-ol) of its food plant (Heed and Kircher, 1965). Cholesterol was not found in the tissues of these flies, which contained mostly lathosterol with a little 7-dehydrocholesterol (Goodnight and Kircher, 1971). Shortly thereafter, a 7-dehydrosterol was found essential for another insect, the ambrosia beetle *Xyleborus ferrugineus;* this wood-tunneling beetle, which carries a symbiotic fungus whose mycelium provides its sole food, requires 7-dehydrocholesterol or ergosterol in order to undergo pupation, although cholesterol is adequate for larval growth (Chu *et al.*, 1970; Norris and Chu, 1971). Complete development occurs with either 7-dehydrocholesterol or ergosterol alone in the diet, consonant with ergosterol being the only sterol of the beetle's fungus food. Thus, for *Xyleborus,* the essential metabolic sterol is ergosterol, a phytosterol, whereas cholesterol and apparently also lanosterol, hitherto nonutilized by several other insects tested, e.g., the khapra beetle (Agarwal, 1970), are sparing sterols, a reversal of the usual order of things.

Many phytophagous insects such as the silkworm are able to utilize plant sterols, commonly β-sitosterol, more efficiently than cholesterol, and recent reports suggest that β-sitosterol rather than cholesterol is the primary essential sterol for a moth, *Crambus trisectus* (Dupnik and Kamm, 1970), and a weevil, *Hylobius pales* (Richmond and Thomas, 1975). Caution is required in interpreting the results for β-sitosterol, since the possibility that this sterol may function as a phagostimulant is not excluded, especially as it is known to be a phagostimulant for the silkworm (Hamamura, 1970). However, in work with the cabbage fly *Hylemya brassicae,* using a meridic diet that supported full development without added sterol (presumably because of carryover of egg reserves, probably including phytosterol derived from the previous generation), the fact that minute doses of cholesterol were lethal at the second instar moult (Dambre-Raes, 1976) indicates a lack of cholesterol utilization and suggests that instances of phytosterol rather than cholesterol being essential may be less unusual than heretofore seemed the case. Certainly, the simple paradigm of 15 years back based on the centrality of cholesterol for both metabolic and structural functions is no longer adequate to embrace the many novel cases of sterol utilization and metabolism unraveled in recent years.

Before considering some of the recent findings, it will be useful to summarize what are now believed to be the main metabolic steps whereby phytosterols are dealkylated to cholesterol in the many typical phytophagous and omnivorous insects in which cholesterol does play a central physiological role. Much of this information was gleaned from studies of *Manduca sexta,* a large moth whose larvae equally readily utilize cholesterol or the common phytosterols (β-sitosterol, stigmasterol, campesterol, fucosterol, brassicasterol, 24-methylenecholesterol, and many others); essentially similar pathways have been found in other Lepidoptera, locusts, cockroaches, beetles, a thysanuran, etc., as summarized in the several reviews of insect sterol metabolism cited above.

In *Manduca,* β-sitosterol is converted via fucosterol, fucosterol-24,28-epoxide, and desmosterol to cholesterol; campesterol proceeds via 24-methylenecholesterol to desmosterol and thence to cholesterol; and stigmasterol goes via Δ5,22,24-cholestatrienol and desmosterol to cholesterol. The convergence of these pathways via desmosterol accounts for the large amount of cholesterol (an "animal" sterol) present in *Manduca* to be used physiologically as such for structural functions and also available for 7-dehydrogenation to 7-dehydrocholesterol en route to the ecdysones. An essentially similar pattern has been found in many other phytophagous and omnivorous insects, and in what follows, we are concerned mainly with divergences from it. The cereal-infesting beetle, *Tribolium confusum,* whose tissue sterols are unusual in including about 50% 7-dehydrocholesterol rather than the small or trace amount commonly found in other insects, exhibits variant pathways in addition to the foregoing (Svoboda *et al.*, 1972): β-sitosterol proceeds as for *Manduca* to desmosterol, then may go via Δ5,7,24-cholestatrienol to 7-dehydrocholesterol as the primary end metabolite rather than cholesterol; and for stigmasterol, in addition to the *Manduca* pathway to desmosterol and then as for β-sitosterol above, an alternative loop is available from Δ5,22,24-cholestatrienol to Δ5,7,22,24-cholestatetraenol to Δ5,7,24-cholestatrienol to 7-dehydrocholesterol, again the principal end metabolite. Cholesterol is found in *Tribolium* tissues in rather lesser amounts than 7-dehydrocholesterol, the two sterols being interconvertibly in balance. Interestingly, these pathways from phytosterols to 7-dehydrocholesterol without the intermediate occurrence of cholesterol point to the possibility that the ecdysogenic metabolic sterol function may be accomplished in the absence of cholesterol.

Not all insects that grow well with phytosterols dealkylate them to cholesterol. This was first shown in the housefly, larvae of which develop to apparently normal adults with only β-sitosterol or campesterol in the diet equally as well as with cholesterol, although the latter is selectively absorbed from a dietary mixture of all three; however, adult females fail to produce viable eggs unless some cholesterol is obtained from food, because no cholesterol is formed from phytosterols. In this insect, cholesterol is apparently necessary only as a precursor for ecdysone, as suggested by the complete adequacy of a combination of the sparing sterol, cholestanol, and 7-dehydrocholesterol (Kaplanis *et al.*, 1965; Robbins, 1963). The khapra beetle, *Trogoderma granarium,* also does not dealkylate phytosterols to cholesterol and, like the housefly, tends to selectively absorb cholesterol and campesterol rather than sitosterol, the latter being the main sterol of its diet (Svoboda *et al.*, 1979). This inability to dealkylate phytosterols is surprising in a beetle related to the typical zoophagous dermestids mentioned above, from which it differs in having adopted a primary use of sparing sterols as the norm, doubtless as a necessary correlate of its evolution to a vegetable-feeding status.

Little, if any, conversion of phytosterols to cholesterol occurs in the milkweed bug, *Oncopeltus fasciatus* (Svoboda *et al.*, 1977), although the small amounts of cholesterol present in its tissues through selective absorption of the traces of cholesterol detectable in the experimental food, sunflower seed, would be adequate to provide sufficient ecdysone precursor. Four to 10% of C_{30} sterol was also present, absorbed from the 19% of C_{30} sterol in the seed, but no traces of saturated or Δ7- and Δ5,7-sterols were detected, and the tissue sterol composition of the bug was otherwise virtually a direct reflection of the proportions of β-sitosterol, campesterol, and stigmasterol in the sunflower seed. *Oncopeltus* is unique among insects studied with respect to moulting hormone chemistry in having mainly a C_{28} hormone, makisterone A, rather than the more usual C_{27} ecdysone series. When reared on its natural food, milkweed seed, cholesterol was scarcely detected, although other sterols were essentially similar, hinting that this insect may dispense with C_{27} ecdysones and rely on makisterone derived from campesterol incorporated unchanged from its food.

In contrast to the foregoing insects, the Mexican bean beetle *Epilachna varivastis* exhibits phytosterol transformations but of a totally different kind from those so far discussed (Svoboda *et al.*, 1974, 1975b). Its tissue sterols are characterized by a preponderance of stanols and Δ7-stenols variously derived by reduction of the corresponding Δ5-sterols ingested from the food plant and from dealkylation of some of the resulting C_{28} and C_{29} stanols to C_{27} cholestanol; also, by some undefined route, a substantial conversion of C_{27}, C_{28}, and C_{29} Δ5-sterols to lathosterol (Δ7-cholestenol) occurs. Lathosterol could be an important intermediate for ecdysone synthesis, but soybean leaves contain low levels of cholesterol which is substantially concentrated in the beetle sterols and possibly available for ecdysone production via 7-dehydrocholesterol, although this latter sterol was not detected. Since *Epilachna* is in a family of beetles, the Coccinellidae, most of which are predators, it was of interest to compare its sterol pattern with that of a typical predatory species, *Coccinella septempunctata*. The predator sterols were comprised primarily of cholesterol (50%), substantial campesterol, sitosterol, and stigmasterol (probably derived directly from phytophagous prey), appreciable 7-dehydrocholesterol, no lathosterol, and very little saturated sterol, a pattern with scant resemblance to that of its phytophagous cousin.

As a result of this recent efflorescence in the study of insect sterol metabolism, the simple rationalization into zoophagous and phytophagous types which held the field two decades ago no longer seems particularly helpful. On the other hand, the idea of metabolic and structural sterols retains its explanatory power. It still seems valid to equate metabolic function with ecdysogenesis, although one certainly should not exclude other yet to be discovered metabolic functions, but it can no longer be assumed that cholesterol is a necessary gate through which sterols must pass en route to becoming ecdysones. On this route, the $\Delta 7$ structure is a unifying requisite, and the foregoing examples suggest that there are alternatives to the classical cholesterol to 7-dehydrocholesterol sequence. As for structural functions, which account for the physical bulk of the insect dietary sterol requirement, a diversity of sterols seems able to satisfy these needs, often after metabolic modification but also apparently unmodified. At present, there seems to be no simple pattern to be perceived in this beyond the truism that narrow, especially monophagous dietaries impose specialized sterol requirements and/or unusual metabolic transformations. In the context of the preceding section, this constitutes yet another facet of allelochemic insect/host plant interactions.

4.5. Fat-Soluble Vitamins

In contrast to the steady progress in understanding sterol and fatty acid requirements, the status in insect nutrition of other fat-soluble growth factors required by vertebrates remains very unclear. This is a frustrating field of work, for one thing certainly established is that no insects studied have a pressing need for vitamins A, D, E, or K for growth over one larval cycle. Hence, the further search for signs of deficiency entails the prospect of pursuing studies beyond larval growth into adult reproductive functions (often requiring different diets and feeding situations in holometabolous insects) and possibly through subsequent generations. Few investigators have been inclined to extend study to such arduous lengths, and thus, little new information has accrued since the most recent reviews of this topic (House, 1974; Dadd, 1973, 1977a). Our discussion is largely a summary from these reviews supplemented with specific references insofar as a few recent studies amplify certain points.

In vertebrates, a lack of dietary vitamin A (retinol/retinal) or suitable precursor carotenes, which no animals are known to synthesize, retards growth, causes various gross lesions of mucus membranes, and impairs vision, this latter because the visual pigments, retinenes, comprise retinal complexed with certain proteins, the opsins. It is now well established that insect visual pigments are also retinenes (Goldsmith and Bernard, 1974; White, 1978), and severe dietary deprivation of vitamin A or carotene has been shown to impair the light response in several species, with concomitant microlesions of the ommatidial retinula cells.

Attempts to detect a wider function of vitamin A in terms of abnormalities of growth and development are of uncertain interpretation. Many insects owe their usual color to dietary carotenoids, which often provide yellow carotenoprotein or green "insectoverdin" pigments, the latter being mixtures of carotenoproteins with protein-complexed bile pigments (Rowell, 1971); if reared with carotene-deficient food, such pigments are absent or diminished, or only the bile pigment component is formed, causing an unusual bluish hue. Generally, such pigment-deficient insects grow and function normally, an exception being the locusts *Locusta migratoria* and *Schistocerca gregaria,* in which deprivation of dietary carotene over two generations retarded growth and caused anomalous color and behavior in the second generation (Dadd, 1961). Slightly retarded larval growth in the absence of vitamin A was also detected in a fly, *Agria affinis* (House, 1966), and without carotene in semisynthetic diets, larvae of the moth *Plodia interpunctella* died in the first instar, apparently without feeding, thus indicating a phagostimulant role for carotene or its breakdown products (Morere, 1971b). One cannot unequivocally ascribe any of these few recorded growth effects to a direct nutritional need for vitamin A beyond its essentiality for vision, since all might very well be consequent on behavioral abnormalities resulting from poor sight.

Vitamin D (calciferol) has frequently been studied in insects as a possible substitute for cholesterol, always with entirely negative results. The few cases in which it has been tested in otherwise complete diets have provided no indication of need, perhaps not surprisingly, given the vertebrate function of vitamin D, mobilization of calcium for bone formation, and the lack of bone in insects.

Vitamin E (as α-tocopherol) was first shown to affect insect growth with flour moths of the genus *Ephestia* (Fraenkel and Blewett, 1946b, 1947). It was considered to act by antioxidant protection of essential polyunsaturated fatty acids, since other antioxidants such as ascorbic acid or propyl gallate were equally effective. Since then, α-tocopherol has been included in the formulation of many synthetic diets, especially where essential fatty acids were called for, and it was recently shown to protect arachidonic acid in mosquito diets, a function subserved equally well by ascorbyl palmitate or propyl gallate (Dadd and Kleinjan, 1979b). With arachidonic acid protected by ascorbate, no additional benefit has yet been found for the simultaneous inclusion of α-tocopherol (R. H. Dadd, unpublished data), suggesting its role to be purely antioxidant protection. However, a recent study of fatty acid requirements in a butterfly, *Pieris brassicae,* discusses metabolic data suggesting an endogenous influence of

α-tocopherol on tissue polyunsaturated fatty acid conservation as distinct from external protection of fatty acids in the diet (Turunen, 1976).

A specific nutritional role for vitamin E has, however, been demonstrated in a few insects, primarily in connection with adult reproductive function, reminiscent of the antisterility factor status of the vitamin for mammals. α-Tocopherol improved fecundity in certain moths and beetles, and in the fly *Agria affinis*, it slightly speeded larval growth and was essential for proper embryonation of eggs (House, 1966). On the other hand, its effect on the cricket *Acheta domestica* is most notable in males, which fail to develop viable sperm without dietary α-tocopherol as larvae, although both males and females are smaller in its absence (Meikle and McFarlane, 1965; McFarlane, 1972). As in vertebrates, the physiological action of vitamin E seems bound up with its redox properties, since it interacts nutritionally with copper and is involved via regulation of the melanin-producing phenyloxidase system in cricket nutritional albinism (McFarlane, 1974).

Vitamin K (phytoquinones in plants and menaquinones in bacteria) is required by vertebrates primarily for its role in prothrombinogenesis and thus would be expected to be unnecessary for animals lacking this type of blood-clotting mechanism. Occasionally, vitamin K, usually as menadione, has been tested with insects for a possible growth-stimulatory effect, until very recently with entirely negative results. However, McFarlane (1976) tested vitamin K with crickets as a possible substitute for vitamin E because of the close similarity between the side chains of the two molecules; he found that it stimulated growth but was without effect on male sterility and suggests that it acts in this insect to spare vitamin E (McFarlane, 1977).

It is evident that, where required, the main effect of vitamin E in insects is not expressed until the adult reproductive stage is reached, and this may be why a need for it has been observed so sporadically. More multigenerational studies may well show the requirement to be widespread.

4.6. Sugar and Carbohydrate Utilization

Insects differ greatly in their requirements for carbohydrates. Some species require none in their diets, and those that do demonstrate no great specificity in the requirement. The determination of carbohydrate requirements and utilization is complicated by the fact that certain sugars, such as glucose and sucrose, are potent phagostimulants, whereas others, such as xylose, galactose, mannose, and sorbose, inhibit feeding (see Dadd, 1977a). Utilizable carbohydrates that are not phagostimulatory may not be ingested in sufficient quantities to support growth or survival. Utilization also depends on the ability of a species to digest complex carbohydrates into diffusable and absorbable forms.

Some insect larvae that do not require dietary carbohydrates are able to break down dietary protein or lipid to satisfy their energy needs; these include the housefly *Musca domestica* (Brookes and Fraenkel, 1958), the blowfly *Phormia regina* (Cheldelin and Newberg, 1959), the screwworm *Cochliomyia hominivorax* (Gingrich, 1964), the fleshfly *Calliphora erythrocephala* (Sedee, 1958), the clothes moth *Tineola bisselliella* (Fraenkel and Blewett, 1946a), and, surprisingly, some flour beetles *Tribolium confusum* and *Silvanus* (*Oryzaephilus*) *surinamensis* (Fraenkel and Blewett, 1943). Larval growth of the mosquito *Aedes aegypti* was improved if glucose was incorporated into the artificial diet, but its effect was related to the level of amino acids present in the diet (R. H. Dadd and V. P. Sneller, unpublished observations). This species will develop suboptimally without dietary carbohydrate (Akov, 1962; Golberg and De Meillon, 1948a).

Larvae of other species require only moderate amounts of carbohydrate in their diet; concentrations of glucose, sucrose, or fructose between 15 and 30% of the dry weight of the diet being satisfactory. Optimum carbohydrate concentration is quite critical for some species such as the corn borer *Diatraea grandiosella* (Chippendale and Reddy, 1974), the locusts *Locusta migratoria* and *Schistocerca gregaria* (Dadd, 1960), the European corn borer *Ostrinia nubilalis,* the silkworm *Bombyx mori,* and the spruce budworm *Choristoneura fumiferana* (see Chippendale, 1978). Certain insect larvae that normally infest stored products with a high starch content have been shown to require as high as 70% carbohydrate on a dry weight basis in their diets; *Ephestia kuehniella* and *Sitodrepa panicea* fall into this category (Fraenkel and Blewett, 1943).

Tabulations of carbohydrate utilization for growth (Hirano and Ishii, 1957; Dadd, 1977a) indicate that the hexoses, glucose and fructose, were well utilized by the majority of the insects tested critically, as were the disaccharides sucrose and maltose and the trisaccharide melezitose. Pentoses and sorbose were shown to be of little nutritive value, and the utilization of galactose, mannose, oligosaccharides, and polysaccharides varied greatly depending on the species studied. Mannose was considered to be toxic to *Tenebrio molitor* (Fraenkel, 1955) and to adult bees (Vogel, 1931). The adults of many species of Diptera require only carbohydrate and water for maintenance. Because of these simple requirements, they have been used to study the effects of various carbohydrates on longevity.

As with the studies dealing with larval growth, the pentoses and sorbose are not utilized. There are good utilization of glucose and fructose and species-variable and often indeterminate responses to mannose and galactose. The variability in the responses to oligosaccharides and polysaccharides depends on the range of digestive enzymes possessed by the various species tested (see Dadd, 1973).

4.7. Minerals

The difficulties of precisely controlling the content of inorganic ions in synthetic diets and in formulating diets deficient in trace minerals are formidable.

Mineral reserves can be carried over from the egg, and many dietary constituents are contaminated with minerals. Ultrapure diets such as those used to detect the requirements for vanadium and tin in the rat (Schwarz and Milne, 1971) have not yet been applied to insects. Because of these difficulties, the determination of mineral requirements is the least investigated area in insect nutrition. Synthetic diets for insects often include a salt mixture developed for mammals. Such mixtures contain more calcium, sodium, and cholride than is required by insects which do not need to maintain endoskeletons made of bone. Salt mixtures based on the ash content of natural food have supported good growth in the European cornborer *Ostrinia nubialis* (Beck *et al.*, 1968), larvae of the fly *Agria affinis* (= *A. housei*) (House and Barlow, 1965), and larvae of the red-banded leafroller *Argyrotaenia velutinana* (Rock *et al.*, 1964).

There is general agreement that insects require potassium, phosphate, and magnesium (Dadd, 1973, 1977a; House, 1974) in substantial amounts. Iron, zinc, manganese, and copper have been shown to be essential nutrients in the aphid *Myzus persicae* (Dadd, 1967), the aphid *Aphis fabea* (Dadd and Krieger, 1967), the silkworm *Bombyx mori* (Horie *et al.*, 1967; Ito and Niimura, 1966), the confused flour beetle *Tribolium confusum* (Medici and Taylor, 1966, 1967), and larvae of the mosquito *Culex pipiens* (R. H. Dadd, unpublished data). Because these ions are cofactors in enzymatic reactions that presumably occur in all insects, they are probably universally required. Dadd (1977a) speculates that with the possible exception of iodine, a necessity for thyroid function unique to vertebrates, insects probably require all the ions necessary for basic animal physiological functions.

Trace metals may affect symbiotes which are necessary for the survival of certain insect species. The cockroach *Blattella germanica* requires symbiotes that are affected by a delicate balance of calcium, zinc, and manganese (Brooks, 1960; Gordon, 1959; Henry and Block, 1961), and trace metals may affect the symbiotes in aphids (see Dadd, 1977a). The form in which trace metals are presented in liquid diets can affect their availability. The aphid *Myzus persicae* could be continuously reared when trace metals were provided as chelates of sodium EDTA, whereas metal salts were ineffective (Dadd, 1968). If ascorbic or citric acid is incorporated into these liquid diets, trace metals supplied as chloride salts are effective, but this effectiveness is probably affected by chelation reactions occurring during diet formulation (Mittler, 1976).

4.8. Water Balance

Insects, like other animals, require water; however, little quantitative information is available on the intake and utilization of water by insects (Waldbauer, 1968). Water balance in insects and mites has recently been reviewed (Arlian and Veselica, 1979). Wharton and Arlian (1972) note that water is unique when compared with other major nutrients in that it is sufficiently volatile to be exchanged between the air and the organism. It is the most concentrated nutrient; in actively metabolizing insect tissues, more than 99 of every 100 molecules is water. Insects must maintain their body water content and water concentration within tolerable limits. Terrestrial species constantly lose body water which must be replaced, and aquatic species constantly gain water which must be eliminated.

Because of their small size, insects have a high surface-area-to-volume ratio. This enhances the loss of water to the atmosphere. The insect exoskeleton with its waxy coating is extremely impermeable to water, but the walls of the trachea, where gas exchange takes place, are more permeable to water than O_2 and CO_2 (Waggoner, 1967). Wharton and Arlian (1972) report that the time required for one-half of the water content to be exchanged with the ambient air varied from 312 hr for the larvae of the beetle *Tenebrio molitor,* which normally infest dry stored products, to 120 hr for the hemipteran *Graphosoma lineatum.* Feeding rates and the efficiency of food conversion may be affected by an insect's drive to regain lost water (Waldbauer, 1968). Under natural conditions, water is obtained by drinking or in the plant or animal tissues ingested as food. Under artificial rearing conditions, water is supplied either *ad libitum* or as a large constituent of the synthetic diet.

Insects that eat extremely dry food obtain metabolic water from the oxidation of carbohydrate and fat (Fraenkel and Blewett, 1944; Leclercq, 1948; Murray, 1968). Under these conditions, the composition of the diet and the relative humidity of the atmosphere is critical. The growth of larvae of *Tenebrio molitor* resembles the development of bird and insect eggs in that a narrow range of water balance must be maintained. Too much or too little metabolic water can interfere with normal growth and development (Machin, 1975; Murray, 1968). Many insects are adapted to take water from unsaturated ambient air. Atmospheric absorption in *T. molitor* larvae and in the firebrat *Thermobia domestica* takes place in the rectum (Noble-Nesbitt, 1970a,b). Nymphs and adult females of the desert cockroach *Arenivaga investigata* absorb water vapor from unsaturated atmospheres above 82.5% RH through two bladderlike extensions of the hypopharynx which are protruded from the mouth (O'Donnell, 1978). This is another example of how the nutrition of insects differs from that of vertebrates.

5. Concluding Remarks

Information about insect nutrition will continue to come from a broad spectrum of studies ranging from those using natural foods to those using precisely controlled diets and axenic rearing conditions. Studies on basic descriptive nutrition providing information on species of economic importance will continue. There are still many taxonomic groups and categories of feeding habits that have not been investigated. Satisfactory diets and rearing methods have been developed for plant sucking insects such as aphids, leaf hoppers, and lygaeid bugs only in the last decade (see Dadd, 1973). The nutritional requirements of predatory insects and insects that are internal parasites are almost completely unknown.

Nutritional studies to extend our knowledge of insects' ecological ranges and their metabolism will continue to develop. Ecological nutrition attempts to relate what an insect requires to what the natural environment can provide as food. This approach requires a knowledge of the insects' feeding behavior, nutritional needs, and digestive and absorptive abilities coupled with knowledge of the chemistry of the potential natural foods and how these foods can match the insects' requirements, since both vary at different seasons and stages of development. Nutritional studies involving metabolic processes and how these are controlled in multicellular organisms will probably require more sophisticated chemical and physical techniques, particularly those involving isotopes, coupled perhaps with the use of mutants that can contribute to our knowledge about how genes act in integrated organisms to influence metabolic control (see Sang, 1972).

ACKNOWLEDGMENTS. We are deeply indebted to Ms. Rosemary Hewson Tanner for her expert editorial assistance.

References

Agarwal, H. C., 1970, Sterol requirements of the beetle *Trogoderma, J. Insect Physiol.* **16**:2023.

Akov, S., 1962, A qualitative study of the nutritional requirements of *Aedes aegypti* L. larvae, *J. Insect Physiol.* **8**:319.

Alfin-Slater, R. B., and Aftergood, L., 1971, Physiological functions of essential fatty acids, *Prog. Biochem. Pharmacol.* **6**:214.

Applebaum, S. W., and Lubin, Y., 1967, The comparative effects of vitamin deficiency on development and on adult fecundity of *Tribolium castaneum, Entomol. Exp. Appl.* **10**:23.

Arai, N., and Ito, T., 1967, Nutrition of the silkworm *Bombyx mori* XVI. Quantitative requirements for essential amino acids, *Bull. Sericult. Exp. Sta. Tokyo* **21**:373.

Arlian, L. G., and Veselica, M. M., 1979, Water balance in insects and mites, *Comp. Biochem. Physiol.* **64A**:191.

Auclair, J. L., 1963, Aphid feeding and nutrition, *Annu. Rev. Entomol.* **8**:439.

Auclair, J. L., 1969, Nutrition of plant-sucking insects on chemically defined diets, *Entomol. Exp. Appl.* **12**:623.

Barlow, J. S., 1962, Pyridoxine requirements of *Agria affinis* (Fall.), *Nature,* **196**:193.

Barrett, F. M., and Friend, W. G., 1970, Uric acid synthesis in *Rhodnius prolixus, J. Insect Physiol.* **16**:121.

Beck, S. D., 1965, Resistance of plants to insects, *Annu. Rev. Entomol.* **10**:207.

Beck, S. D., and Chippendale, G. M., 1968, Environmental and behavioural aspects of the mass rearing of plant-feeding Lepidopterans, in: *Radiation, Radioisotopes, and Rearing Methods in the Control of Insect Pests,* pp. 19–30, IAEA, Vienna.

Beck, S. D., Chippendale, G. M., and Swinton, D. E., 1968, Nutrition of the European corn borer,

Ostrinia nubilalis. VI. A larval rearing medium without crude plant fractions. *Ann. Entomol. Soc. Am.* **61**:459.

Bodnaryk, R. P., and Levenbook, L., 1969, The role of β-alanyl-L-tyrosine (sarcophagine) in puparium formation in the fleshfly, *Sarcophaga bullata, Comp. Biochem. Physiol.* **30**:909.

Bongers, U., 1970, Aspects of host-plant relationship of the Colorado potato beetle, *Meded. Landbouw. Wag.* **70**(10):1.

Brookes, V. J., and Fraenkel, G., 1958, The nutrition of the larva of the housefly, *Musca domestica* L., *Physiol. Zool.* **31**:208.

Brooks, M. A., 1960, Some dietary factors that affect ovarial transmission of symbiotes, *Proc. Helminth. Soc. Wash.* **27**:212.

Brooks, M. A., 1964, Symbiotes and the nutrition of medically important insects, *Bull. WHO* **31**:555.

Brust, M., and Fraenkel, G., 1955, The nutritional requirements of larvae of the blowfly, *Phormia regina* Meig, *Physiol. Zool.* **28**:186.

Bursell, E., 1967, The excretion of nitrogen in insects, *Adv. Insect Physiol.* **4**:33.

Butler, C. G., 1943, Bee paralysis May-sickness, *Bee World* **24**:3.

Chauvin, R., 1956, *Physiologie de l'Insecte. Le Comportement, les Grandes Fonctions, Ecophysiologie,* Institute Nationale des Recherches Agronomique, Paris.

Cheldelin, V. H., and Newburgh, R. W., 1959, Nutritional studies on the blowfly, *Ann. N.Y. Acad. Sci.* **77**:373.

Chippendale, G. M., 1978, The functions of carbohydrates in insect life processes, in: *Biochemistry of Insects* (M. Rockstein, ed.), pp. 2-54, Academic Press, New York.

Chippendale, G. M., and Beck, S. D., 1964, Nutrition of the European corn borer, *Ostrinia nubilalis* (Hubn.). V. Ascorbic acid as the corn leaf factor, *Entomol. Exp. Appl.* **7**:241.

Chippendale, G. M., and Beck, S. D., 1968, Biochemical requirements for mass rearing plant-feeding lepidopterans, in: *Radiation, Radioisotopes, and Rearing Methods in the Control of Insect Pests,* pp. 31-39, IAEA, Vienna.

Chippendale, G. M., and Reddy, G. P. V., 1974, Dietary carbohydrates: Role in feeding behavior and growth of the southwestern corn borer, *Diatraea grandiosella, J. Insect Physiol.* **20**:751.

Chippendale, G. M., Beck, S. D., and Strong, F. M., 1965, Nutrition of the cabbage looper, *Trichoplusia ni* (Hubn.). I. Some requirements for larval growth and wing development, *J. Insect Physiol.* **11**:211.

Chu, H.-M., Norris, D. M., and Kok, L. T., 1970, Pupation requirement of the beetle, *Xyleborus ferrugineus:* Sterols other than cholesterol, *J. Insect Physiol.* **16**:1379.

Clark, A. J., and Bloch, K., 1959, Function of sterols in *Dermestes vulpinus, J. Biol. Chem.* **234**:2583.

Clayton, R. B., 1964, The utilization of sterols by insects, *J. Lipid Res.* **5**:3.

Clements, A. N., 1963, *The Physiology of Mosquitoes,* Pergamon Press, Oxford.

Cooke, J., and Sang, J. H., 1970, Utilization of sterols by larvae of *Drosophila melanogaster, J. Insect Physiol.* **16**:801.

Dadd, R. H., 1960, The nutritional requirements of locusts. III. Carbohydrate requirements and utilization, *J. Insect Physiol.* **5**:301.

Dadd, R. H., 1961, Observations on the effects of carotene on the growth and pigmentation of locusts, *Bull. Entomol. Res.* **52**:63.

Dadd, R. H., 1963, Feeding behaviour and nutrition in grasshoppers and locusts, *Adv. Insect Physiol.* **1**:47.

Dadd, R. H., 1964, A study of carbohydrate and lipid nutrition in the wax moth, *Galleria mellonella* (L.), using partially synthetic diets, *J. Insect Physiol.* **10**:161.

Dadd, R. H., 1967, Improvement of synthetic diet for the aphid *Myzus persicae* using plant juices, nucleic acids, or trace metals, *J. Insect Physiol.* **13**:763.

Dadd, R. H., 1968, Problems connected with inorganic compounds of aqueous diets, *Bull. Entomol. Soc. Am.* **14**:22.

Dadd, R. H., 1970a, Arthropod nutrition, *Chem. Zool.* **5**:35.

Dadd, R. H., 1970b, Digestion in insects, *Chem. Zool.* **5**:117.

Dadd, R. H., 1973, Insect nutrition: Current developments and metabolic implications, *Annu. Rev. Entomol.* **18**:381.

Dadd, R. H., 1977a, Qualitative requirements and utilization of nutrients: Insects, in: *CRC Handbook Series in Nutrition and Food* (M. Rechcigl, Jr., ed.), pp. 305–346, CRC Press, Cleveland.

Dadd, R. H., 1977b, Qualitative requirements and utilization of nutrients: arthropods, in: *CRC Handbook Series in Nutrition and Food* (M. Rechcigl, Jr., ed.), pp. 347–352, CRC Press, Cleveland.

Dadd, R. H., 1978, Amino acid requirements of the mosquito *Culex pipiens:* Asparagine essential, *J. Insect Physiol.* **24**:25.

Dadd, R. H., 1979, Nucleotide, nucleoside and nutritional requirements of the mosquito *Culex pipiens, J. Insect Physiol.* **25**:353.

Dadd, R. H., 1980, Essential fatty acids for the mosquito *Culex pipiens, J. Nutr.* **110**:1152.

Dadd, R. H., 1981, Essential fatty acids for mosquitoes, other insects, and vertebrates, in: *Current Topics in Insect Endocrinology and Nutrition* (G. Baskaran, S. Friedman, and J. G. Rodriguez, eds.), pp. 189–214, Plenum Press, New York.

Dadd, R. H., and Kleinjan, J. E., 1977, Dietary nucleotide requirements of the mosquito, *Culex pipiens, J. Insect Physiol.* **23**:333.

Dadd, R. H., and Kleinjan, J. E., 1979a, Essential fatty acid for the mosquito *Culex pipiens:* Arachidonic acid, *J. Insect Physiol.* **25**:495.

Dadd, R. H., and Kleinjan, J. E., 1979b, Vitamin E, ascorbyl palmitate, or propyl gallate protect arachidonic acid in synthetic diets for mosquitos, *Entomol. Exp. Appl.* **26**:222.

Dadd, R. H., and Krieger, D. L., 1967, Continuous rearing of aphids of the *Aphis fabea* complex on sterile synthetic diet, *J. Econ. Entomol.* **60**:1512.

Dadd, R. H., and Krieger, D. L., 1968, Dietary amino acid requirements of the aphid *Myzus persicae, J. Insect Physiol.* **14**:741.

Dadd, R. H., Krieger, D. L., and Mittler, T. E., 1967, Studies on the artificial feeding of the aphid *Myzus persicae* (Sulzer)—IV. Requirements for water-soluble vitamins and ascorbic acid, *J. Insect Physiol.* **13**:249.

Dadd, R. H., Friend, W. G., and Kleinjan, J. E., 1980, Arachidonic acid requirement for two species of *Culiseta* reared on synthetic diet, *Can. J. Zool.* **58**:1845.

Dambre-Raes, H., 1976, The effect of dietary cholesterol on the development of *Hylemya brassicae, J. Insect Physiol.* **22**:1287.

David, J., 1967, Methods d'evaluation des besoins nutritionnels des insectes eleves sur milieux artificiels, *Appl. Nutr. Aliment.* **21**:25.

Davis, G. R. F., 1964, The importance of carnitine in the diet of larvae of *Oryzaephilus surinamensis* (L) (Coleoptera: Silvanidae), *Adv. Insect Physiol.* **72**:70.

Davis, G. R. F., 1966, Replacement of RNA in the diet of *Oryzaephilus surinamensis* L. (Coleoptera: Silvanidae) by purines, pyrimidines, and ribose, *Can. J. Zool.* **44**:781.

Davis, G. R. F., 1968, Phagostimulation and consideration of its role in artificial diets, *Bull. Entomol. Soc. Can.* **14**:27.

Davis, G. R. F., 1971, Protein nutrition of *Tenebrio molitor* L. V. Amino acid mixtures as replacements for protein of the artificial diet, *Adv. Insect Physiol.* **79**:11.

Davis, G. R. F., and Sosulski, F. W., 1973, Improvement of basic diet for use in determining the nutritional value of proteins with larvae of *Tenebrio molitor* L., *Adv. Insect Physiol.* **81**:495.

Destephano, D. B., and Brady, V. E., 1977, Prostaglandin and prostaglandin synthetase in the cricket, *Acheta domesticus, J. Insect Physiol.* **23**:905.

Downer, R. G. H., 1978, Functional role of lipids in insects, in: *Biochemistry of Insects* (M. Rockstein, ed.), pp. 57–92, Academic Press, New York.

Dupnik, T. D., and Kamm, J. A., 1970, Development of an artificial diet for *Crambus trisectus, J. Econ. Entomol.* **63**:1578.

Earle, N. W., Slatten, B., and Burks, M. L., 1967, Essential fatty acids in the diet of the boll weevil, *Anthonomus grandis* (Boheman) (Coleoptera: Curculionidae), *J. Insect Physiol.* **13**:187.

Ehrhardt, P., 1968a, Der Vitaminbedarf einer siebrohrensargenden Aphide *Neomyzus circumflexus* Buckt, *Z. Vergl. Physiol.* **60**:416.

Ehrhardt, P., 1968b, Einfluss von Ernahrungsfaktoren auf die Entwicklung von Safte saugenden Insekten unter besonderer Berucksichtigung von Symbioten, *Z. Parasitenkd.* **31**:38.

Ehrhardt, P., 1968c, Nachweis einer durch symbiotische Mikroorganismen bewirkten Sterinsynthese in kunstlich ernahten Aphiden (Homoptera, Rhynchota, Insecta), *Experientia* **24**:82.

Ehrhardt, P., 1969, Die Rolle von Methionin, Cystein, Cystin, und Sulfat bei der kunstlichen Ernahrung von *Neomyzus (Aulacorthum) circumflexus* Buckt, *Biol. Zentralbl.* **88**:335.

Falk, D. R., and Nash, D., 1972, The search for auxotrophic mutants in *Drosophila melanogaster*, in: *Insect and Mite Nutrition* (J. G. Rodriguez, ed.), pp. 19–31, North-Holland, Amsterdam.

Falk, D. R., and Nash, D., 1974, Sex-linked auxotrophic and putative auxotrophic mutants of *Drosophila melanogaster*, *Genetics* **76**:755.

Fast, P. G., 1964, Insect lipids: A review, *Mem Entomol. Soc. Can.* **37**:1.

Fast, P. G., 1970, Insect lipids, *Prog. Chem. Fats Other Lipids* **11**:181.

Fraenkel, G., 1955, Inhibitory effects of sugars on the growth of the mealworm *Tenebrio molitor* L., *J. Cell. Comp. Physiol.* **45**:393.

Fraenkel, G., 1958, The effect of zinc and potassium in the nutrition of *Tenebrio molitor*, with observations on the expression of a carnitine deficiency, *J. Nutr.* **65**:361.

Fraenkel, G., 1959a, A historical and comparative survey of the dietary requirements of insects, *Ann. N.Y. Acad. Sci.* **77**:267.

Fraenkel, G., 1959b, The raison d'etre of the secondary plant substances, *Science* **129**:1466.

Fraenkel, G., 1969, Evaluation of our thoughts on secondary plant substances, *Entomol. Exp. Appl.* **12**:473.

Fraenkel, G., and Blewett, M., 1943, The basic food requirements of several insects, *J. Exp. Biol.* **20**:28.

Fraenkel, G., and Blewett, M., 1944, The utilisation of metabolic water in insects, *Bull. Entomol. Res.* **35**:127.

Fraenkel, G., and Blewett, M., 1946a, The dietetics of the clothes moth, *Tineola bisselliella* Hum, *J. Exp. Biol.* **22**:156.

Fraenkel, G., and Blewett, M., 1946b, Linolenic acid, vitamin E and other fat-soluble substances in the nutrition of certain insects, *J. Exp. Biol.* **22**:172.

Fraenkel, G., and Blewett, M., 1947, Linoleic acid and arachidonic acid in the metabolism of the insects *Ephestia kuehniella* and *Tenebrio molitor*, *Biochem. J.* **41**:475.

Friend, W. G., 1958, Nutritional requirements of phytophagous insects, *Annu. Rev. Entomol.* **3**:57.

Friend, W. G., 1968, The nutritional requirements of Diptera, in: *Radiation, Radioisotopes, and Rearing Methods in the Control of Insect Pests*, pp. 41–57, IAEA, Vienna.

Friend, W. G., 1978, Physical factors affecting the feeding responses of *Culiseta inornata* to ATP, sucrose, and blood, *Ann. Entomol. Soc. Am.* **71**:935.

Friend, W. G., 1981, Diet destination in *Culiseta inornata:* Effects of feeding conditions on the response to ATP and sucrose, *Ann. Entomol. Soc. Am.* **74**:151.

Friend, W. G., and Smith, J. J. B., 1977, Factors affecting feeding by bloodsucking insects, *Annu. Rev. Entomol.* **22**:309.

Galford, J. R., 1972, Some basic nutritional requirements of the smaller European elm bark beetle larvae, *J. Econ. Entomol.* **65**:681.

Galun, R., 1966, Feeding stimulants of the rat flea *Xenopsylla cheopis* Roth, *Life Sci.* **5**:1335.

Galun, R., 1977, Responses of blood-sucking arthropods to vertebrate hosts, in: *Chemical Control of Insect Behavior, Theory and Application* (H. H. Shorey, J. J. McKelvey, Jr., eds.), pp. 103–115, John Wiley & Sons, New York.

Geer, B. W., 1963, A ribonucleic acid-protein relationship in *Drosophila* nutrition, *J. Exp. Zool.* **154**:353.

Geer, B. W., 1964, Inheritance of the dietary ribonucleic acid requirement of *Drosophila melanogaster, Genetics* **49**:787.

Geer, B. W., and Vovis, G. F., 1965, The effects of choline and related compounds on the growth and development of *Drosophila melanogaster, J. Exp. Zool.* **158**:223.

Geer, B. W., Vovis, G. F., and Yund, M. A., 1968, Choline activity during the development of *Drosophila melanogaster, Physiol. Zool.* **41**:280.

Geer, B. W., Dolph, W. W., Maguire, J. A., and Dates, R. J., 1971, The metabolism of dietary carnitine in *Drosophila melanogaster, J. Exp. Zool.* **176**:445.

Gilbert, L. I., 1967, Lipid metabolism and function in insects, *Adv. Insect Physiol.* **4**:69.

Gilmour, D., 1963, *The Biochemistry of Insects,* Academic Press, New York.

Gingrich, R. E., 1964, Nutritional studies on screwworm larvae with chemically defined media, *Ann. Entomol. Soc. Am.* **57**:351.

Golberg, L., and De Meillon, B., 1948a, The nutrition of the larva of *Aedes aegypti* Linnaeus. 3. Lipid requirements, *Biochem. J.* **43**:372.

Golberg, L., and De Meillon, B., 1948b, The nutrition of the larva of *Aedes aegypti* Linnaeus. 4. Protein and amino-acid requirements, *Biochem. J.* **43**:379.

Goldsmith, T. H., and Bernard, G. D., 1974, The visual system of insects, in; *Physiology of Insecta,* Vol. 2 (M. Rockstein, ed.), pp. 165–272, Academic Press, New York.

Goodnight, K. C., and Kircher, H. W., 1971, Metabolism of lathosterol by *Drosophila pachea, Lipids* **6**:166.

Gordon, H. T., 1959, Minimal nutritional requirements of the german roach, *Blattella germanica* L., *Ann. N.Y. Acad. Sci.* **77**:290.

Gordon, H. T., 1961, Nutritional factors in insect resistance to chemicals, *Annu. Rev. Entomol.* **6**:27.

Gordon, H. T., 1968, Quantitative aspects of insect nutrition, *Am. Zool.* **8**:131.

Gordon, H. T., 1972, Interpretations of insect quantitive nutrition, in: *Insect and Mite Nutrition* (J. G. Rodriguez, ed.), pp. 73–106, North Holland, Amsterdam.

Guarnieri, M., and Johnson, R. M., 1970, The essential fatty acids, *Adv. Lipid Res.* **8**:115.

Hagen, K. S., and Tassan, R. L., 1966, Artificial diet for *Chrysopa carnea* Stephens, in: *Ecology of Aphidophagous Insects* (I. Hodek, ed.), pp. 83–87, Academia, Dr. W. Junk, The Hague.

Hamamura, Y., 1970, The substances that control the feeding behavior and growth of the silkworm *Bombyx mori* L., in: *Control of Insect Behaviour by Natural Products* (D. L. Wood, R. M. Silverstein, and M. Nakajima, eds.), pp. 55–80, Academic Press, New York.

Harborne, J. B., 1978, *Biochemical Aspects of Plant and Animal Coevolution,* Academic Press, New York.

Haydak, M. H., 1970, Honey bee nutrition, *Annu. Rev. Entomol.* **15**:143.

Heed, W. B., and Kircher, H. W., 1965, Unique sterol in the ecology and nutrition of *Drosophila pachea, Science* **149**:758.

Hegdekar, B. M., 1970, Amino acid analogues as inhibitors of insect reproduction, *J. Econ. Entomol.* **63**:1950.

Henry, S. M., 1962, The significance of microorganisms in the nutrition of insects, *Trans. N.Y. Acad. Sci. Ser. II* **24**:676.

Henry, S. M., and Block, R. J., 1961, The sulfur metabolism of insects. VI. Metabolism of the sulfur amino acids and related compounds in the german cockroach, *Blattella germanica* (L.), *Contrib. Boyce Thomson Inst.* **21**:129.

Hinton, T., 1956, The effects of arginine, ornithine and citrulline on the growth of *Drosophila, Arch. Biochem. Biophys.* **62**:78.

Hinton, T., 1959, Miscellaneous nutritional variations, environmental and genetic, in *Drosophila, Ann. N.Y. Acad. Sci.* **77**:366.

Hinton, T., Noyes, T. D., and Ellis, J., 1951, Amino acids and growth factors in a chemically defined medium for *Drosophila, Physiol. Zool.* **24**:335.

Hirano, C., and Ishii, S., 1957, Nutritive values of carbohydrates for the growth of larvae of the rice stem borer; *Chilo suppressalis* Walker, *Bull. Natl. Inst. Agric. Sci.* **7**:90.

Hobson, R. P., 1935, On a fat soluble factor required by blowfly larvae. II. Identity of the growth factor with cholesterol, *Biochem. J.* **29**:2023.

Hogan, G. R., 1972, Development of *Tribolium castaneum* (Coleoptera: Tenebrionidae) in media supplemented with pyrimidines and purines, *Ann. Entomol. Soc. Am.* **65**:631.

Holman, R. T., 1977, The deficiency of essential fatty acids, in: *Polyunsaturated Fatty Acid* (W. H. Kunau, R. T. Holman, eds.), pp. 163–182, American Oil Chemists Society, Champaign.

Horie, Y., Watanabe, K., and Ito, T., 1966, Nutrition of the silkworm, *Bombyx mori*. XIV. Further studies on the requirements for B vitamins, *Bull. Sericult. Exp. Sta. Tokyo* **20**:393.

Horie, Y., Watanabe, K., and Ito, T., 1967, Nutrition of the silkworm *Bombyx mori*. XVIII. Quantitative requirements for potassium, phosphorus, magnesium, and zinc, *Bull. Sericult. Exp. Sta. Tokyo* **22**:181.

House, H. L., 1949, Nutritional studies with *Blattella germanica* (L.) reared under aseptic conditions. III. Five essential amino acids, *Can. Entomol.* **81**:133.

House, H. L., 1954a, Nutritional studies with *Pseudosarcophaga affinis* (Fall.), a dipterous parasite of the spruce budworm, *Choristoneura fumiferana* (Clem.). I. A chemically defined medium and aseptic-culture technique, *Can. J. Zool.* **32**:331.

House, H. L., 1954b, Nutritional studies with *Pseudosarcophaga affinis* (Fall.), a dipterous parasite of the spruce budworm, *Choristoneura fumiferana* (Clem.). II. Effects of eleven vitamins on growth, *Can. J. Zool.* **32**:342.

House, H. L., 1958, Nutritional requirements of insects associated with animal parasitism, *Exp. Parasitol.* **7**:555.

House, H. L., 1961, Insect nutrition, *Annu. Rev. Entomol.* **6**:13.

House, H. L., 1962, Insect nutrition, *Annu. Rev. Biochem.* **31**:653.

House, H. L., 1963, Nutritional diseases, in: *Insect Pathology*, Vol. I (E. A. Steinhaus, ed.), pp. 133–160, Academic Press, New York.

House, H. L., 1964, Effects of dietetic nucleic acids and components on growth of *Agria affinis* (Fallen) (Diptera: Sarcophagidae), *Can. J. Zool.* **42**:801.

House, H. L., 1965a, Insect nutrition, in: *The Physiology of Insecta*, Vol. 2 (M. Rockstein, ed.), pp. 769–813, Academic Press, New York.

House, H. L., 1965b, Digestion, in: *The Physiology of Insecta*, Vol. 2 (M. Rockstein, ed.), pp. 815–858, Academic Press, New York.

House, H. L., 1965c, Effects of low levels of the nutrient content of a food and of nutrient imbalance on the feeding and the nutrition of a phytophagous larva, *Celerio euphorbiae* (Linnaeus) (Lepidoptera: Sphingidae), *Can. Entomol.* **97**:62.

House, H. L., 1966, Effects of vitamins E and A on growth and development, and the necessity of vitamin E for reproduction in the parasitiod *Agria affinis* (Fallen), *J. Insect Physiol.* **12**:409.

House, H. L., 1969, Effects of different proportions of nutrients on insects, *Entomol. Exp. Appl.* **12**:659.

House, H. L., 1974, Nutrition, in: *The Physiology of Insecta* (M. Rockstein, ed.), pp. 1–62, Academic Press.

House, H. L., and Barlow, J. S., 1957, New equipment for rearing small numbers of *Pseudosarcophaga affinis* (Fall.) (Diptera: Sarcophagidae) for experimental purposes, *Can. Entomol.* **89**:145.

House, H. L., and Barlow, J. S., 1965, The effects of a new salt mixture developed for *Agria affinis* on the growth rate, body weight, and protein content of the larvae, *J. Insect Physiol.* **11**:915.

House, H. L., Singh, P., and Batsch, W. W., 1971, Artificial diets for insects: A compilation of references with abstracts, *Inform. Bull. Res. Inst. Can. Dept. Agric.*, Belleville, No. 7.

Inokuchi, T., Horie, Y., and Ito, T., 1967, Nutrition of the silkworm *Bombyx mori*. XIX. Effects of omission of essential amino acids in each of the larval instars, *Bull. Sericult. Exp. Sta. Tokyo* **22**:195.

Inokuchi, T., Horie, Y., and Ito, T., 1969, Urea cycle in the silkworm, *Bombyx mori, Biochem. Biophys. Res. Commun.* **35**:783.

Ito, T., 1967, Nutritional requirements of the silkworm, *Bombyx mori* L., *Proc. Jpn. Acad.* **43**:57.

Ito, T., and Arai, N., 1965, Nutrition of the silkworm, *Bombyx mori*. IX. Further studies on the nutritive effects of ascorbic acid, *Bull. Sericult. Exp. Sta. Tokyo* **20**:1.

Ito, T., and Arai, N., 1966, Nutrition of the silkworm, *Bombyx mori*. XI. Requirements for aspartic and glutamic acids, *J. Insect Physiol.* **12**:861.

Ito, T., and Niimura, M., 1966, Nutrition of the silkworm, *Bombyx mori*. XII. Nutritive effects of minerals, *Bull. Sericult. Exp. Sta. Tokyo* **20**:361.

Kaplanis, J. N., Robbins, W. E., Monroe, R. E., Shortino, T. J., and Thompson, M. J., 1965, The utilization and fate of β-sitosterol in the larva of the housefly, *Musca domestica* L., *J. Insect Physiol.* **11**:251.

Kasting, R., and McGinnis, A. J., 1958, The use of glucose labeled with carbon-14 to determine the amino acids essential for an insect, *Nature* **182**:1380.

Kasting, R., and McGinnis, A. J., 1960, Use of glutamic acid-u-C14 to determine nutritionally essential amino acids for larvae of the blowfly, *Phormia regina, Can. J. Biochem. Physiol.* **38**:1229.

Kasting, R., and McGinnis, A. J., 1967, An artificial diet and some growth factor requirements for the pale western cutworm, *Can. J. Zool.* **45**:787.

Kircher, H. W., and Gray, M. A., 1978, Cholestanol-cholesterol utilization by axenic *Drosophila melanogaster, J. Insect Physiol.* **24**:555.

Lafon, M., 1938, Le besoin qualitatif d'azote chez *Drosophila melanogaster* Meig, *C. R. Acad. Sci. [D] (Paris)* **207**:306.

Lake, P., and Friend, W. G., 1968, The use of artifical diets to determine some of the effects of *Nocardia rhodnii* on the development of *Rhodnius prolixus, J. Insect Physiol.* **14**:543.

Lands, W. E. M., Martin, E. H., and Crawford, C. G., 1977, Functions of polyunsaturated fatty acids: Biosynthesis of prostaglandins, in: *Polyunsaturated Fatty Acid* (W. H. Kunau and R. T. Holman, eds.), pp. 193–228, American Oil Chemists Society, Champaign.

Lea, A. O., and Delong, D. M., 1958, Studies on the nutrition of *Aedes aegypti* larvae, *Proc. Int. Cong. Entomol.* **10**:299.

Lea, A. O., Dimond, J. B., and DeLong, D. M., 1956, A chemically defined medium for rearing *Aedes aegypti* larvae, *J. Econ. Entomol.* **49**:313.

Leclercq, J., 1948, Contribution a l'etude du metabolisme de l'eau chez la larve de *Tenebrio molitor* L., *Arch. Int. Physiol.* **55**:412.

Leclercq, J., and Lopez-Francos, L., 1966, Nutrition protidique chez *Tenebrio molitor* L. VII. Nouveaux essais de remplacement de la caseine par des preparations d'acides amines, *Adv. Insect Physiol.* **74**:397.

Leclercq, J., and Lopez-Francos, L., 1967, Nutrition protidique chez *Tenebrio molitor* L. VIII. Sur la valeur nutritive des fractions de la caseine, *Adv. Insect Physiol.* **75**:89.

Levinson, H. Z., and Gothilf, S., 1965, A semisynthetic diet for axenic growth of the carob moth *Ectomyelois ceratoniae* (Zell.), *Riv. Parasitol.* **26**:19.

Levinson, H. Z., and Navon, A., 1969, Ascorbic acid and unsaturated fatty acids in the nutrition of the Egyptian cotton leafworm, *Prodenia litura, J. Insect Physiol.* **15**:591.

Levinson, Z. H., 1962, The function of dietary sterols in phytophagous insects, *J. Insect Physiol.* **8**:191.

Lipke, H., and Fraenkel, G., 1956, Insect nutrition, *Annu. Rev. Entomol.* **1**:17.

Machin, J., 1975, Water balance in *Tenebrio molitor,* L. larvae; the effect of atmospheric water absorption, *J. Comp. Physiol.* **101**:121.

Maddrell, S. H. P., 1972, The functioning of insect Malpighian tubules, in: *Role of Membranes in Secretory Processes,* pp. 338–356, North-Holland, Amsterdam.

Markkula, M., and Laurema, S., 1967, The effect of amino acids, vitamins, and trace elements on the development of *Acyrthosiphon pisum* (Harris), *Ann. Agric. Fenn.* **6**:77.

McFarlane, S. E., 1972, Vitamin E, tocopherol quinone and selenium in the diet of the house cricket, *Acheta domesticus* L., *Israel J. Entomol.* **7**:7.

McFarlane, S. E., 1974, The function of copper in the house cricket and the relation of copper to vitamin E, *Can. Entomol.* **106**:441.

McFarlane, S. E., 1976, Vitamin K: A growth factor for the house cricket (Orthoptera: Gryllidae), *Can. Entomol.* **108**:391.

McFarlane, S. E., 1977, Vitamins E and K in relation to growth of the house cricket (Orthoptera: Gryllidae), *Can. Entomol.* **109**:329.

Medici, J. C., and Taylor, M. W., 1966, Mineral requirements of the confused flour beetle, *Tribolium confusum* (DuVal), *J. Nutr.* **88**:181.

Medici, J. C., and Taylor, M. W., 1967, Interrelationships among copper, zinc, and cadmium in the diet of the confused flour beetle, *J. Nutr.* **93**:307.

Meikle, J. E. S., and McFarlane, S. E., 1965, The role of lipid in the nutrition of the house cricket, *Acheta domesticus* L., *Can. J. Zool.* **45**:87.

Mews, A. R., Langley, P. A., Pimley, R. W., and Flood, M. E. T., 1977, Large-scale rearing of tsetse flies (*Glossina* spp.) in the absence of a living host, *Bull. Entomol. Res.* **67**:119.

Michelbacher, A. E., Hoskins, W. M., and Herms, W. B., 1932, The nutrition of fleshfly larvae *Lucilia sericata* (Meig.). I. The adequacy of sterile synthetic diets, *J. Exp. Zool.* **64**:109.

Mittler, T. E., 1971, Dietary amino acid requirements of the aphid *Myzus persicae* affected by antibiotic uptake, *J. Nutr.* **101**:1023.

Mittler, T. E., 1972, Interactions between dietary components, in: *Insect and Mite Nutrition* (J. G. Rodriguez, ed.), pp. 211-223, North-Holland, Amsterdam.

Mittler, T.E., 1976, Ascorbic acid and other chelating agents in the trace mineral nutrition of the aphid *Myzus persicae* on artificial diets, *Entomol. Exp. Appl.* **20**:81.

Mittler, T. E., Tsitsipis, J. A., and Kleinjan, J. E., 1970, Utilization of dehydroascorbic acid and some related compounds by the aphid *Myzus persicae* feeding on an improved diet, *J. Insect Physiol.* **16**:2315.

Morere, J.-L., 1971a, Remplacement de l'huile de germe de mais par des acides gras dans l'alimentation de type meridique de *Plodia interpunctella* (Hbn.) (Lepidotere—Pyralidae), *C. R. Acad. Sci. [D] (Paris)* **272**:133.

Morere, J.-L. 1971b, Le carotene: Substance indispensable pour la nutrition de *Plodia interpunctella* (Lep. Pyralidae), *C. R. Acad. Sci. [D] (Paris)* **272**:2229.

Morrison, P. E., and Davies, D. M., 1964, Repeated ovarian cycles with ribonucleic acid in the diet of adult house flies, *Nature* **201**:948.

Murray, D. R. P., 1968, The importance of water in the normal growth of larvae of *Tenebrio molitor*, *Entomol. Exp. Appl.* **11**:149.

Naton, E., 1967, Die Auswirkungen der Carnitinblockade durch γ-Butyrobetain bei den Larven von *Tribolium destructor* Uyttenb. und einigen enderen Vorratsschadlingen, *J. Stored Products Res.* **3**:49.

Nayar, J. K., 1964, The nutritional requirements of grasshoppers. I. Rearing of the grasshopper *Melanoplus bivittatus* (Say) on a completely defined synthetic diet and some effects of different concentrations of B-vitamin mixture, linoleic acid and β-carotene, *Can. J. Zool.* **42**:11.

Naylor, A. F., 1963, Glutamic and aspartic acids and sucrose in the diet of the flour beetle *Tribolium confusum*, *Can. J. Zool.* **41**:1127.

Newburg, D. S., and Fillios, L. C., 1979, A requirement for dietary asparagine in pregnant rats, *J. Nutr.* **109**:2190.

Newburg, D. S., Frankel, D. L., and Fillios, L. C., 1975, An asparagine requirement in young rats fed the dietary combination of aspartic acid, glutamine, and glutamic acid, *J. Nutr.* **105**:356.

Noble-Nesbitt, J., 1970a, Water uptake from sub-saturated atmospheres: Its site in insects, *Nature* **225**:753.

Noble-Nesbitt, J., 1970b, Water balance in firebrat *Thermobia domestica*, *J. Exp. Biol.* **52**:193.

Noda, H., Wada, K., and Saito, T., 1979, Sterols in *Laodelphax striatellus* with special reference to the intracellular yeastlike symbiotes as a sterol source, *J. Insect Physiol.* **25**:443.

Norby, S., 1971, A specific requirement for pyrimidine in rudimentary mutants of *Drosophila melanogaster, Hereditas* **66**:205.

Norris, D. M., and Chu, H.-M., 1971, Maternal *Xyleborus ferrugineus* transmission of sterol or sterol-dependent metabolites necessary for progeny pupation, *J. Insect Physiol.* **17**:1741.

Nunakata, K., 1970, Insect antifeedants in plants, in: *Control of Insect Behaviour by Natural Products* (D. L. Wood, R. M. Silverstein, and M. Nakajima, eds.), pp. 179–187, Academic Press, New York.

O'Donnell, M. J., 1978, The site of water vapour absorption in *Arenivaga investigata,* in: *Comparative Physiology—Water, Ions and Fluid Mechanics* (K. S. Schmidt-Nielsen, L. Bolis, and S. H. P. Maddrell, eds.), pp. 115–120, Cambridge University Press, Cambridge.

Ouye, M. T., and Vanderzant, E. S., 1964, B-vitamin requirements of the pink bollworm, *J. Econ. Entomol.* **57**:427.

Pant, N. C., Gupta, P., and Nayar, J. K., 1960, Physiology of intracellular symbiotes of *Stegobium paniceum* L. with special reference to amino acid requirements of the host, *Experientia* **15**:311.

Pausch, R. D., and Fraenkel, G., 1966, The nutrition of the larva of the oriental rat flea *Xenopsylla cheopis* (Rothschild), *Physiol. Zool.* **39**:202.

Reddy, G. P. V., and Chippendale, G. M., 1972, Observations on the nutritional requirements of the southwestern corn borer, *Diatraea grandiosella, Entomol. Exp. Appl.* **15**:51.

Retnakaren, A., and Beck, S. D., 1968, Amino acid requirements and sulphur amino acid metabolism in the pea aphid, *Acyrthosiphon pisum* (Harris), *Comp. Biochem. Physiol.* **24**:611.

Richmond, J. A., and Thomas, H. A., 1975, *Hylobius pales:* Effect of dietary sterols on development and on sterol content of somatic tissue, *Ann. Entomol. Soc. Am.* **68**:329.

Rivers, J. P. W., Sinclair, A. J., and Crawford, M. A., 1975, Inability of the cat to desaturate essential fatty acids, *Nature* **258**:171.

Robbins, W. E., 1963, Studies on the utilization, metabolism and function of sterols in the house-fly, *Musca domestica,* in: *Radiation and Radioisotopes Applied to Insects of Agricultural Importance,* pp. 269–280, IAEA, Vienna.

Robbins, W. E., Kaplanis, J. N., Svoboda, J. A., and Thompson, M. J., 1971, Steroid metabolism in insects, *Annu. Rev. Entomol.* **16**:53.

Rock, G. C., 1967, Aseptic rearing of the codling moth on synthetic diets: Ascorbic acid and fatty acid requirements, *J. Econ. Entomol.* **60**:1002.

Rock, G. C., and King, K. W., 1967a, Qualitative amino acid requirements of the red-banded leafroller, *Argyrotaenia velutinana, J. Insect Physiol.* **13**:59.

Rock, G. C., and King, K. W., 1967b, Quantitative amino acid requirements of the red-banded leafroller, *Argyrotaenia velutinana, J. Insect Physiol.* **13**:175.

Rock, G. C., Glass, E. H., and Patton, R. L., 1964, Axenic rearing of the redbanded leafroller, *Argyrotaenia velutinana,* on meridic diets, *Ann. Entomol. Soc. Am.* **57**:617.

Rock, G. C., Patton, R. L., and Glass, E. H., 1965, Studies on the fatty acid requirements of *Argyrotaenia velutinana* (Walker), *J. Insect Physiol.* **11**:91.

Rock, G. C., Khan, A., and Hodgson, E., 1975, The nutritional value of seven D-amino acids and α-ketoacids for *Argyrotaenia velutinana, Heliothis zea,* and *Phormia regina, J. Insect Physiol.* **21**:693.

Rockstein, M., 1978, *Biochemistry of Insects,* Academic Press, New York.

Rodriguez, J. G., 1966, Axenic arthropoda: Current status of research and further possibilities, *Ann. N.Y. Acad. Sci.* **139**:53.

Rowell, C. H. F., 1971, The variable colouration of the acridoid grasshoppers, *Adv. Insect Physiol.* **8**:146.

Royes, W. V., and Robertson, F. W., 1964, The nutritional requirements and growth relations of different species of *Drosophila, J. Exp. Zool.* **156**:105.

Sang, J. H., 1956, The quantitative nutritional requirements of *Drosophila melanogaster, J. Exp. Biol.* **33**:45.
Sang, J. H., 1959, Circumstances affecting the nutritional requirements of *Drosophila melanogaster, Ann. N.Y. Acad. Sci.* **77**:352.
Sang, J. H., 1962, Relationships between protein supplies and B-vitamin requirements, in axenically cultured *Drosophila, J. Nutr.* **77**:355.
Sang, J. H., 1972, The use of mutants in nutritional research, in: *Insect and Mite Nutrition* (J. G. Rodriguez, ed.), pp. 9–17, North-Holland, Amsterdam.
Schoonhoven, L. M., 1968, Chemosensory bases of host plant selection, *Annu. Rev. Entomol.* **13**:115.
Schwarz, K., and Milne, D. B., 1971, Growth effects of vanadium in the rat, *Science* **174**:426.
Sedee, D. J. W., 1958, Dietetic requirements and intermediary protein metabolism of an insect (*Calliphora erythrocephala* Meig.), *Entomol. Exp. Appl.* **1**:38.
Settya, B. N. Y., and Ramaiah, T. R., 1979, Isolation and identification of prostaglandins from the reproductive organs of male silkmoth *Bombyx mori* L., *Insect Biochem.* **9**:613.
Settya, B. N. Y., and Ramaiah, T. R., 1980, Effects of prostaglandins and inhibitors of prostaglandin biosynthesis on oviposition in the silkmoth *Bombyx mori* L., *J. Exp. Biol.* **13**:539.
Sharma, G. K., Hodgson, E., and Rock, G. C., 1972, Nutrition and metabolism of sulphur amino acids in *Argyrotaenia velutinana* larvae, *J. Insect Physiol.* **18**:9.
Singh, K. R. P., and Brown, A. W. A., 1957, Nutritional requirements of *Aedes aegypti* L., *J. Insect Physiol.* **1**:199.
Singh, P., 1972, Bibliography of artificial diets for insects and mites, N.Z. *Dept. Sci. Indust. Res.* **209**:15.
Singh, P., 1977, *Artificial Diets for Insects, Mites, and Spiders,* Plenum Press, New York.
Sivapalan, P., and Gnanapragasam, N. C., 1979, The influence of linoleic acid and linolenic acid on adult moth emergence of *Homona coffearia* from meridic diets *in vitro, J. Insect Physiol.* **25**:393.
Smith, B. C., 1965, Effects of food on the longevity, fecundity, and development of adult coccinellids (Coleoptera: Coccinellidae), *Can. Entomol.* **97**:910.
Smith, J. J. B., and Friend, W. G., 1976, Further studies on potencies of nucleotides as gorging stimuli during feeding in *Rhodnius prolixus, J. Insect Physiol.* **22**:607.
Sneller, V. P., and Dadd, R. H., 1981, Lecithin in synthetic larval diet for *Aedes aegypti* improves larval and adult performance, *Entomol. Exp. Appl.* **29**:9.
Sprecher, H., 1977, Biosynthesis of polyunsaturated fatty acids and its regulation, in: *Polyunsaturated Fatty Acid* (W. H. Kunau and R. T. Holman, eds.), pp. 1–18, American Oil Chemists Society, Champaign.
Stanley-Samuelson, D. W., 1980, Long-chain polyunsaturated fatty acids in whole-animal and specific-tissue extracts of insects, Abstract #577, *Am. Zool.* **20**:832.
Stanley-Samuelson, D. W., and Dadd, R. H., 1981, Arachidonic acid and other tissue fatty acids of *Culex pipiens* reared with various concentrations of dietary arachidonic acid, *J. Insect Physiol.*, in press.
Svoboda, J. A., Robbins, W. E., Cohen, C. F., and Shortino, T. J., 1972, Phytosterol utilization and metabolism in insects: Recent studies with *Tribolium confusum,* in: *Insect and Mite Nutrition* (J. G. Rodriguez, ed.), pp. 505–516, North-Holland, Amsterdam.
Svoboda, J. A., Thompson, M. J., Elden, T. C., and Robbins, W. E., 1974, Unusual composition of sterols in a phytophagous insect, the mexican bean beetle reared on soybean plants, *Lipids* **9**:752.
Svoboda, J. A., Kaplanis, J. N., Robbins, W. E., and Thompson, M. J., 1975a, Recent developments in insect steroid metabolism, *Annu. Rev. Entomol.* **20**:205.
Svoboda, J. A., Thompson, M. J., Robbins, W. F., and Elden, T. C., 1975b, Unique pathways of sterol metabolism in the Mexican bean beetle, a plant-feeding insect, *Lipids* **10**:524.

Svoboda, J. A., Dutky, S. R., Robbins, W. E., and Kaplanis, J. N., 1977, Sterol composition and phytosterol utilization and metabolism in the milkweed bug, *Lipids* **12**:318.

Svoboda, J. A., Thompson, M. J., Robbins, W. E., and Kaplanis, J. N., 1978, Insect steroid metabolism, *Lipids* **13**:742.

Svoboda, J. A., Nair, A. M. G., Agarwal, H. C., and Robbins, W. E., 1979, The sterols of the khapra beetle, *Trogoderma granarium* Everts, *Experientia* **35**:1454.

Thompson, M. J., Svoboda, J. A., Kaplanis, J. N., and Robbins, W. E., 1972, Metabolic pathways of steroids in insects, *Annu. Rev. Entomol.* **16**:53.

Thorsteinson, A. J., 1960, Host selection in phytophagous insects, *Annu. Rev. Entomol.* **5**:193.

Tinoco, J., Babcock, R., Hincenbergs, I., Medwakowski, B., Miljanich, P., and Williams, M. A., 1979, Linolenic acid deficiency, *Lipids* **14**:166.

Trager, W., 1947, Insect nutrition, *Biol. Rev.* **22**:148.

Trager, W., 1953, Nutrition, in: *Insect Physiology* (K. D. Roeder, ed.), pp. 350–386, John Wiley & Sons, New York.

Treherne, J. E., 1967, Gut absorption, *Annu. Rev. Entomol.* **12**:43.

Turner, R. B., 1971, Dietary amino acid requirements of the cotton aphid *Aphis gossypii:* The sulphur-containing amino acids, *J. Insect Physiol.* **17**:2451.

Turunen, S., 1974, Polyunsaturated fatty acids in the nutrition of *Pieris brassicae* (Lepidoptera), *Ann. Zool. Fenn.* **11**:300.

Turunen, S., 1976, Vitamin E: Effect on lipid synthesis and accumulation of linolenate in *Pieris brassicae, Ann. Zool. Fenn.* **13**:148.

Uvarov, B. P., 1928, Insect nutrition and metabolism, *Trans. Entomol. Soc. Lond.* **76**:255.

Vanderzant, E. S., 1963, Nutrition of the boll weevil larva, *J. Econ. Entomol.* **56**:357.

Vanderzant, E. S., 1966, Defined diets for phytophagous insects, in: *Insect Colonization and Mass Production* (C. N. Smith, ed.), pp. 273–303, Academic Press, New York.

Vanderzant, E. S., 1968, Dietary requirements for the bollworm, *Heliothis zea,* for lipids, choline, and inositol and the effects of fats and fatty acids on the composition of the body fat, *Ann. Entomol. Soc. Am.* **61**:120.

Vanderzant, E. S., 1969a, Physical aspects of artificial diets, *Entomol. Exp. Appl.* **12**:642.

Vanderzant, E. S., 1969b, An artificial diet for larvae and adults of *Chrysopa carnea,* an insect predator of crop pests, *J. Econ. Entomol.* **62**:256.

Vanderzant, E. S., and Chremos, J. H., 1971, Dietary requirements of the boll weevil for arginine and the effect of arginine analogues on growth and on the composition of the body amino acids, *Ann. Entomol. Soc. Am.* **64**:480.

Vanderzant, E. S., and Richardson, C. D., 1963, Ascorbic acid in the nutrition of plant feeding insects, *Science* **140**:989.

Vanderzant, E. S., and Richardson, C. D., 1964, Nutrition of the adult boll weevil: Lipid requirements, *J. Insect Physiol.* **10**:267.

Vanderzant, E. S., Kerur, D., and Reiser, R., 1957, The role of dietary fatty acids in the development of the pink bollworm, *J. Econ. Entomol.* **50**:606.

Vanderzant, E. S., Pool, M. C., and Richardson, C. D., 1962, The role of ascorbic acid in the nutrition of three cotton insects, *J. Insect Physiol.* **8**:287.

Vinson, S. B., 1967, Effect of several nutrients on DDT resistance in tobacco hornworm, *J. Econ. Entomol.* **60**:565.

Vogel, B., 1931, Uber die Beziehungen swischen Sussgeschmack und Nahrwert von Zuckern und Zuckeralkoholen bei der Honigbiene, *Z. Vergl. Physiol.* **14**:273.

Vyse, E. R., and Sang, J. H., 1971, A purine and pyrimidine mutant of *Drosophila melanogaster, Genet. Res.* **18**:117.

Waggoner, P. E., 1967, Moisture loss through the boundary layer, *Biometeorology* **3**:41.

Waldbauer, G. P., 1962, The growth and reproduction of maxillectomized tobacco hornworms feeding on normally rejected non-solanaceous plants, *Entomol. Exp. Appl.* **5**:147.

Waldbauer, G. P., 1964, The consumption, digestion and utilization of solanaceous and non-solanaceous plants by larvae of the tobacco hornworm, *Protoparce sexta* (Johan.), *Entomol. Exp. Appl.* **7**:253.

Waldbauer, G. P., 1968, The consumption and utilization of food by insects, *Adv. Insect Physiol.* **5**:229.

Wardojo, S., 1969, Some factors relating to the larval growth of the Colorado potato beetle, *Leptinotarsa decemlineata* Say, on artificial diets, *Meded. Landbouw. Wag.* **69**:1.

Wharton, G. W., and Arlian, L. G., 1972, Utilization of water by terrestrial mites and insects, in: *Insect and Mite Nutrition* (J. G. Rodriguez, ed.), pp. 153–165, North-Holland, Amsterdam.

White, R. H., 1978, Insect visual pigments, *Adv. Insect Physiol.* **13**:35.

Wigglesworth, V. B., 1972, *The Principles of Insect Physiology,* Seventh Edition, Methuen, London.

Wood, D. L., Silverstein, R. M., and Nakajima, M., 1970, *Control of Insect Behaviour by Natural Products,* Academic Press, New York.

Wyatt, G. R., 1967, The biochemistry of sugars and polysaccharides in insects, *Adv. Insect Physiol.* **4**:287.

Yasgan, S., 1972, A chemically defined synthetic diet and larval nutritional requirements of the endoparasitoid *Itoplectis conquistor* (Hymenoptera), *J. Insect Physiol.* **18**:2123.

Yokoyama, T., 1963, Sericulture, *Annu. Rev. Entomol.* **8**:287.

Chapter 9

The Nutrient Requirements of Cultured Mammalian Cells

William J. Bettger and Richard G. Ham

1. Introduction

1.1. Definitions

This chapter seeks to bridge a gap in current research by analyzing the nutrient requirements of cultured mammalian cells from a perspective that is usually applied only to nutritional studies with intact animals. The combined experience of the authors includes both whole-animal nutrition and the growth requirements of cultured cells. In order to make our presentation clear to members of both disciplines, it is necessary to begin by comparing a number of concepts and specific terms from the two fields.

1.1.1. Growth and Cellular Multiplication

In whole-animal nutrition, the critical test of nutritional adequacy is usually whether a diet will support normal growth and development, which often has more stringent requirements than does maintenance of a healthy adult. Growth refers to a net increase in size and mass of the animal, which for any given organ system typically proceeds in three precisely regulated stages: (1) hyperplasia (an increase in cell number); (2) hyperplasia with hypertrophy (an increase in cell

---

William J. Bettger and Richard G. Ham • Department of Molecular, Cellular and Developmental Biology, University of Colorado, Boulder, Colorado 80309. Present Address for William J. Bettger: Department of Biochemistry, St. Louis University School of Medicine, St. Louis, Missouri 63104.

size); and (3) hypertrophy alone. Hyperplasia, or cellular multiplication, represents the critical stage in development and occurs in different organs for varying periods of time, including some where cellular replacement continues throughout the life-span (Enesco and Leblond, 1962; Winick and Noble, 1965). Malnutrition during the period of cellular multiplication typically results in decreased cell number, which cannot be reversed later by adequate nutrient intake, whereas malnutrition after the period of cellular multiplication results in decreased cell size, which generally can be reversed (Winick and Noble, 1966; Pike and Brown, 1975).

In cell culture, the objective is usually to achieve an increase in cell number with no net change in the size or mass of individual cells. This is sometimes loosely referred to as cellular "growth," but it is preferable to use the terms "multiplication" or "proliferation" to distinguish it from situations where cellular size actually does change in culture. The multiplication of animal cells in culture offers a unique model system for the study of hyperplasia, provided that the right cell type and the right type of assay are employed, as will be discussed below.

1.1.2. Normal and Transformed Cells

As used in this chapter, the term "normal" refers only to cells that do not differ in any significant way from cells found in healthy intact animals. By definition, a normal cell has an unaltered euploid karyotype (usually diploid) and does not form a tumor when injected into a nude mouse or other test animal. Most types of normal cells are also dependent on anchorage to a suitable substrate for multiplication *in vitro* and cease multiplication when they become crowded (density-dependent inhibition of multiplication). In addition, they typically undergo cellular senescence and permanently withdraw from multiplication after a finite number of doublings in culture.

At the other extreme, fully malignant cells that form tumors in appropriate test animals typically exhibit a set of properties that are collectively referred to as "transformed." These include loss of anchorage dependence, loss of density-dependent inhibition of multiplication, aneuploidy, and, in most cases, immortality (formation of permanent cell lines that continue to multiply indefinitely without cellular senescence). Many cell lines in culture possess properties that are intermediate between fully normal and fully transformed. In this review, cells that exhibit any transformed properties are considered not to be normal. It should be noted, however, that certain properties commonly associated with transformation are normal for certain kinds of cells (e.g., lack of anchorage dependence in chondrocyte and leukocyte cultures).

The history of mammalian cell culture has been heavily influenced by permanent cell lines such as mouse L and HeLa. Such lines exhibit many transformed properties. In addition, they appear to have undergone extensive evolution in culture, including simplification of their growth requirements (Ham, 1974a, 1981). Since such cells no longer reliably reflect the requirements of normal cells in the intact organism, relatively little will be said about them in this chapter. However, it should be noted that many widely used culture media were originally developed to satisfy the requirements of permanent cell lines rather than normal cells. These media and the cell lines that were used for their development have been reviewed in detail in other publications (Higuchi, 1973; Katsuta and Takaoka, 1973; Ham and McKeehan, 1979).

Once such medium is of interest to this review, however. In 1959, Eagle published a ''minimum essential medium'' (MEM) which contained, in a well-balanced formulation, only those nutrients that he could show to be needed for multiplication of mouse L and HeLa cells in the presence of a small amount of dialyzed serum (Eagle, 1959). Although there are some exceptions, the 29 components of MEM (glucose, 13 amino acids, choline, inositol, six vitamins, and seven major inorganic ions) are widely accepted as essential nutrients for most cultured cells and are often used as the starting point for more extended studies of cellular growth requirements (Ham, 1981). It should also be emphasized that MEM is not a true ''minimum essential'' nutrient mixture in that it was designed for use with dialyzed serum and does not contain certain nutrients that tend to remain bound to dialyzed serum macromolecules (e.g., biotin, vitamin B_{12}, iron, and other inorganic trace elements).

In recent years, cell culture technology has progressed to a point where a variety of normal cell types can now be grown *in vitro* using well-defined techniques (Jakoby and Pastan, 1979; Harris *et al.*, 1980). This chapter will focus primarily on the nutrient requirements of normal cells and the relationship between these requirements and those of intact animals of the same species.

Although each cell type exhibits some degree of individuality in its cycle of cell division because of the complex interplay of environmental signals with its genome, the overall division process in mammalian cells can be generalized into four major parts, designated as M (mitosis), G_1 (''gap'' between M and S), S (DNA synthesis), and G_2 (''gap'' between S and M). In addition, non-transformed cells can withdraw into a quiescent state of indefinite length designated G_0. Each stage of the cell cycle is defined by a series of biochemical and cytological events, which have recently been reviewed (Pardee *et al.*, 1978). Multiplication of cells *in vivo* appears to involve similar stages.

1.1.3. Nutrients and Growth Requirements

''Nutrient'' is a highly restrictive term which, in this chapter, will be used only to refer to a chemical substance that is taken into a cell and utilized as a substrate in biosynthesis or energy metabolism, as a catalyst in one of these processes, or as a structural component of a cellular organelle. Substances (and conditions) that promote cellular multiplication by other means are specifically excluded from the definition of a nutrient but are included in the much broader term "growth requirement," which will be used in this chapter to refer to virtually anything (including nutrients) that has a positive effect on cellular multiplication. For example, normal cells typically require hormones and "growth factors" that appear to control multiplication by interactions with stereospecific receptors in a manner that is usually considered regulatory rather than nutritional. In addition to nutrients, hormones, and growth factors, the total set of growth requirements for normal cells encompasses diverse variables such as temperature, pH, osmolarity, quantitative balance, and the nature of the surface the cells are grown on. Pharmacological agents are also sometimes added to currently available media to stimulate multiplication.

The nonnutrient growth requirements of cultured normal cells are described only superficially in this chapter, which focuses primarily on cellular nutrient requirements. However, the broader set of growth requirements has been discussed in detail in other recent reviews (Ham and McKeehan, 1978a,b, 1979; Ham, 1981; Barnes and Sato, 1980), and several recent volumes have been devoted primarily to hormones and growth factors (Sato and Ross, 1979; Jiminez de Asua *et al.*, 1980; Fox, 1981). For purposes of this writing, it is sufficient to say that adequate cellular nutrition is an obligatory prerequisite for continued action of the nonnutrient factors that stimulate multiplication in normal cells.

1.2. Types of Cellular Multiplication Assays

There exist a variety of cellular multiplication assays whose major differences are the initial cellular population density and the total amount of replication that is involved. At one extreme, a nearly confluent monolayer of cells that has entered G_0 because of nutrient deprivation may be tested for its ability to undergo a single round of DNA synthesis in response to addition of a specific nutrient. At the other extreme, in the clonal growth assay, single isolated cells are required to multiply sufficiently to form discretely separated colonies, each of which is a clone of genetically identical cells.

Clonal growth is the most stringent assay for adequacy and freedom from toxicity of a culture medium (Ham, 1980b). It starts with a very small cellular inoculum that is unable to "condition" the medium and requires each cell to undergo multiple rounds of proliferation (a minimum of six and usually 10–12 sequential doublings) such that any nutrients inadvertently introduced with the cellular inoculum are quickly exhausted. Techniques and special requirements for clonal growth have been discussed in detail elsewhere (Ham, 1972, 1974b, 1981; Ham and McKeehan, 1978a,b, 1979). As used in this chapter, the term "clonal growth" will be restricted to a maximum inoculum of 2000 cells per ml of medium in a culture dish large enough so that each surviving cell can form a discrete colony.

1.3. Statement of Intent

The objective of this chapter is to analyze the nutrient requirements of cultured cells in a manner that is meaningful to studies of whole-animal nutrition. An accurate comparison of the requirements of cultured cells to those of healthy intact animals makes it necessary to consider only cultured cells whose requirements have not deviated significantly from those of normal cells in the animal. We have therefore focused on normal cells in this chapter and have excluded data from transformed cells whenever possible. In addition, data based on clonal growth assays have been given the most weight, since such assays are the only procedures that can accurately define nutrient requirements of single cells without interference caused by conditioning of the medium and carryover of nutrients from the inoculum, as discussed above. To keep the review as simple as possible, we have also restricted it to mammals. The limited data available on growth requirements of avian cells generally corroborate the mammalian data, with differences only in details.

These restrictions leave a rather limited body of data to work with for several reasons: (1) the number of different types of normal mammalian cells that have been grown under clonal conditions is quite small; (2) clonal growth of normal cells in media without complex undefined supplements has been achieved only very recently, and thus far, only for a very few cell types (Table I); and (3) a rigorous analysis of nutrient requirements has not yet been carried to completion for any type of normal mammalian cell in a fully defined medium.

In spite of these limitations, there is sufficient information available to construct an overview of the nutritional requirements of normal mammalian cells in culture that both demonstrates the value of such studies and points to the need for their expansion in the future. By drawing direct comparisons between whole-animal nutrition and cellular nutrition whenever possible, we seek specifically to encourage more nutritionists to utilize cultured cells as experimental tools, particularly in studies of human nutrition, where direct experimentation is often not possible.

2. Nutrition *in Vivo* and *in Vitro*

2.1. Historical Separation

The history of the development of nutrition as a science and current knowledge of the nutritional requirements for animals including man have been extensively reviewed (Pike and Brown, 1975; Goodhart and Shils, 1980). However, the science of nutrition has not readily encompassed the nutrition of mammalian cells in culture, which has been, and to some extent remains, a neglected problem in modern biology and nutrition (Ham, 1974a). There exists no unified textbook on the nutrition of mammalian cells. The closest current approximations

Table I. Growth of Normal Mammalian Cells in Defined Media

Cell type and reference	Medium	Supplements replacing serum	Assay system
Human fibroblasts (fetal lung and neonatal foreskin) Bettger *et al.* (1981)	MCDB 110	Insulin, EGF, dexamethasone, liposomes (made from mixed phospholipids, cholesterol, sphingomyelin, vitamin E), prostaglandins E_1, $F_{2\alpha}$, phosphoenolpyruvate, glutathione(SH), dithiothreitol[a]	Clonal, 1000 cells/ 60-mm dish, 12–14 days
Human keratinocytes (neonatal foreskin) Tsao *et al.* (1981)	MCDB 152	Insulin, EGF, hydrocortisone, progesterone, ethanolamine, phosphoethanolamine, transferrin	Clonal, 1000 cells/ 60-mm dish, 10–14 days
Rabbit chondrocytes (elastic cartilage of the ear) Jennings and Ham (1981)	MCDB 110	Insulin, EGF, liposomes (same as for human fibroblasts)	Clonal, 1000 cells/ 60-mm dish, 10–14 days
Mouse kidney epithelial cells Taub and Sato (1979)	DME : F12 (1 : 1)	Insulin, hydrocortisone, triiodothryonine, PGE_1, transferrin	5000 cells/ 35-mm dish, 10 days
Rabbit corneal epithelium and endothelium Chan and Haschke (1981)	DME	Insulin, hydrocortisone, progesterone, putrescine, selenium, transferrin (attached with serum)	Monolayer (epithelium) Explants (endothelium)
Mouse Sertoli cells Mather and Sato (1979)	DME : F12 (1 : 1)	Insulin, EGF, follicle-stimulating hormone, somatomedin C, growth hormone, transferrin	Monolayer
Human prostatic epithelium Chaproniere-Rickenberg and Webber (1980)	RPMI 1640	EGF, dexamethasone, putrescine (or spermidine), polyvinylpyrollidone, transferrin, vitamin A	Outgrowth from explanted acini
Rat ovarian follicle line RF-1 Orly and Sato (1979)	DME : F12 (1 : 1)	Insulin, hydrocortisone, transferrin, fibronectin	13,000 cells/ 35-mm dish
Rat thyroid line FRTL Ambesi-Impiombato *et al.* (1980)	Modified F12	Insulin, hydrocortisone, transferrin, somatostatin, glycyl-histidyl-lysine, thyrotropin (attached with low serum)	2×10^5 cells/ 60-mm dish

[a]The supplement for human diploid fibroblasts is added in three parts: A, prostaglandins, PEP, glutathione, and dithiothreitol; B, the liposome mixture; and C, insulin, EGF, and dexamethasone.

are multiauthor research monographs that deal with selected topics in depth but fail to achieve a comprehensive or consistent coverage of the entire field (Rothblat and Cristofalo, 1972; Katsuta, 1978; Waymouth *et al.*, 1981). All of the older reviews of cellular nutrition reflect the fact that until quite recently nearly all studies of cellular growth requirements were done with permanent cell lines (Morgan, 1958; Levintow and Eagle, 1961; Waymouth, 1965; Higuchi, 1973;

Katsuta and Takaoka, 1973), and even relatively recent reviews freely mix plentiful data from transformed cells with the limited data available from normal cells (Ham and McKeehan 1978b, 1979; Rizzino *et al.*, 1979; Ham, 1981).

Most cell culturists are reluctant to apply the principles of animal nutrition studies rigorously and consistently to their research, and most nutritionists are reluctant to use cells in culture to study nutritional principles. This may be partially because normal cells in culture require more than simple nutrients to grow optimally, as discussed above. Part of the problem also lies in the fact that relatively few investigators have sufficiently diverse backgrounds and experience to deal effectively with the complex three-way interactions that occur among nutrients, regulatory factors, and cultured normal cells. It is primarily in laboratories that employ a team approach, with experts from diverse areas of cell biology, biochemistry, and nutrition, that such studies are being pursued successfully.

2.2. Total Parenteral Nutrition and Cellular Nutrition

The advent of a new branch of nutrition, called total parenteral nutrition (TPN) (Goodhart and Shils, 1980), appears to be drawing the disciplines of nutrition and cell culture closer together. In particular, TPN makes it easier for nutritionists to think of "food" as a solution that bypasses gut absorption and becomes directly available to cells *in vivo*. From there it is only a short conceptual leap to an interest in cell culture media.

Table II compares a typical complete TPN formula for humans with two recently developed defined media for clonal growth of normal human fibroblasts. The first, designated MCDB 110 + ABC, supports rapid clonal growth of human diploid fibroblasts (Bettger *et al.*, 1981). It consists of a defined nutrient medium (MCDB 110) plus three complex supplements (A, B, C) that contain additional nutrients as well as hormones and growth factors (cf. Table I). The second, designated MCDB 111, is an experimental formulation that contains only true nutrients and supports suboptimal multiplication of human diploid fibroblasts in the total absence of exogenously added proteins or hormones. It consists of MCDB 110 plus additional nutrients, some of which are also in supplements A and B.

Comparison of the TPN solution to these culture media clearly illustrates the difference between nutrition of cells *in vivo* and *in vitro*. The basic difference is that cells in culture cannot simply be "fed" in the sense of introducing a daily allowance of essential nutrients into an extracellular environment that is precisely controlled by its own homeostatic systems. Cells *in vitro* do not have the regulating effect of liver and kidneys for control of extracellular concentrations of essential nutrients and toxic waste products. The objective of a TPN solution is to provide all essential nutrients for the entire organism together with sufficient energy (calories) to maintain a normal level of metabolism. This is typically achieved by supplying a recommended daily allowance (RDA) of each essential

Table II. Comparison of a Total Parenteral Nutrition (TPN) Formula with Defined Media for Human Fibroblasts

Nutrient	TPN formula[a]		Fibroblast defined media[b]	
	mg/liter	% available energy	mg/liter	% available energy
Crystalline amino acids	3.6×10^4	10%	816.7	48%
L-Isoleucine	2,478		3.9	
L-Leucine	3,234		13.1	
L-Lysine	3,234		36.6	
L-Methionine	1,890		4.5	
L-Phenylalanine	2,016		4.6	
L-Threonine	1,428		11.9	
L-Tryptophan	546		2.1	
L-Valine	2,352		11.7	
L-Alanine	2,520		8.9	
L-Arginine	1,302		210.7	
L-Histidine	1,008		20.9	
L-Proline	3,990		34.5	
L-Serine	2,100		10.5	
Glycine	7,518		22.5	
L-Cysteine	84		8.8	
L-Aspartate	—		13.3	
L-Asparagine	—		15.0	
L-Glutamate	—		14.7	
L-Glutamine	—		365.0	
L-Tyrosine	—		3.5	
Glucose	2.5×10^5	69%	730	43%
Lipid	3.36×10^4	21%	10	1%
Soybean oil	3×10^4		—	
Egg phosphatides	3.6×10^3		—	
Soybean lecithin	—		6	
Cholesterol	—		3	
Sphingomyelin	—		1	
Other organic compounds			143.3	8%
Adenine	—		1.35	
Choline chloride	—		13.9	
Inositol	—		18.0	
Putrescine	—		1.6×10^{-4}	
Sodium pyruvate	—		110	
Thymidine	—		7.2×10^{-2}	
Phosphoenol pyruvate	—		1.8	
Ethanolamine	—		0.6[c]	
Taurine	—		1.25[c]	
Major inorganic salts				
K^+	585		195	
Na^+	689		2668	
Mg^{2+}	97		24.3	
Ca^{2+}	200		40.0	
Cl^-	779		3929	
P(i)	927		93.0	
S(i)	—		32.0	

(continued)

Table II. (*Continued*)

Nutrient	TPN formula[a] mg/liter	TPN formula[a] % available energy	Fibroblast defined media[b] mg/liter	Fibroblast defined media[b] % available energy
Vitamins				
Vit C	500		3.6 (Ascorbate-2-SO_4)[c]	
Vit A	10,000 IU/liter		—	
Vit D	1,000 IU/liter		—	
Vit E	51 IU/liter		0.26 IU/liter	
Thiamine	50		0.34	
Riboflavin	10		0.11	
Pyridoxine	15		6.2×10^{-2}	
Niacinamide	100		6.1	
Dexpanthenol	25		0.23 (pantothenic acid)	
Vit B_{12}	5×10^{-3}		0.13	
Folic acid	0.2		6.0×10^{-4} (folinic acid)	
Biotin	3×10^{-2}		7.3×10^{-3}	
Vit K	1.8×10^{-1}		—	
Trace elements				
I	5×10^{-2}		—	
Fe	4		0.28	
Zn	7.5		3.2×10^{-2}	
Cu	1.6		6.3×10^{-5}	
Cr	7×10^{-3}		—	
Mn	0.5		5.5×10^{-5}	
Se	0.125[e]		2.4×10^{-3}	
Mo	—		9.6×10^{-5}	
V	—		2.5×10^{-4}	
Ni	—		2.9×10^{-5}	
Sn	—		5.9×10^{-5}	
Si	—		1.4×10^{-2}	
Nonnutrient components of fibroblast defined media				
Phenol red (pH indicator)	—		1.2	
HEPES (buffer)			7.1×10^{3}	
Insulin	—		9.5×10^{-4}[d]	
Epidermal growth factor	—		3.0×10^{-5}[d]	
Dexamethasone	—		2.0×10^{-4}[d]	
Prostaglandin E_1	—		9.0×10^{-6}[d]	
Prostaglandin $F_{2\alpha}$	—		7.1×10^{-4}[d]	
Glutathione (reduced)	—		2.0×10^{-4}[d]	
Dithiothreitol	—		1.0×10^{-3}[d]	

[a]The TPN formula used in this table is a composite, based on Schneider and Buchwald (1979), Meng (1977), and American Medical Association (1979).
[b]Except where indicated otherwise, the values given apply both to MCDB 110 plus supplements as shown in Table I (Bettger *et al.*, 1981), and to MCDB 111, which is an experimental formulation designed to support as much multiplication of human diploid fibroblasts as possible in the complete absence of proteins or other nonnutrient multiplication-promoting substances (W. Bettger, unpublished observations).
[c]This component is present in MCDB 111 but absent from MCDB 110 plus supplements.
[d]This component is present in MCDB 110 plus supplements but is absent from MCDB 111.
[e]Present as a containment (Smith and Goos, 1980).

nutrient and then bringing the solution to isotonicity with energy-providing compounds. Such solutions have high concentrations of glucose and amino acids and relatively low levels of the major ions. Cell culture media, on the other hand, must start with concentrations of major ions that simulate mammalian extracellular fluid. The concentrations of energy-providing compounds that can be supplied must therefore be kept much lower than in TPN solutions to avoid hypertonicity. Toxic effects of high concentrations of vitamins and trace minerals must also be carefully avoided.

Thus, although the objective in both cases is complete cellular nutrition, the strategies that must be followed are rather different. Cultured cells must actually live and multiply in their medium as it is formulated, whereas a TPN solution is introduced at a controlled rate into a more complex homeostatic system that is self-correcting within reasonable limits. For cultured cells, satisfaction of all nutritional needs must be integrated into an extracellular milieu that also satisfies all of their physiological needs. The current challenge in cell culture is to accomplish both while at the same time freeing the culture medium completely from undefined additives such as serum and also reducing the concentrations of defined protein growth factors and hormones to levels that are as low as possible.

2.3. Cell Type Specificity of Requirements

A major complication in the study of cellular nutrition is the immense diversity that is involved. The nutrient requirements of intact animals differ quantitatively and to a lesser degree qualitatively from one species to another (Pike and Brown, 1975). In addition, nutritional requirements also change with the developmental age of the animal (Pike and Brown, 1975; Goodhart and Shils, 1980). Species differences are also observed in the growth requirements of cultured cells. In addition, the diversity is further compounded by the fact that each cell type from a given species tends to have a different set of qualitative and quantitative nutritional requirements (Ham and McKeehan, 1978b, 1979; Ham, 1980, 1981). This is an expected result in view of functional specialization of cells, including metabolic cooperation in which organs such as the liver and kidneys carry out many biochemical functions for the entire body. However, since there are at least 100 different cell types of potential interest in the mammalian body, the complexity that could be encountered in studies of nutrient requirements of mammalian cells in culture is increased at least 100-fold over that of the intact organism.

3. Nutrients

In this section, requirements of intact mammals and cultured normal cells for specific nutrients are compared systematically. Organic nutrients have been subdivided into five categories (amino acids, carbohydrates, lipids, vitamins, and other ogranic nutrients including "vitaminlike" factors) which are discussed in that sequence, followed by major inorganic ions and trace elements. Within each group, the discussion begins with a brief resume of whole animal requirements followed by a more detailed description of the requirements of normal cells in culture.

3.1. Amino Acids

3.1.1. Requirements of Intact Animals

The amino acid requirements of mammals can be subdivided into two classes: (1) the requirement for essentail (indispensable) amino acids and (2) the requirement for a nontoxic source of utilizable nitrogen. The essential amino acids are those that cannot be synthesized by the aminal at rates sufficient for protein synthesis and other biochemical functions. For most mammalian species including man, eight L-amino acids, isoleucine, leucine, lysine, methionine, phenylalanine, threonine, tryptophan, and valine, are strictly essential (Munro, 1969). Two other amino acids, arginine and histidine, vary in their essentiality from species to species and also with the age of the animal. In addition, the nonprotein amino acid taurine has been shown to be essential for the growing cat (Knopf *et al.*, 1978).

The remaining protein amino acids are classified as "nonessential." They are readily synthesized by the intact animal when an adequate pool of utilizable nitrogen and other essential precursors are available. However, there exist within the intact animal major tissue-specific differences in the synthesis and metabolism of certain amino acids. Thus, for example, it is reasonable to expect that cultured cells derived from organs other than the liver will require arginine and tyrosine. There is also evidence for age-specific differences in the amino acid requirements of intact animals. For example, in the human infant, methionine meets the requirements for all of the sulfur amino acids; but, in the human fetus, the enzyme cystathionase (EC 4.2.1.15) is absent, which makes cyst(e)ine an essential amino acid that must be supplied from maternal sources (Gaull *et al.*, 1972). Since many cultured normal cells are derived from fetal sources, they may in some cases reflect the special nutrient needs of fetal tissues. Amino acid metabolism in mammals has been extensively reviewed (Munro, 1969; Felig, 1975).

3.1.2. Requirements of Cultured Cells

The distinction between "essential" and "nonessential" amino acids is difficult to make precisely for cultured cells for several reasons.

1. Biosynthetic pathways for amino acids that are nonessential for the intact

animal operate with varying degrees of efficiency in various types of cells. Because of the lack of a liver and absence of metabolic interactions with the rest of the body, cultured cells typically require more kinds of "essential" amino acids than whole animals of the same species. For example, media lacking arginine and/or tyrosine are often used to select against growth of other cell types in attempts to culture functional liver parenchymal cells.

2. "Nonessential" amino acids that are made in sufficient amounts to satisfy individual cellular needs frequently leak from the cells into the extracellular space. If the volume of medium per cell is large, as in a clonal growth assay, such loss may keep the intracellular concentration of certain amino acids too low to support adequate protein synthesis for multiplication. It therefore becomes necessary to supply such amino acids from an external source for clonal growth but not when a larger cellular inoculum is used. In the latter case, the extracellular concentration of metabolically produced amino acids rises to a level that allows sufficient intracellular accumulation for normal multiplication (Eagle and Piez, 1962; Ham, 1972, 1974b, 1981). Serine is nearly always needed for clonal growth, and glycine and asparagine are also common "population-dependent" requirements.

3. In addition to the essentiality of amino acids as building blocks for protein synthesis, certain amino acids appear to have other key roles in multiplication of cells in culture, especially under serum-free conditions. Glutamine appears to have particularly major roles in many metabolic pathways. An adequate extracellular concentration of glutamine is needed for proper utilization of glucose as an energy source by many types of cells (Zielke *et al.*, 1978), and it has been proposed that glutamine may be the primary energy-producing compound in some cases (Reitzer *et al.*, 1979). In some circumstances, the amount of glutamine needed for multiplication may be increased by the chemical lability of glutamine.

An excess of cysteine or cystine has been shown to be quite inhibitory to clonal growth of human diploid fibroblasts (Ham *et al.*, 1977), although the mechanisms that are involved have not been specifically identified. Cysteine and cystine may artificially affect the multiplication potential of culture media by their effects on extracellular protein-bound sulfhydryl groups which influence cellular multiplication (Corfield and Hay, 1978). Alternatively, cysteine and histidine both bind trace metals tightly, and excesses could reduce trace metal availability. On the other hand, mixed complexes of amino acids appear to serve as a major source of trace minerals in serum-free cultures (Perrin and Agarwal, 1973).

A discussion of essential and nonessential amino acids for cultured cells usually begins with the 13 amino acids in Eagle's MEM [arginine, cyst(e)ine, glutamine, histidine, isoleucine, leucine, lysine, methionine, phenylalanine, threonine, tryptophan, tyrosine, and valine]. Although there are exceptions, these 13 amino acids are commonly regarded as "essential" both for permanent lines and for normal cells. Reliable data are available, however, only for a few types of normal cells. The metabolism of these 13 amino acids in cultured cells has been reviewed in detail, including extensive discussion of special conditions under which some types of cells can be grown without certain of them (Eagle and Levintow, 1965; Patterson, 1972).

The current trend in the design of cell culture media is toward complete media in which all 20 of the amino acids that are involved in protein synthesis are supplied in carefully balanced concentrations to minimize cellular biosynthetic loads. Quantitative balance of amino acids is of major importance, both in whole-animal nutrition (Harper, 1964) and in the design of culture media (Ham, 1974a, 1981). However, optimum balance relationships vary widely from one type of cultured cell to another and do not appear to match precisely either the amino acid profiles of intracellular proteins or those of "optimal" dietary proteins.

The amino acid profiles of well-balanced dietary proteins are more homogeneous than those of culture media that have been optimized for clonal growth of normal cells or even those of MEM supplemented with "nonessential" amino acids for clonal growth (Table III). Culture media are generally characterized by very high concentrations of glutamine and arginine and relatively low concentrations of many of the amino acids that are essential for whole animals, particularly the sulfur amino acids, the aromatic amino acids, and isoleucine.

3.2. Carbohydrates

3.2.1. Carbohydrates in Animal Nutrition

In animal nutrition, digestible carbohydrates are important energy-yielding substrates and influence the taste of animal and human diets. All carbohydrates found in cells can be synthesized by the cells' metabolic machinery, and thus, none are required in the strict sense of the word. However, there is a need for a bulk energy source in animal diets, and carbohydrates generally provide a large portion of this requirement. In the average human diet in the United States, carbohydrates make up approximately 50% of the derivable energy (Williams, 1977). Also, dietary carbohydrate is required for maximal efficiency of the energy-utilization system in animals. Specifically, carbohydrate has a sparing effect on the protein requirement and helps to eliminate the toxic waste products of β-oxidation of fat. Thus, although specific carbohydrates are not absolute requirements, carbohydrates as a class are clearly beneficial in animal nutrition.

3.2.2. Carbohydrates in Cell Culture Media

Glucose is the commonly provided source of energy for cells in culture. However, for most cells, it appears that any of several other metabolizable

Table III. Comparison of Amino Acid Compositions of Dietary Proteins and Culture Media[a]

Amino acid	Whole egg[b]	Soybean meal[b]	MCDB 110[c]	MCDB 152[c]	F12 : DME[c] (1 : 1)	MEM + NEAA[c]
Alanine	—[d]	4.5	1.0	0.6	0.4	1.0
Arginine	6.7	7.0	19.2	11.0	11.5	11.5
Aspartic acid + asparagine[e]	8.2	8.3	2.9	1.1	1.2	2.9
Cysteine[f]	2.3	1.2[g]	0.7	1.8	3.4	2.7
Glutamic acid + glutamine[e]	12.6	18.5	41.8	56.2	35.2	33.7
Glycine	3.6	3.8	2.5	0.5	1.8	0.8
Histidine	2.4	2.5	1.7	0.8	2.2	3.4
Isoleucine	6.9	5.8	0.4	0.1	5.1	5.8
Leucine	9.4	7.6	1.4	4.1	5.6	5.8
Lysine	8.0	6.6	3.2	0.9	6.9	6.4
Methionine	3.3	1.1[g]	0.5	0.3	1.6	1.6
Phenylalanine	5.8	4.8	0.5	0.3	2.4	3.6
Proline	4.5	5.0	3.8	2.2	1.6	1.3
Serine	7.8	5.6	1.2	4.1	2.5	1.2
Threonine	5.0	3.9	1.3	0.8	5.1	5.2
Tryptophan	1.6	1.2	0.2	0.2	0.9	1.1
Tyrosine	4.1	3.2	0.6	0.2	3.7.	4.0
Valine	7.4	5.2	1.3	2.2	5.0	5.1
Total amino acid (mg/liter)	—	—	765.7	1385.0	1028.0	884.1

[a] All amino acid compositions are given as grams of amino acid per 16 g total protein nitrogen. For an idealized protein that is exactly 16% nitrogen, these values are equivalent to weight percentage for each amino acid. Values for cell culture media have been calculated from the total amino acid content of the media.
[b] Values for whole egg and soybean meal are from Block and Weiss (1956).
[c] MCDB 110 is optimized for human diploid fibroblasts (Bettger *et al.*, 1981); MCDB 152 is optimized for human keratinocytes (Tsao *et al.*, 1981); a 1 : 1 mixture of F12 (Ham, 1965) and DME (Morton, 1970) is widely used as a base for hormone- and growth factor-supplemented media (Barnes and Sato, 1980); MEM (shown here with an optional supplement of "non-essential" amino acids) is typical of media used with permanent cell lines (Eagle, 1959).
[d] Not included in source reference.
[e] Since special assays are needed to measure asparagine and glutamine without loss of amide nitrogen, and asparagine and glutamine are not essential for whole animals, data are normally reported as "aspartic acid plus asparagine" and "glutamic acid plus glutamine." The same has been done here for cell culture media, but with the understanding that glutamine is essential for most types of cells and that asparagine is often beneficial.
[f] The values reported are for the total of cystine plus cysteine, all reduced to cysteine.
[g] As a dietary protein, soybean meal is considered to be deficient in sulfur amino acids.

monosaccharides will support good growth, examples being D-mannose, D-fructose, D-galactose, and D-xylose (Burns *et al.*, 1976; Schwartz and Johnson, 1976; Demetrakopoulos *et al.*, 1977). Some cells require the presence of pyruvate to utilize an alternative carbohydrate to glucose (Burns *et al.*, 1976). Sugars other than glucose will cause metabolic alterations in the cell that may or may not affect their multiplicative ability. Galactose and pyruvate are sometimes used in place of glucose specifically to reduce acid production (Leibovitz, 1963; Waymouth, 1978). Pyruvate is also often needed on a population-dependent basis in the presence of glucose, as is discussed in Section 3.5.

It would appear that a source of carbohydrate is essential for growth of mammalian cells in culture, since neither amino acids nor fats can readily be used either as the sole energy source or as substrates to build up a sufficient pool of intracellular carbohydrate intermediates. The profile of potential energy-providing compounds of most cell culture media is vastly different from that of an animal diet. In an animal diet, 10–20% of the calories come from amino acids, 40–60% from carbohydrate, and 20–50% from fat. In cell culture media, a much higher proportion of the potentially available calories are in the form of amino acids ($\approx$50%), with less than 1% of the available energy in the form of fat (Table II). However, it must be recognized that culture media are normally not completely utilized and that metabolic energy is selectively derived from certain components of the media, especially glucose and glutamine. This is particularly true in the clonal growth assay where the total cellular population is far below the maximum that the medium could support.

3.3. Lipids

3.3.1. Lipids in Animal Nutrition

The high energy-yielding capacity of lipids generally increases the caloric density of natural foods and TPN solutions. Dietary lipid makes up some 40% of the calories in the average American diet. However, animals and man can generally perform at maximal efficiency on a diet that is fat-free except for minimal amounts of the essential lipids (essential fatty acids and lipid-soluble vitamins). The essential fatty acids are members of the linoleic (18:2ω6) acid family (Holman, 1968) and perhaps, in some cases, the linolenic (18:3ω6) acid family (Tinoco *et al.*, 1978). These fatty acids cannot be synthesized by the cellular machinery and are intimately involved in the maintenance of normal membrane composition and function. In addition, the linoleic acid family of fatty acids are precursors of prostaglandins, "local" hormonelike compounds that possess diverse biological activity (Samuelsson *et al.*, 1978). Dietary essential fatty acid deficiency leads to a complex series of symptoms resulting in death of the animal (Holman, 1968). The lipid-soluble vitamins are discussed under vitamins.

3.3.2. Lipids in Cell Culture Media

Lipids are frequently not included in the defined portions of cell culture media; however, large amounts of bound lipids are contained in the serum supplements that are usually used with such media (Spector, 1972). The lipid in serum is preferentially utilized by cells in culture, with suppression of *de novo* lipid biosynthesis when exogenous lipids are available. In fact, most cell types will alter the lipid composition of their membranes in response to the lipids in the extracellular environment (Spector, 1972). The question of whether normal diploid mammalian cells in culture "require" any lipid as a nutrient remains somewhat controversial. Much of the controversy stems from the finding that many transformed and highly adapted cell lines, such as mouse L and Erlich ascites carcinoma cells, continue to multiply indefinitely without any lipids in their media. Cells grown in this manner contain no $\omega 6$ series fatty acids in their membranes and exhibit no apparent deleterious effects. However, there is increasing evidence that certain lipids will stimulate growth of both normal and transformed cells. The lipids that have been proposed to stimulate growth are cholesterol, free linoleic acid and its metabolites, free oleic acid, and phospholipids containing linoleic acid (Spector, 1972; King and Spector, 1981; Chen and Kandutsch, 1981; McKeehan and Ham, 1978; Iscove and Melchers, 1978; Holmes *et al.*, 1969). It is conceivable that some cell types in culture would require cholesterol as a nutrient, since a large portion of cholesterol biosynthesis is localized in the liver. One might also expect that linoleic acid would be required for the growth of normal, low-passage mammalian cells that have not undergone nutritional adaptation in culture. It has been shown that cells produce prostaglandins *in vitro* and that exogenous prostaglandins have potent effects on cellular multiplication (Thomas *et al.*, 1974; Jiminez de Asua *et al.*, 1975; Cornwell *et al.*, 1979; Gorman *et al.*, 1979; Zalin, 1979; Bettger and Ham, 1981).

One of the problems of establishing the ''requirements'' for lipid nutrients in cell culture has been the chemistry of the compounds themselves. *In vivo,* lipids are presented to cells as water-soluble lipoproteins, ranging from chylomicrons, with a very high lipid-to-protein ratio, to albumin, with a low lipid-to-protein ratio (Smith *et al.*, 1978). Much of the lipid uptake of cells *in vivo* is receptor mediated and is a highly regulated process. *In vitro,* lipid has been added to cell culture media (1) as lipoprotein, (2) solubilized with detergents, (3) in a nonpolar solvent, and (4) as liposomes. In the development of maximally defined culture media, the use of protein supplements as nutrient carriers (e.g., transferrin for iron, albumin for free fatty acids, β-lipoprotein for cholesterol) is undesirable. Furthermore, the addition of lipids to culture media in absolute ethanol or other relatively nonpolar solvents provides a pharmacological effect of solvent on the cells that is unwarranted and also fails to insure that the lipids will remain in solution after they are introduced into the medium. The use of liposomes to make lipids soluble in culture media without toxicity seems to be the most promising method at present. The interaction of liposomes with mammalian cells has been extensively reviewed (Pagano and Weinstein, 1978).

In this laboratory, mixtures of lipids in the form of liposomes have yielded major growth responses in serum-free clonal growth assays with several types of normal and nontransformed cells including human lung and foreskin fibroblasts (Bettger *et al.*, 1981), rabbit chondrocytes, mouse 3T3 cells, and human keratinocytes (unpublished observations). Although the optimum composition differs somewhat from one cell type to another, the most effective liposomes typically contain a mixture of phospholipids, cholesterol, and sphingomyelin that roughly approximates the composition of intracellular membranes (Bettger *et al.*, 1981). In theory, these liposomes could stimulate cellular multiplication by a variety of mechanisms including: (1) providing specific essential lipids such as linoleic acid or cholesterol; (2) having a sparing effect on the biosynthetic load of the cell; (3) supplying energy precursors to the cell; (4) increasing the rate of movement of other nutrients into the cell coincidental with their own uptake; or (5) a surface-action effect that triggers specific responses at the cell surface. At this point, it is not known by which one or more of these mechanisms complex liposomes stimulate clonal growth of normal diploid cells. However, our data suggest that the inclusion of lipids in the medium may be obligatory for a maximal rate of clonal growth of normal diploid cells in serum-free, hormone-supplemented media.

3.4. Vitamins

3.4.1. Requirements of Intact Animals

Mammals in general require 12 vitamins, including the four fat-soluble vitamins (A, D, E, and K) and the eight members of the B complex (thiamine, riboflavin, niacin, pyridoxine, pantothenic acid, folacin, vitamin B_{12}, and biotin). In addition, primates, guinea pigs, and the flying mammals, all of which lack the enzyme L-gluconolactone oxidase, require vitamin C. The B vitamins function as cofactors for specific enzymes, and their deficiency can result in death of the animal. Vitamin C functions as an oxygen acceptor in several mixed-function oxidase systems. The fat-soluble vitamins perform diverse functions which are beginning to be understood as having very widespread importance to the body as a whole. Vitamin K is a cofactor in the vitamin K-dependent carboxylase system in liver and other cells (Olson and Suttie, 1977). Vitamin A, in addition to its role in the visual process and in protecting epithelial membranes, has an unidentified role in promoting cell division (Zile *et al.*, 19 9). Vitamin D has been shown to form 1,25-dihydroxycholecalciferol, a hormone with a variety of target tissues (DeLuca, 1980). Vitamin E is thought to function (1) as a nonspecific antioxidant, (2) as a specific structural stabilizer of biomembranes, and (3) as a modulator of the prostaglandin synthetase complex (DeDuve and Hayaishi, 1978).

3.4.2. Requirements of Cultured Cells

Most modern cell culture media include all the B vitamins, variably contain vitamin C, and do not contain the vitamins A, D, E, and K. Normal diploid cells in culture exhibit requirements for the B vitamins (Ham, 1981). Addition to the culture medium of certain compounds that are utilized at the substrate level may have a sparing effect on the requirement for vitamins that function catalytically in the biosynthesis of such compounds. Thus, the addition of purines, pyrimidines, and glycine may reduce the requirement for folic acid (McKeehan *et al.*, 1977), and exogenous fatty acids may reduce the requirement for biotin (Messmer and Young, 1977). Some cells have shown growth responses to vitamin C. However, this vitamin is very labile under cell culture conditions, so failure to obtain a response and the occurrence of inhibitory responses are difficult to interpret (Feng *et al.*, 1977; Koch and Biaglow, 1978; Rowe *et al.*, 1977).

A common dogma among cell culturists is that the fat-soluble vitamins are needed by intact animals only for highly specialized functions that are not directly related to most types of cultured cells. However, this conclusion may not be valid in view of the assay systems that have been used and the limited number of cell types that have been studied. Reports concerning vitamin E have been inconsistent (Smith, 1981), but there is evidence suggesting that it may be beneficial to some systems under some conditions (Bettger and Ham, 1981). Vitamin A has been shown to promote multiplication of epidermal cells (Sporn *et al.*, 1973; Christophers, 1974). Vitamins D and K have not been shown to be required by cells in culture. However, it is doubtful that 1,25-dihydroxycholecalciferol, the active form of vitamin D, has been adequately tested.

A comparison of vitamin concentrations in defined media developed for normal mammalian cells and the concentration range in human serum (Table IV) reveals some widespread discrepancies. The common nutritional practice of pro-

Table IV. A Comparison of Vitamin Levels in Defined Media for Normal Mammalian Cells with the Normal Concentration Range in Human Serum[a]

Vitamin	Human serum[b]	MCDB 110	MCDB 152	F12 : DME (1 : 1)
Biotin	$3.8\text{-}6.7 \times 10^{-8}$	3×10^{-8}	6×10^{-8}	1.5×10^{-8}
Folate/folinate	7.9×10^{-8}	1×10^{-9}	1.8×10^{-6}	6×10^{-6}
Niacin/niacinamide	$3.4\text{-}6.8 \times 10^{-5}$	5×10^{-5}	3×10^{-7}	1.6×10^{-5}
D-Pantothenate	$8.6\text{-}14.6 \times 10^{-7}$	1×10^{-6}	1×10^{-6}	9×10^{-6}
Pyridoxine	6.6×10^{-7}	3×10^{-7}	3×10^{-7}	1.0×10^{-5}
Riboflavin	1.7×10^{-7}	3×10^{-7}	1×10^{-7}	6×10^{-7}
Thiamin	3.8×10^{-8}	1×10^{-6}	1×10^{-6}	7×10^{-6}
Vitamin B_{12}	3.6×10^{-10}	1×10^{-7}	3×10^{-7}	5×10^{-7}
Ascorbate	$2.8\text{-}5.6 \times 10^{-5}$	—[c]	—	—
Vitamin A	1000-3000 IU/liter	—	—	—
Vitamin D	660-1650 IU/liter	—	—	—
Vitamin E	11-17 IU/liter	—[d]	—	—
Vitamin K	?	—	—	—

[a]Concentrations are in moles per liter except where indicated otherwise.
[b]Concentration ranges for human serum have been converted to moles per liter from weight ranges compiled by Kutsky (1973).
[c]MCDB 111 (Table II) contains 1×10^{-5} M ascorbate-2-sulfate.
[d]MCDB 110 supplemented for defined medium growth of human diploid fibroblasts and MCDB 111 (Table II) both contain 5.6×10^{-7} M vitamin E (0.26 IU/liter).

viding all vitamins at a substantial excess of their required concentration is not generally utilized in cell culture; many levels of the vitamins are below their *in vivo* concentrations and well below their toxicity levels. Quite conspicuously, vitamins C, A, D, E, and K are not present in the culture media, at least partially because of methodological difficulties in their use. (Note that vitamin E is included in the liposome supplement for human diploid fibroblasts in Table II.)

The comparison points out a potential shortcoming of current cell culture media and suggests some possible changes for the future. (1) All vitamins required by the parent animal of the cell type involved should be employed in the nutrient medium at least until it can be demonstrated unequivocally that they are not beneficial. (This would universally include the active forms of vitamins A, D, E, and K.) (2) Vitamins should be incorporated at concentrations well above the level of apparent minimal requirements but always below the level of toxicity. (3) If a vitamin shows a major toxic response at very low concentration, alternate forms of the vitamin should be employed that can be used at higher concentrations without a toxic effect. (4) Any vitamin that shows a major growth response at a level vastly different from that found in the serum of the same species should be examined to determine if it is truly functioning in its *in vivo* biochemical role.

3.4.3. "Vitaminlike" Substances

The question of what constitutes a true vitamin has always been a difficult one, both in whole-animal nutrition and in cell culture. The usual test for water-soluble vitamins is whether a cofactor role can be demonstrated. Lipoic acid, which functions as a true vitamin for certain organisms but has not been shown to be required from exogenous sources by mammals (Pike and Brown, 1975), was included in medium F10 on the basis of a possible marginally beneficial effect (Ham, 1963) and has since been included in many other media without ever being shown to be clearly needed.

Ubiquinone is not usually classified as a true vitamin since it is beneficial to intact animals only under special conditions such as vitamin E deficiency, severe nutritional anemia, and genetically determined dystrophy in mice (Pike and Brown, 1975). However, its catalytic role in mitochondrial electron transport is consistent with the designation of "vitamin." Ubiquinone is seldom tested in cell culture systems because of its lipid-soluble nature, and there is no convincing evidence that it is beneficial to cultured cells.

Choline and inositol have stimulatory effects that are sometimes referred to as "vitaminlike" both in whole animal nutrition and in nutrition of cultured cells. However, a closer examination of their biochemical roles indicates that these compounds function at the substrate, rather than the catalytic, level. They are therefore, discussed among other organic nutrients in Section 3.5.

3.5. Other Organic Nutrients

3.5.1. Requirements of Whole Animals

When provided with adequate sources of calories, amino acids, vitamins, and essential lipids as discussed above, intact mammals exhibit few other requirements for organic nutrients. Choline and *myo*-inositol, both of which are structural components of phospholipids, are required by some species (Pike and Brown, 1975). Typically, these are the only "other organic" nutrients required by healthy intact animals provided with adequate amounts of the nutrients that have been discussed previously.

3.5.2. Requirements of Cultured Cells

The greatest divergence between the nutrient requirements of whole animals and those of cultured cells occurs in the category of other organic nutrients. Cultured normal cells tend to require a number of organic compounds that are not required by intact animals. In general, these compounds satisfy all of the criteria for true nutrients, although in some cases, certain compounds may also have regulatory roles, as will be discussed below. Most of these requirements are thought to reflect biochemical specialization of various cell types within the intact animal and the fact that cooperative interactions with specialized metabolic tissues such as the liver and the kidneys are absent in cell culture. In many cases, the requirements of the cultured cells are population-dependent, as discussed in Section 3.1. Thus, certain intermediates with limited rates of synthesis leak out of the individual cells. Crowded cultures build up substantial extracellular concentrations of these intermediates and, by equilibration, also raise their intracellular concentrations to levels that are sufficient to support intermediary metabolism and biosynthesis, whereas sparse cultures cannot do so because of near-infinite extracellular dilution.

(a) Choline, myo-*Inositol, and Ethanolamine.* Choline and *myo*-inositol are both included in Eagles's MEM, and both are quite generally considered to be required for cellular multiplication, although in some permanent cell lines, the requirement for *myo*-inositol is population dependent. Phosphoethanolamine has been shown to be the component of pituitary extract that stimulates growth of a rat mammary tumor line (Kano-Sueoka *et al.*, 1979), and more recently ethanolamine has been shown to have a similar growth-promoting effect with a higher specific activity (T. Kano-Sueoka, personal communication). Ethanolamine and/or phosphoethanolamine are also required for clonal growth of normal human epidermal keratinocytes in a defined medium (Tsao *et al.*, 1981). Like choline and *myo*-inositol, ethanolamine is a structural component of phospholipids, although on the basis of the small amount needed for growth of rat mammary tumor lines, there remains some question as to whether the

multiplication-promoting effect may be primarily regulatory rather than at the substrate level (T. Kano-Sueoka, personal communication).

(b) Intermediates of Energy Metabolism. Cells grown at low cellular density generally require pyruvate or other 2-oxocarboxylic acids (Neuman and McCoy, 1958; Eagle, 1959; Ham, 1962, 1974b). These metabolic intermediates accumulate in media as "conditioning" factors when cells are inoculated at high densities. Similar intermediates also accumulate in the blood plasma of intact animals. The amount of 2-oxocarboxylic acid required for clonal growth rises sharply as the amount of serum supplement is reduced, and a key regulatory role for the 2-oxocarboxylic acids in cellular multiplication has been suggested (McKeehan and McKeehan, 1979; Groelke *et al.*, 1979). Acetate has been incorporated into several cell culture media (Ham and McKeehan, 1979), and acetate was found to stimulate the clonal growth of normal human keratinocytes grown with minimal amounts of serum protein (Peehl and Ham, 1980). An end product of intermediary metabolism, carbon dioxide is also needed on a population-dependent basis, as is discussed in Section 3.8.2.

(c)Polyamines. One of the first events in cells that are stimulated to multiply is an increase in ornithine decarboxylase activity which results in the synthesis of putrescine. A number of cases are known in which the addition of putrescine, spermidine, or spermine from exogenous sources is beneficial to cellular multiplication (Ham, 1964; Clo *et al.*, 1976, 1979; Tabor and Tabor, 1976). Unusually high concentrations of putrescine have recently been reported to favor selective growth of human bronchial epithelial cells in primary culture (Stoner *et al.*, 1980). Spermine and spermidine are converted by enzymes found in bovine serum to products that have been reported to be selectively toxic to fibroblasts in cultures of human prostatic epithelium (Webber and Chaproniere-Rickenberg, 1980). Except in special cases such as this, only putrescine should be used in media containing any type of bovine serum.

(d) Nucleic Acid Components. Many types of cultured cells respond positively to the presence of purine bases and/or pyrimidine nucleosides in the culture medium. Human diploid fibroblasts exhibit the best clonal growth when adenine and thymidine are included in their culture medium (McKeehan *et al.*, 1977). Human epidermal keratinocytes require a higher level of adenine than human fibroblasts (Peehl and Ham, 1980). Many types of normal cells in culture are inefficient in their utilization of folic or folinic acid, so biosynthesis of purines, thymidine, and glycine is often rate limiting for multiplication if these substrates are not provided from an external source (Neugut and Weinstein, 1979; Ham, 1981). Although not strictly required, these compounds used judiciously can significantly improve cellular multiplication rates.

The most effective forms for normal cells appear to be adenine and thymidine. Hypoxanthine is incorporated into many classical media (Ham and McKeehan, 1979), but detailed comparisons suggest that adenine is more effective for normal cells than any other purines, purine nucleosides, or purine nucleotides, either singly or in combinations. Free pyrimidines are not utilized effectively by cultured cells, and thymidine is more effective than thymidine nucleotides or other pyrimidine nucleosides.

3.6. Major Ions

3.6.1. Requirements of Intact Animals

Mammals require a dietary source of the minerals calcium, phosphorus, magnesium, sodium, chlorine, potassium, and sulfur. The major portion of sulfur is bound to organic material, and the dietary requirement is generally met by the sulfur in the amino acids methionine, cysteine, and cystine. The other elements generally occur in the diet in ionic form. The concentrations of these ions in extracellular fluids are precisely regulated over a wide range of dietary intakes by the kidney and other tissues. This indicates a profound importance for a constant extracellular level of these ions in the maintenance of normal cellular function. The intracellular concentrations of these ions are also highly regulated, largely by membrane-bound, ATP-driven pumps. A sudden surge of one or more of these ions in or out of cells may be a basic metabolic signal, even in cells without electrically excitable membranes (Kaplan, 1978). The ions phosphate, calcium, and magnesium exhibit major interactions in terms of absorption. Any change in the homeostasis of any one generally results in changes in the others. Sodium, potassium, and chloride are highly permeable to cells, and their gradients between intra- and extracellular space are maintained by a highly regulated pump that operates at the expense of cellular energy (Dahl and Hokin, 1974).

3.6.2. Requirements of Cultured Cells

The major ions, sodium, potassium, calcium, magnesium, chloride, and phosphate, are all required by cells in culture. Sulfur is also required in an organic bound form. There are reports of marginal benefit from the addition of inorganic sulfate to culture media (McKeehan *et al.*, 1977), but these need more study before they can be viewed as conclusive. The major inorganic ions are required both as nutrients and for other roles such as the maintenance of osmolarity, pH, and membrane potential. The nonnutrient aspects of the requirements for the major inorganic ions have been extensively reviewed (Ham, 1981). Phosphorus, calcium, magnesium, potassium, and sodium appear to have key regulatory roles in cell division (Leffert, 1980). It is not clear if the intracellular concentrations of these ions are the critical factors in determining their regulatory roles in cell division or if extracellular concentrations have separate and independent regulatory functions. However, many hormone effects appear to mediated by changes in the intracellular concentrations of these ions.

Table V. Ratios of Major Inorganic Ions in Culture Media Used for Clonal Growth Assays

Culture medium	Cell type	Na^+/K^+	Ca^{2+}/Mg^{2+}	Ca^{2+}/PO_4
MCDB 110	Human fibroblasts, rabbit chondrocytes	22	1.0	0.33
MCDB 152	Human keratinocytes	100	0.05	0.015
F12/DME (1 : 1)	Mouse kidney epithelia	37	1.5	1.1
F12	Rabbit choroid cells	50	0.5	0.33
Human plasma[a]	——	29	1.7	1.2

[a]Ratios calculated from values for individual ions given by Guyton (1971).

Manipulation of the ratios of the extracellular concentrations of these major regulatory ions can significantly alter growth and differentiation of various types of cultured cells (Rubin and Koide, 1976; Hennings *et al.*, 1980; Waymouth, 1981; Ham, 1981). Thus, a wide range of concentrations of these ions is seen in culture media that are optimized for specific cell types. Examples are shown in Table V. Sodium-to-potassium ratios range from 22 to 100, calcium-to-magnesium levels from 0.05 to 1.5, and calcium-to-phosphate levels from 0.015 to 1.1 in various culture media.

3.7. Trace Elements

3.7.1. Requirements of Intact Animals

The trace element requirements of mammals are less well established than those for other nutrient classes because of methodological difficulties in experimentally producing deficiencies of some trace elements. The trace mineral requirements of animals have been reviewed in detail (Underwood, 1977). Those elements generally recognized as required are iron, copper, molybdenum, cobalt, manganese, zinc, chromium, iodine, and selenium. The elements nickel, fluorine, vanadium, and silicon have been shown to be required in some species when raised under extreme conditions to prevent contamination. The elements arsenic and tin are suspected to be required. The trace minerals generally function as cofactors in specific enzymes, such as iron in catalase, copper in cytochrome C oxidase, zinc in carbonic anhydrase, and selenium in glutathione peroxidase. Iodine functions as a part of the hormone thyroxine and is thought to have no catalytically related activity. Complex interrelationships are seen among many of the trace minerals (i.e., Fe and Cu, Zn and Cu, Cu and Mo). The transition metals compete for almost all metal-binding ligands according to their individual affinities.

3.7.2. Requirements of Cultured Cells

Although all of these trace elements could theoretically be required by cells in culture, growth responses to many of them are not likely to be detected unless a major effort is made to eliminate contaminating levels from the chemicals, water, glassware, and sterilizing filters used in medium preparation. In current culture systems, strong growth responses can often be seen for iron, zinc, and selenium, which are generally assumed to be required by a variety of cell types (Ham, 1981). The insolubility of ferric iron has led many investigators to use transferrin in defined media (Barnes and Sato, 1980). However, experience in our laboratory and others has shown that freshly dissolved ferrous iron used in conjunction with an optimized balance relationship to other trace elements will fully satisfy the iron requirements of a variety of cell types. Effects of copper and manganese are generally marginal. Four major aspects of the culture system can affect the apparent trace element requirements: (1) the background level of contamination, (2) the bioavailability of the form of the element that has been added to the medium, (3) the relative balance of the trace minerals in the medium, and (4) the cell type. At present, a wide range of kinds and amounts of trace element supplementation is found among various culture media (Ham and McKeehan, 1979). Major differences are seen between plasma concentrations of trace elements in mammals and those in culture media (Table VI).

Table VI. Comparison of Trace Elements Concentration in Serum to Concentrations Supplied in Various Culture Media

Element	Human serum[a] (μg/liter)	MCDB 110 (μg/liter)	MCDB 152 (μg/liter)	F12/DME (1 : 1) (μg/liter)
Iron	1200	279	84[b]	98
Zinc	1000	32.5	228[b]	98
Copper	1100	0.063	0.69	0.32
Manganese	2–10	0.055	0.055	—
Selenium	110	2.4	2.4	—[c]
Molybdenum	5–15	0.096	0.096	—
Nickel	1–4	0.029	0.029	—
Silicon	1000	14	14	—
Tin	<0.09	0.059	0.059	—
Vanadium	<10	0.25	0.25	—
Chromium	7–15	—	—	—
Cobalt	1–15	5.9[d]	17.7[d]	29.4[d]
Iodine (i)	1–6	—	—	—

[a] Values from Underwood (1977).
[b] Recent data indicate that growth of human keratinocytes under defined conditions can be improved by increasing iron and reducing zinc in MCDB 152 to levels comparable to those in MCDB 110.
[c] Selenium is frequently added to DME:F12 when it is used for serum-free growth of cells with hormone and growth factor supplements (Barnes and Sato, 1980).
[d] Supplied as vitamin B_{12}.

There have been recent reports of positive growth responses to the addition of cadmium to culture media for some cell types (Barnes and Sato, 1980). Cadmium accumulates in animal tissues, but it is generally viewed as highly toxic in excess and probably not essential (Underwood, 1977). The observed stimulation emphasizes the difficulty of establishing true trace element requirements of cells in culture. A heavy metal, such as cadmium, may stimulate cells by one or more of a multitude of mechanisms including the following: (1) it could be a required nutrient (i.e., a cofactor in a specific cadmium-metalloenzyme); (2) it could act as a semispecific sulfhydryl reagent, reacting with membrane sulfhydryl groups that control membrane permeability to a variety of organic and inorganic compounds; (3) it could compete with other trace metals for nonspecific binding sites and thus release required trace metals, such as Zn, Cu, or Fe, which may be growth limiting, making them more available for cellular utilization; (4) it could nonspecifically stimulate the synthesis of a metallothioneinlike protein that has multitude of effects on trace mineral metabolism and possibly other cellular functions; and (5) it could act as a toxin that, at threshold concentrations, could elicit a positive growth response from the cells.

Thus, barring the discovery of a specific cofactor role for cadmium, it would be very difficult to extrapolate from cell culture results to a definitive whole-animal requirement for cadmium. This points out the limitation of using cellular multiplication as the sole criterion for a nutritional requirement. However, at the same time, the existence of a responsive cell culture system greatly facilitates the search for a specific biochemical role of cadmium, if one exists. If found, such a role, together with the cellular multiplication response, could point the way toward similar studies in intact animals. In addition, cell culture provides a means for extrapolating animal trace element requirements to humans in the absence of long-term depletion experiments which cannot ethically be performed with young growing human subjects.

Classical cell culture media are notoriously deficient in inorganic trace elements (Ham and McKeehan, 1979). However, the choice of many workers who are currently developing low-serum or serum-free media for normal cells is to supply in the culture medium most or all of the trace minerals that are known to be required by whole animals (Barnes and Sato, 1980; Ham, 1981). This approach is considered desirable even when a specific response cannot be demonstrated, since it buffers the system against inconsistency resulting from possible variable levels of contamination and also recognizes that the trace elements are probably required by the cells in question even though specific responses cannot be demonstrated. The trace minerals for which a definitive growth response cannot be demonstrated are incorporated at levels well below the levels that yield a toxic response. This is likely to result in listed trace mineral concentrations that are severalfold lower than the concentrations that are actually in the media because of contamination.

3.8. Water, Osmolarity, pH, Buffers, Carbon Dioxide, Oxygen

3.8.1. Requirements of Whole Animals

The largest single nutrient in whole animals is water, which constitutes some 70% of body weight. Water, in addition to being the major solvent in living systems, is a nutrient by the definition given in this review. It is a structural component of membranes and a cosubstrate for a multitude of enzymatic reactions. The extent of dilution of body fluids with water is precisely controlled by homeostatic mechanisms to maintain a constant osmolarity. The relative preponderance of ionization products of water (H^+, OH^-) is also precisely controlled at a physiological pH value. This is achieved by a variety of metabolic buffering mechanisms including the ratio of carbon dioxide to bicarbonate ion.

Carbon dioxide is produced in excess and kept at a regulated extracellular level by respiratory control mechanisms. Although commonly viewed as a waste product, carbon dioxide is also a metabolic intermediate that participates in a variety of biosynthetic reactions, include synthesis of purines, pyrimidines, fatty acids, and intermediates of the citric acid cycle, which must be replenished as they are utilized as substrates in other biosynthetic reactions.

Oxygen is not generally classified as a nutrient in whole animals, since it is not taken in through the digestive system. However, it functions as a nutrient on the cellular level. An animal will die more quickly from a lack of oxygen than from a deficiency of any nutrient, including water. Oxygen is required in a multitude of enzymatic reactions and is the terminal electron acceptor in the mitochondrial electron transport chain. It also functions as a direct substrate in mixed-function oxidases. For example, the hydroxyl oxygen of hydroxyproline in collagen is derived from respiratory oxygen.

3.8.2. Requirements of Cultured Cells

Cells in culture require a water-based medium, and water is the most predominant nutrient by weight, as it is in the whole animal. Water performs many if not all of the same functions it does in animals. The osmolarity range for optimal cellular multiplication is relatively narrow and differs from cell type to cell type (Waymouth, 1970; Ham and McKeehan, 1978b). Precise control of incubator humidification is very important to prevent evaporation of water from culture media and to maintain correct osmolarity of media in long-term experiments. In Table VII, the concentrations of the major osmotically active substances in various culture media are compared to plasma, interstitial fluid, and intracellular fluid. In culture media, there is considerably more contribution to total osmolarity of the medium by amino acids, pyruvate, acetate, chloride, and organic buffers such as HEPES than there is in interstitial fluid. In culture media, Ca^{2+}, Mg^{2+}, creatine, lactate, protein, and urea represent smaller contributions

Table VII. Comparison of Major Osmotically Active Substances in Extracellular and Intracellular Fluids *in Vivo* and in Various Cell Culture Media[a]

	Plasma[b]	Interstitial[b] fluid	Intracellular[b] fluid	MCDB 110	MCDB 152	F12/DME (1:1)
Na^+	144	137	10	115	150	155
K^+	5	4.7	141	5	1.5	4.2
Ca^{2+}	2.5	2.4	—	1	0.03	1.05
Mg^{2+}	1.5	1.4	31	1	0.6	0.7
Cl^-	107	112.7	4	120	130	130
HCO_3^-	27	28.3	10	—	14	29
PO_4^{-3}	2	2	11	3	2	0.96
SO_4^{-2}	0.5	0.5	1	1	0.004	0.4
HEPES	—	—	—	30	28	—
Phosphocreatine	—	—	45	—	—	—
Carnosine	—	—	14	—	—	—
Amino acids	2	2	8	5.4	9.8	7.0
Creatine	0.2	0.2	9	—	—	—
Lactate	1.2	1.2	1.5	—	—	—
ATP	—	—	5	—	—	—
Hexose monophosphate	—	—	3.7	—	—	—
Glucose	5.6	5.6	—	4	6	7.8
Protein	1.2	0.2	4	—	—	—
Urea	4	4	4	—	—	—
Pyruvate	—	—	—	1	0.5	1
Acetate	—	—	—	—	3.7	—
Total	303.7	302.2	302.2	286.4	346.1	337.1

[a] All values are in mOsmoles per liter of water.
[b] From Guyton (1971).

to total osmolarity than they do in interstitial fluid. This is intriguing, since cells *in vivo* grow in interstitial fluid. However, one must keep in mind that microenvironments within tissues may more readily reflect the composition of "optimized" cell culture media as compared to plasma or interstitial fluid which tends to represent an "average" compostion. Also, in many cases, the rate of multiplication of cells in optimized media *in vitro* is substantially greater than the multiplication rate for the same types of cells *in vivo*.

The pH of the culture medium must be maintained over a very narrow range. The optimum pH varies with the cell type and with the type of medium used but typically extends over only a few tenths of a pH unit. Older culture media generally employed a relatively physiological CO_2/bicarbonate system together with phosphate. The current trend favors use of synthetic organic buffers such as HEPES for more precise control (Eagle, 1973; Ham, 1981). For clonal cultures, however, it remains necessary to utilize a carbon dioxide-enriched atmosphere (usually 2–5%) in the cell culture incubator. Quite apart from the buffering capacity of CO_2 and bicarbonate, it is necessary to maintain by equilibration an intracellular level that is adequate for normal levels of biosynthesis (Ham, 1974b). Cells in culture require oxygen as a nutrient, with gaseous O_2 in ing dissolved in the nutrient medium (Ham and McKeehan, 1979). Most cells do better at partial pressures of oxygen that are lower than those in the ambient atmosphere (Taylor *et al.*, 1974; McKeehan *et al.*, 1977). However, it is usually possible to obtain adequate growth at ambient oxygen levels provided that a well-balanced medium with an adequate level of selenium is employed (McKeehan *et al.*, 1977; Ham and McKeehan, 1979). In comparing results from different laboratories, it may be necessary to take into account the effect of altitude on ambient oxygen tension.

4. Nutritional Studies *in Vitro:* Summary and Prognosis

4.1. Current Status

At the present time, rapid progress is being made in the development of defined media for normal diploid mammalian cells. Each time it becomes possible to grow another type of normal cell in a defined medium, it also becomes possible to analyze in precise detail the total set of growth requirements, both nutrient and nonnutrient, for that type of cell. This in turn opens the way for detailed studies at the cellular level of responses to all nutrients that are required by whole animals of the species from which the cell is derived. In addition, the cellular studies may identify substances that need to be examined as possible nutrients at the whole-animal level. An example is cadmium. Although many alternative explanations are possible, as discussed above, the fact that cadmium stimulates cellular multiplication under certain conditions makes it desirable to examine in greater detail the possibility that cadmium, which is known to accumulate in biological systems, may actually function as an essential trace element.

Table VIII summarizes the current status of studies on the responses of normal cells in culture to nutrients that are required at the whole-animal level and also of studies on cellular nutrients that have not been shown to be required by whole animals. In interpreting Table VIII, the reader must keep in mind that there is substantial variation from one cell type to another and also that detailed studies of growth requirements of normal cells in defined media are still in early stages of development. Thus, certain types of nutrients have not yet been fully sorted out into categories of "essential," "stimulatory," and "having little or no effect" at the cellular level.

In vitro studies of requirements for amino acids, carbohydrates, and most of the B vitamins have been relatively thorough. Areas where additional study is

Table VIII. Summary and Comparison of Nutrient Requirements of Intact Mammals and Cultured Mammalian Cells[a]

Nutrient	Whole mammal response[b]	Response of cultured cells
Amino acids		
Alanine	N	Normally not required
Arginine	S,D	Required, with possible exception of liver cells
Asparagine	N	Often required on a population-dependent basis
Aspartic Acid	N	Normally not required
Cyst(e)ine	P,D	Required by normal cells; some permanent lines are capable of a low level of synthesis
Glutamine	N	Generally required, although some cells can grow in crowded cultures with glutamic acid instead
Glutamic acid	N	Normally not required
Glycine	N	Often required on a population-dependent basis; also needed if folic acid metabolism is impaired
Histidine	S,D	Always required
Isoleucine	R	Always required
Leucine	R	Always required
Lysine	R	Always required
Methionine	R	Always required
Phenylalanine	R	Always required
Proline	N	Normally not required[c]
Serine	N	Frequently required on a population-dependent basis
Threonine	R	Always required
Tryptophan	R	Always required, excess often very inhibitory
Tyrosine	P	Required, with possible exception of liver cells
Valine	R	Always required
Carbohydrates		
Glucose[d]	B	Required, can generally be replaced with other closely related sugars
Lipids		
Linoleic acid	R	Probably required by normal cells, but data not conclusive
Linolenic acid	U	Current data are inadequate to evaluate possible requirement by normal cells
Oleic acid	N	Stimulatory to some cell types; data inadequate to generalize
Cholesterol	N	Major growth-promoting effect; may be essential for some types of normal cells
Sphingomyelin	N	Moderate growth stimulation for some types of cells; data inadequate to generalize
Phospholipids	N	Clearly beneficial for certain normal cells; possible requirement not clearly distinguished from fatty acids
Vitamins		
Biotin	R	Probably required by all cells; requirement difficult to demonstrate when serum protein is present[e]

(*continued*)

Table VIII. (*Continued*)

Nutrient	Whole mammal response[b]	Response of cultured cells
Vitamins (*continued*)		
Folic acid/folinic acid	R	Required by all cells; requirement partially replaced by purines, thymidine, glycine
Lipoic acid	N	Included in some media, probably not required
Niacin	R[f]	Always required
Pantothenic acid	R	Always required
Pyridoxine	R	Always required; requirement partially replaced by nonessential amino acids
Riboflavin	R	Always required
Thiamin	R	Always required
Vitamin A	R	Stimulatory to some cell types; apparently not required for many, but needs further study
Vitamin B_{12}	R	Probably required by all cells; requirement difficult to demonstrate when serum protein is present[e]; end product substitution may be possible
Vitamin C	S	Some reports of benefit; instability in culture medium makes precise study difficult; many cell types apparently able to multiply in absence of added vitamin C
Vitamin D	R	Generally considered not to be required, but needs further study in highly defined media for normal cells
Vitamin E	R	Data are controversial; apparently beneficial to some cell types under some circumstances; needs further study in highly defined media for normal cells
Vitamin K	R	Generally considered not to be required, but needs further study in highly defined media for normal cells
Other organic nutrients		
Acetate	N	Beneficial for some cell types; further study needed
Choline	S	Generally required
Ethanolamine	N	Required by some types of cells; further study is needed under conditions where contribution from phospholipids is precisely controlled
myo-Inositol	S	Often required, sometimes on a population-dependent basis
2-Oxocarboxylic acids	N	Pyruvate or related compounds often required on a population-dependent basis; may also have a regulatory role
Polyamines	N	Putrescine is sometimes beneficial; in absence of bovine serum, spermine or spermidine can also be used; high levels may be selective against fibroblastic overgrowth; role may be primarily regulatory
Purines	N	Adenine or hypoxanthine frequently stimulatory to normal cells; essential if folic acid metabolism is impaired

(*continued*)

Table VIII. (*Continued*)

Nutrient	Whole mammal response[b]	Response of cultured cells
Other organic nutrients (*continued*)		
Thymidine	N	Frequently beneficial to normal cells; essential if folic acid metabolism is impaired
Major inorganic ions		
Calcium	R	Probably always required; certain cells such as keratinocytes require only low levels; may have regulatory roles
Magnesium	R	Always required; may have regulatory roles
Potassium	R	Always required
Sodium	R	Always required
Chloride	R	Always required
Phosphate	R	Always required
Sulfate	N	Possible slight beneficial effect on certain normal cells; more study needed
Arsenic	U	Not adequately studied; no current evidence for requirement
Cadmium	N(?)	Reported to be beneficial to some cell types; further study is needed
Chromium	R	Not adequately studied; no current evidence for requirement
Cobalt	R	Probably required by all cells as component of vitamin B_{12}; no clear evidence for or against other roles
Copper	R	Clean background difficult to obtain; probably generally required; needs further study
Fluorine	U	Not adequately studied; no current evidence for requirement
Iodine	R	Thyroid hormone reported beneficial to some cell types; no clear evidence for any other roles
Iron	R	Probably required by all cells; low solubility of ferric iron makes detailed studies difficult; often supplied as transferrin complex
Manganese	R	Several reports of moderate stimulation, but needs further study with cleaner background
Molybdenum	R	Scattered reports of benefit, but needs further study with cleaner background
Nickel	U	Not adequately studied; no current evidence for requirement
Selenium	R	Clearly required by a number of different cell types
Silicon	U	Not adequately studied; no current evidence for requirement
Tin	U	Not adequately studied; no current evidence for requirement
Vanadium	U	Limited data suggest possible slight benefit; needs further study with cleaner background

(*continued*)

Table VIII. (*Continued*)

Nutrient	Whole mammal response[b]	Response of cultured cells
Major inorganic ions (*continued*)		
Zinc	R	Clearly required by a number of different cell types
Oxygen and water		
Oxygen	R	Required by all cell types adequately examined; reduced partial pressure sometimes beneficial
Water	R	Absolute requirement

[a] This table summarizes the current cell culture status of all known nutrient requirements of whole mammals and also lists major cell culture nutrients that are not known to be required by intact mammals.
[b] The responses of whole mammals to the nutrients have been coded as follows (note that more than one letter may be used for the same nutrient in some cases): B, beneficial, but not strictly essential; D, requirement varies with developmental stage; N, nonessential; P, not strictly required, but will spare the requirement for an essential precursor (cysteine for methionine, tyrosine for phenylalanine); R, essential nutrient required by mammals in general; S, required by some species but not others; U, uncertain status—limited studies indicate a requirement but are inadequate to generalize to other species.
[c] Certain widely studied Chinese hamster ovary lines (CHO, $CHOK_1$) require proline because of mutation.
[d] Various other sugars can replace glucose, both for whole animals and for cultured cells (cf. Section 3.2).
[e] Because of the presence of binding proteins for biotin and vitamin B_{12}, serum will generally supply sufficient amounts of these vitamins even after extensive dialysis.
[f] For rats, the requirement for niacin can be replaced by a high level of dietary tryptophan.
[g] Because of the presence of trace elements as contaminants in basal media, many of the potential trace element requirements of cultured cells have not been adequately analyzed.

clearly needed include lipids, fat-soluble vitamins, vitamin C, biotin, vitamin B_{12}, trace elements, and the interrelationship among energy-producing substrates. Recent data are making it increasingly clear that lipids are of major importance in cellular nutrition, but it is not yet clear which (if any) specific compounds are absolutely required. Results with lipid-soluble vitamins have been highly variable, probably because of technical difficulties related to their lipid-soluble nature. Rapid degradation of vitamin C in standard culture media makes it necessary to seek special forms that are more stable and yet retain their biological activity. Biotin and vitamin B_{12} both bind tightly to dialyzed serum and can be studied adequately only under serum-free conditions which have just been achieved for the cell types of greatest interest. Widespread contamination of culture systems with trace elements makes necessary exhaustive purifications before detailed studies of trace element requirements can be undertaken. Trace elements are also quite toxic in excess and tend to interact with many other components of the culture medium. Patterns of energy metabolism in cultured cells and the profile of potential energy-producing compounds in culture media are sufficiently different from those that occur *in vivo* to warrant further investigation of the optimal distribution of energy yield from amino acids, carbohydrates, and lipids as well as from specific compounds within those major classes for cultured cells.

To summarize, the tools are now available for studying nutrition at the cellular level in precise detail, but the number of unanswered questions concerning cellular nutrition remains very large.

4.2. Prognosis

Despite the fact that tools for detailed nutritional studies with normal cells in culture are now available, the studies themselves are progressing rather slowly. Most of the laboratory groups that are currently developing defined media for normal cells are primarily interested in regulatory factors and control of the cell cycle rather than in nutrition per se. Although the media that they are developing make detailed nutritional studies with normal cells fully feasible, those studies are generally not being undertaken because a wide range of other projects are of greater interest to the current investigators. Thus, in order for nutritional studies *in vitro* to progress rapidly, it will be necessary for investigators whose primary interest is in nutrition to begin utilizing the newly developed defined medium systems.

The ability to grow normal mammalian cells in a defined nutrient broth supplemented with a small number of purified hormones and growth factors has opened the doors for a whole new type of nutritional study. Just as the advent of purified diets caused a boom in whole-animal nutrition studies, the development of defined media for normal cells can be expected to do the same for cellular nutrition studies. The tools are available now. The only remaining question is how rapidly they will be utilized. This chapter attempts to provide the background and encouragement for such endeavors.

References

Ambesi-Impiombato, F. S., Parks, L. A. M., and Coon, H. G., 1980, Culture of hormone-dependent functional epithelial cells from rat thyroids, *Proc. Natl. Acad. Sci. U.S.A.* **77**:3455.

American Medical Association, 1979, Guidelines for essential trace element preparations for parenteral use, *J. Parenteral Enteral Nutr.* **3**:263.

Barnes, D., and Sato, G., 1980, Methods for growth of cultured cells in serum-free medium, *Anal. Biochem.* **102**:255.

Bettger, W. J., and Ham, R. G., 1981, Effects of non-steroidal anti-inflammatory agents and antioxidants on the clonal growth of human diploid fibroblasts, *Prog. Lipid Res.*, in press.

Bettger, W. J., Boyce, S. T., Walthall, B. J., and Ham, R. G., 1981, Rapid clonal growth and serial passage of human diploid fibroblasts in a lipid-enriched, synthetic medium supplemented with EGF, insulin and dexamethasone, *Proc. Natl. Acad. Sci. U.S.A.*, in press.

Block, R. J., and Weiss, K. W., 1956, *Amino Acid Handbook*, Charles C. Thomas, Springfield, Illinois.

Burns, R. L., Rosenberger, P. G., and Klebe, R. J., 1976, Carbohydrate perferences of mammalian cells, *J. Cell. Physiol.* **88**:307.

Chan, K. Y., and Haschke, R. H., 1981, Trophic effect of rabbit corneal cells on trigeminal neurons in coculture, *Invest. Ophthalmol. Vis. Sci.*, in press.

Chapronière-Rickenberg, D., and Webber, M. M., 1980, A synthetic medium for the growth of primary cultures of adult human prostatic epithelium, *In Vitro* **16**:214.

Chen, H. W., and Kandutsch, A. A., 1981, Cholesterol requirement for cell growth: Endogenous synthesis vs. exogenous sources, in: *The Growth Requirements of Vertebrate Cells in Vitro* (C. Waymouth, R. G. Ham, and P. J. Chapple, eds.), Cambridge University Press, New York, in press.

Christophers, E., 1974, Growth stimulation of cultured post embryonic epidermal cells by vitamin A acid, *J. Invest. Dermatol.* **63**:450.

Clo, C., Orlandini, G. C., Casti, A., and Guarnieri, C., 1976, Polyamines as growth stimulating factors in eukaryotic cells, *Ital. J. Biochem.* **25**:94.

Clo, C., Calderera, C. M., Tantini, B., Benalal, D., and Bachrach, U., 1979, Polyamines and cellular adenosine 3′:5′-cyclic monophosphate, *Biochem. J.* **182**:641.

Corfield, V. A., and Hay, R. J., 1978, Effects of cystine or glutamine restriction on human diploid fibroblasts in culture, *In Vitro* **14**:787.

Cornwell, D. G., Huttner, J. J., Milo, G. E., Panganamala, R. V., Sharma, H. M., and Geer, J. C., 1979, Polyunsaturated fatty acids, vitamin E, and the proliferation of aortic smooth muscle cells, *Lipids* **14**:194.

Dahl, J. L., and Hokin, L. E., 1974, The sodium-potassium adenosin triphosphatase, *Annu. Rev. Biochem.* **43**:327.

DeDuve, C., and Hayaishi, O., 1978, *Tocopherol, Oxygen and Biomembranes,* Elsevier/North-Holland, Amsterdam.

DeLuca, H. F., 1980, Some new concepts emanating from a study of the metabolism and function of vitamin D, *Nutr. Rev.* **38**:169.

Demetrakopoulos, G. E., Gonzalez, F., Colofiore, J., and Amos, H., 1977, Growth of chick and mammalian cells on D-xylose, *Exp. Cell Res.* **106**:167.

Eagle, H., 1959, Amino acid metabolism in mammalian cell cultures, *Science* **130**:432

Eagle, H., 1973, The effect of environmental pH on the growth of normal and malignant cells, *J. Cell. Physiol.* **82**:1.

Eagle, H., and Levintow, L., 1965, Amino acid and protein metabolism. I. The metabolic characteristics of serially propagated cells, in: *Cells and Tissues in Culture* (E. N. Willmer, ed.), pp. 277–296, Academic Press, New York.

Eagle, H., and Piez, K., 1962. The population-dependent requirement by cultured mammalian cells for metabolites which they can synthesize, *J. Exp. Med.* **116**:29.

Enesco, M., and Leblond, C. P., 1962, Increase in cell number as a factor in the growth of organs and tissues of the young male rat, *J. Embryol. Exp. Morphol.* **10**:530.

Felig, P., 1975, Amino acid metabolism in man, *Annu. Rev. Biochem.* **44**:933.

Feng, J., Melcher, A. H., Brunette, D. M., and Moe, H. K., 1977, Determination of L-ascorbic acid levels in culture medium: Concentrations in commercial media and maintenance of levels under conditions of organ culture, *In Vitro* **13**:91.

Fox, C. F., 1981, *Control of Cellular Division and Development, 1980 ICN-UCLA Symposium,* Alan R. Liss, New York, in press.

Gaull, G., Sturman, J. A., and Raiha, N. C. R., 1972, Development of mammalian sulfur metabolism: Absence of cystathionase in human fetal tissues, *Pediatr. Res.* **6**:538.

Goodhart, R. S., and Shils, M. E., 1980, *Modern Nutrition in Health and Disease,* Sixth Edition, Henry Kimpton Publishers, London.

Gorman, R. R., Hamilton, R. D., and Hopkins, N. K., 1979, Stimulation of human foreskin fibroblast adenosine 3′:5′-cyclic monophosphate levels by prostacyclin, *J. Biol. Chem.* **254**:1671.

Groelke, J. W., Baseman, J. B., and Amos, H., 1979, Regulation of the $G_1 \rightarrow S$ phase transition in chick embryo fibroblasts with α-keto acids and L-alanine, *J. Cell. Physiol.* **101**:391.

Guyton, A. C., 1971, *Textbook of Medical Physiology,* Fourth Edition, W. B. Saunders, Philadelphia.

Ham, R.G., 1962, Clonal growth of diploid Chinese hamster cells in a synthetic medium supplemented with purified proteins fractions, *Exp. Cell Res.* **28**:489.

Ham, R. G., 1963, An improved nutrient solution for diploid Chinese hamster and human cell lines, *Exp. Cell Res.* **29**:515.

Ham, R. G., 1964, Putrescine and related amines as growth factors for a mammalian cell line, *Biochem. Biophys. Res. Commun.* **14**:34.

Ham, R. G., 1965, Clonal growth of mammalian cells in a chemically defined, synthetic medium, *Proc. Natl. Acad. Sci. U.S.A.* **53**:288.

Ham, R. G., 1972, Cloning of mammalian cells, *Methods Cell Physiol.* **5**:37.

Ham, R. G., 1974a, Nutritional requirements of primary cultures. A neglected problem of modern biology, *In Vitro* **10**:119.

Ham, R. G., 1974b, Unique requirements for clonal growth, *J. Natl. Cancer Inst.* **53**:1459.

Ham, R. G., 1980, Dermal fibroblasts, *Methods Cell Biol.* **21A**:255.

Ham, R. G., 1981, Survival and growth requirements of nontransformed cells, in: *The Handbook of Experimental Pharmacology,* Vol. 57 (R. Baserga, ed.), p. 13, Springer-Verlag, New York.

Ham, R. G., and McKeehan, W. L., 1978a, Development of improved media and culture conditions for clonal growth of normal diploid cells, *In Vitro* **14**:11.

Ham, R. G., and McKeehan, W. L., 1978b, Nutritional requirements for clonal growth of nontransformed cells, in: *Nutritional Requirements of Cultured Cells* (H. Katsuta, ed.), pp. 63–115, University Park Press, Baltimore.

Ham, R. G., and McKeehan, W. L., 1979, Media and growth requirements, *Methods Enzymol.* **58**:44.

Ham, R. G., Hammond, S. L., and Miller, L. L., 1977, Critical adjustment of cysteine and glutamine concentrations for improved clonal growth of WI-38 cells, *In Vitro* **13**:1.

Harper, A. E., 1964, Amino acid toxicities and imbalances, in: *Mammalian Protein Metabolism,* Vol. 2 (H. N. Munro and J. B. Allison, eds.), pp. 87–134, Academic Press, New York.

Harris, C. C., Trump, B. F., and Stoner, G. D., 1980, *Normal Human Tissue and Cell Culture. Methods in Cell Biology,* Vols. 21A and 21B, Academic Press, New York.

Hennings, H., Michael, D., Cheng, C., Steinert, P., Holbrook, K., and Yuspa, S. H., 1980, Calcium regulation of growth and differentiation of mouse epidermal cells in culture, *Cell* **19**:245.

Higuchi, K., 1973, Cultivation of animal cells in chemically defined media, a review, *Adv. Appl. Microbiol.* **16**:111.

Holman, R. T., 1968, Essential fatty acid deficiency, *Prog. Chem. Fats Other Lipids* **9**:279.

Holmes, R., Helms, J., and Mercer, G., 1969, Cholesterol requirement of primary diploid human fibroblasts, *J. Cell Biol.* **42**:262.

Iscove, N. N., and Melchers, F., 1978, Complete replacement of serum by albumin, transferrin, and soybean lipid in cultures of lipopolysaccharide-reactive B lymphocytes, *J. Exp. Med.* **147**:923.

Jakoby, W. B., and Pastan, I. H., 1979, *Cell Culture Methods in Enzymology,* Vol. 58, Academic Press, New York.

Jennings, S. J., and Ham, R. G., 1981, Clonal growth of primary cultures of rabbit chondrocytes in a defined medium, *In Vitro* **17**:238.

Jimenez de Asua, L., Clingan, D., and Rudland, P. S., 1975, Initiation of cell proliferation in cultured mouse fibroblasts by prostaglandin $F_{2\alpha}$, *Proc. Natl. Acad. Sci. U.S.A.* **72**:2724.

Jimenez de Asua, L., Levi-Montalcini, R., Shields, R., and Iacobelli, S., 1980, *Control Mechanisms in Animal Cells: Specific Growth Factors,* Raven Press, New York.

Kano-Sueoka, T., Cohen, D. M., Yamaizuimi, Z., Nishimura, S., Mori, M., and Fujiki, H., 1979, Phosphoethanolamine as a growth factor of a mammary carcinoma cell line of rat, *Proc. Natl. Acad. Sci. U.S.A.* **76**:5741.

Kaplan, J. G., 1978, Membrane cation transport and the control of proliferation of mammalian cells, *Annu. Rev. Physiol.* **40**:19.

Katsuta, H., 1978, *Nutritional Requirements of Cultured Cells,* University Park Press, Baltimore.

Katsuta, H., and Takaoka, T., 1973, Cultivation of cells in protein- and lipid-free synthetic media, *Methods Cell Biol.* **6**:1.

King, M. E., and Spector, A. A., 1981, Lipid metabolism in cultured cells, in: *The Growth Requirements of Vertebrate Cells in Vitro* (C. Waymouth, R. G. Ham, and P. J. Chapple, eds.), Cambridge University Press, New York, in press.

Knopf, K., Sturman, J. A., Armstrong, M., and Hayes, K. C., 1978, Taurine: An essential nutrient for the cat, *J. Nutr.* **108**:773.

Koch, C. J., and Biaglow, J. E., 1978, Toxicity, radiation sensitivity modification, and metabolic effects of dehydroascorbate and ascorbate in mammalian cells, *J. Cell Physiol.* **94**:299.

Kutsky, R. J., 1973, *Handbook of Vitamins and Hormones,* Van Nostrand Reinhold, New York.

Leffert, H. L., 1980, *Growth Regulation by Ion Fluxes, Annals of the New York Academy of Sciences,* Vol. 339, New York Academy of Sciences, New York.

Leivobitz, A., 1963, The growth and maintenance of tissue-cell cultures in free gas exchange with the atmosphere, *Am. J. Hyg.* **78**:173.

Levintow, L., and Eagle, H., 1961, Biochemistry of cultured mammalian cells, *Annu. Rev. Biochem.* **30**:605.

Mather, J. P., and Sato, G. H., 1979, The use of hormone-supplemented serum-free media in primary cultures, *Exp. Cell Res.* **124**:215.

McKeehan, W. L., and Ham, R. G., 1978, Phospholipid vesicles as synthetic, defined carriers for lipids in aqueous culture media, *In Vitro* **14**:353.

McKeehan, W. L., and McKeehan, K. A., 1979, Oxocarboxylic acids, pyridine nucleotide-linked oxidoreductases and serum factors in regulation of cell proliferation, *J. Cell Physiol.* **101**:9.

McKeehan, W. L., McKeehan, K. A., Hammond, S. L., and Ham, R. G., 1977, Improved medium for clonal growth of human diploid fibroblasts at low concentrations of serum protein, *In Vitro* **13**:399.

Meng, H. C., 1977, Parenteral nutrition: Principles, nutrient requirements, techniques, and clinical applications, in: *Nutritional Support of Medical Practice* (H. A. Schneider, C. E. Anderson, and D. B. Cousin, eds.), pp. 152–183, Harper & Row, New York.

Messmer, T. O., and Young, D. V., 1977, The effects of biotin and fatty acids on SV3T3 cell growth in the presence of normal calf serum, *J. Cell. Physiol.* **90**:265.

Morgan, J. F., 1958, Tissue culture nutrition, *Bacteriol. Rev.* **22**:20.

Morton, H. J., 1970, A survey of commercially available tissue culture media, *In Vitro* **6**:89.

Munro, H. N., 1969, Evolution of protein metabolism in mammals, in: *Mammalian Protein Metabolism,* Vol. III (H. N. Munro, ed.), pp. 133–182, Academic Press, New York.

Neugut, A. I., and Weinstein, I. B., 1979, Growth limitation of BHK-21 and its relation to folate metabolism, *In Vitro* **15**:363.

Neuman, R. E., and McCoy, T. A., 1958, Growth-promoting properties of pyruvate, oxalacetate and α-ketoglutarate for isolated Walker carcinosarcoma 256 cells, *Proc. Soc. Exp. Biol. Med.* **98**:303.

Olson, R. E., and Suttie, J. W., 1977, Vitamin K and γ-carboxyglutamate biosynthesis, *Vitam. Horm.* **35**:59.

Orly, J., and Sato, G., 1979, Fibronectin mediates cytokinesis and growth of rat follicular cells in serum-free medium, *Cell* **17**:295.

Pagano, R. E., and Weinstein, J. N., 1978, Interactions of liposomes with mammalian cells, *Annu. Rev. Biophys. Bioeng.* **7**:435.

Pardee, A. B., Dubrow, R., Hamlin, J. L., and Kletzein, R. F., 1978, Animal cell cycle, *Annu. Rev. Biochem.* **47**:715.

Patterson, M. K., Jr., 1972, Uptake and utilization of amino acids by cell cultures, in: *Growth, Nutrition, and Metabolism of Cells in Culture* (G. H. Rothblat and V. J. Cristofalo, eds.), pp. 171–209, Academic Press, New York.

Peehl, D. M., and Ham, R. G., 1980, Clonal growth of human keratinocytes with small amounts of dialyzed serum, *In Vitro* **16**:526.

Perrin, D. D., and Agarwal, R. P., 1973, Multimetal-multiligand equilibria: A model for biological systems, in: *Metal Ions in Biological Systems* (H. Sigel, ed.), pp. 168-206, Marcel Dekker, New York.

Pike, R. L., and Brown, M. L., 1975, *Nutrition: An Integrated Approach,* John Wiley and Sons, New York.

Reitzer, L. J., Wice, B. M., and Kennell, D., 1979, Evidence that glutamine, not sugar, is the major energy source for cultured HeLa cells. *J. Biol. Chem.* **254**:2669.

Rizzino, A., Rizzino, H., and Sato, G., 1979, Defined media and the determination of nutritional and hormonal requirements of mammalian cells in culture, *Nutr. Rev.* **37**:369.

Rothblat, G. H., and Cristofalo, V. J., 1972, *Growth, Nutrition and Metabolism of Cells in Culture,* Three Vols., Academic Press, New York.

Rowe, D. W., Starman, B. J., Fujimoto, W. Y., and Williams, R. H., 1977, Differences in growth response to hydrocortisone and ascorbic acid by human diploid fibroblasts, *In Vitro* **13**:824.

Rubin, H., and Koide, T., 1976, Mutual potentiation by magnesium and calcium of growth in animal cells, *Proc. Natl. Acad. Sci. U.S.A.* **73**:168.

Samuelsson, B., Goldyne, M., Granstrom, E., Hamberg, M., Hammerstrom, S., and Malmstem, C., 1978, Prostaglandins and thromboxanes, *Annu. Rev. Biochem.* **47**:997.

Sato, G. H., and Ross, R., 1979, *Hormones and Cell Culture, Books A and B,* Cold Spring Harbor Laboratory, Cold Spring Harbor, New York.

Schneider, P. D., and Buchwald, H., 1979, Total paranteral nutrition and elemental diets in: *Quick Reference to Clinical Nutrition* (S. L. Halpern, ed.), pp. 295-302, J. B. Lippincott, Philadelphia.

Schwartz, J. P., and Johnson, G. S., 1976, Metabolic effects of glucose deprivation and of various sugars in normal and transformed cell lines, *Arch. Biochem. Biophys.* **173**:237.

Smith, L. C., Pownall, H. J., and Gotto, A. M., Jr., 1978, The plasma lipoproteins: Structure and metabolism, *Annu. Rev. Biochem.* **47**:751.

Smith, J. R., 1981, The fat soluble vitamins, in: *The Growth Requirements of Vertebrate Cells In Vitro* (C. Waymouth, R. G. Ham, and P. J. Chapple, eds.), Cambridge University Press, New York.

Smith, J. L., and Goos, S. M., 1980, Selenium nutritive in total paranteral nutrition: Intake levels, *J. Parenteral Enteral Nutr.* **4**:23.

Spector, A. A., 1972, Fatty acid, glyceride, and phospholipid metabolism, in: *Growth, Nutrition and Metabolism of Cells in Culture,* Vol. 1 (G. H. Rothblat and V. J. Cristofalo, eds.), pp. 257-296, Academic Press, New York.

Sporn, M. B., Dunlop, N. M., and Yuspa, S. H., 1973, Retinyl acetate: Effect on cellular content of RNA in epidermis in cell culture in chemically defined medium, *Science* **182**:722.

Stoner, G. D., Harris, C. C., Myers, G. A., Trump, B. F., and Connor, R. D., 1980, Putrescine stimulates growth of human bronchial epithelial cells in primary culture, *In Vitro* **16**:399.

Tabor, C. W., and Tabor, H., 1976, 1,4-Diaminobutane (putrescine), spermidine and spermine, *Annu. Rev. Biochem.* **45**:285.

Taub, M., and Sato, G. H., 1979, Growth of kidney epithelial cells in hormone-supplemented, serum-free medium, *J. Supramol. Struct.* **11**:207.

Taylor, W. G., Richter, A., Evans, V. J., and Sanford, K. K., 1974, Influence of oxygen and pH on plating efficiency and colony development of WI-38 and vero cells, *Exp. Cell Res.* **86**:152.

Thomas, D. R., Philpott, G. W., and Jaffe, B. M., 1974, The relationship between concentration of prostaglandin E and rates of cell replication, *Exp. Cell Res.* **84**:40.

Tinoco, J., Babcock, R., Hincenbergs, I., Medwadowski, B., and Miljanick, P., 1978, Linoleic acid deficiency: Changes in fatty acid patterns in female and male rats raised on a linolenic acid-deficient diet for two generations, *Lipids* **13**:6.

Tsao, M. C., Walthall, B. J., and Ham, R. G., 1981, Clonal growth of normal human epidermal keratinocytes in a defined medium, *J. Cell Physiol.*, in press.

Underwood, E. J., 1977, *Trace Elements in Human and Animal Nutrition,* Fourth Edition, Academic Press, New York.

Waymouth, C., 1965, Construction and use of synthetic medium, in: *Cells and Tissues in Culture* (E. N. Willmer, ed.), pp. 99–142, Academic Press, New York.

Waymouth, C., 1970, Osmolality of mammalian blood and of media for culture of mammalian cells, *In Vitro* **6**:109.

Waymouth, C., 1978, Studies on chemically defined media and the nutritional requirements of cultures of epithelial cells, in: *Nutritional Requirements of Cultured Cells* (H. Katsuta, ed.), pp. 39–61, University Park Press, Baltimore.

Waymouth, C., 1981, Major ions, buffer systems, pH, osmolality, and water quality, in: *The Growth Requirements of Vertebrate Cells in Vitro* (C. Waymouth, R. G. Ham, and P. J. Chapple, eds.), Cambridge University Press, New York, in press.

Waymouth, C., Ham, R. G., and Chapple, R. J. 1981, *The Growth Requirements of Vertebrate Cells in Vitro,* Cambridge University Press, New York, in press.

Webber, M. M., and Chaproniere-Richenberg, D., 1980, Spermine oxidation products are selectively toxic to fibroblasts in cultures of normal human prostatic epithelium, *Cell Biol. Int. Rep.* **4**:185.

Williams, S. R., 1977, *Nutrition and Diet Therapy,* Third Edition, C. V. Mosby, St. Louis.

Winick, M., and Noble, A., 1965, Quantitative changes in DNA, RNA and protein during prenatal and postnatal growth in the rat, *Dev. Biol.* **12**:451.

Winick, M., and Noble, A., 1966, Cellular response in rats during malnutrition at various ages, *J. Nutr.* **89**:300.

Zalin, R., 1979, The cell cycle, myoblast differentiation and prostaglandin as a developmental signal, *Dev. Biol.* **71**:274.

Zielke, H. R., Ozand, P. T., Tildon, J. T., Sevdalian, D. A. and Cornblath, M., 1978, Reciprocal regulation of glucose and glutamine utilization by cultured human diploid fibroblasts, *J. Cell. Physiol.* **95**:41.

Zile, M. H., Bunge, E. C., and DeLuca, H. F., 1979, On the physiological basis of vitamin A-stimulated growth, *J. Nutr.* **109**:1787.

Chapter 10

Fatty Acid Metabolism in the Neonatal Ruminant

Raymond Clifford Noble and John Herbert Shand

1. Introduction

In spite of the rapid advances that have been made in both biochemical and physiological techniques in recent years, the enclosed environment in which the fetus exists still provides an area of biological investigation that poses many unsolved questions. Without doubt, the relationship between the lipid metabolism of the fetus and that of its mother, in particular the part played by the mother in the supply of lipids during uterine development, is an area of investigation in which extreme contradictions exist. For instance, although fetal lipids may accumulate either by *de novo* synthesis within the fetal tissues themselves or from the maternal circulation through transfer across the placenta, the relative quantitative importance of the two processes in the provision of lipids to the developing fetus still remains an area of considerable doubt. The exceptions to this are, of course, the essential fatty acids for which there is an obligatory placental transfer in all mammalian species. Until recently it was generally accepted that lipids did not cross the placenta in significant amounts. However, the acceptance of the idea that the fetus largely relied upon its own ability to synthesize lipids was questioned with the recognition that the least abundant of the maternal plasma lipid components, the unesterified fatty acids, could provide a major part of the lipid requirement of the fetus.

With the emergence of the fetus to the extrauterine environment, extensive

---

Raymond Clifford Noble and John Herbert Shand • Department of Biochemistry, The Hannah Research Institute, Ayr KA6 5HL, Scotland.

changes are required to be made to its lipid metabolism. As opposed to other species, the neonatal period of the ruminant animal is further complicated by the adaptations required during the first few weeks following birth in order to change from a predominantly monogastric existence into the specialized physiological and biochemical arrangements associated with the ruminant animal.

2. Lipid Composition of Fetal and Neonatal Tissues

2.1. Liver

In spite of the importance of the liver in the regulation of lipid metabolism, analysis of the lipid components of liver tissue has been virtually confined to the adult ruminant, in particular adult sheep, with very few observations on liver tissue from fetal and neonatal animals.

The lipid contents of the livers from newborn calves and lambs appear to be significantly lower than those of the adult. Analyses of the proportions (weight percentage) of the major lipid classes from calves (Poukka, 1966) and newborn lambs (Downing, 1964) have shown that, although triglycerides and phospholipids formed the bulk of the lipid present, triglycerides comprised a far lower percentage of the total lipid in the livers of newborn animals than in adult tissue where these two components were present in approximately equal proportions. In the liver of the newborn ruminant, phospholipids appear to comprise the major lipid fraction and therefore, not surprisingly, have been analyzed in much greater detail than the other lipid components. Within the phospholipids present in the livers of fetal lambs (Downing, 1964), phosphatidyl choline comprised some 47% and phosphatidyl ethanolamine some 23% of the total phospholipids present; sphingomyelin (9%) comprised the largest proportion of the minor phospholipids which also included phosphatidyl inositol, phosphatidyl serine, and cardiolipin. Compared with the adult tissue, the phospholipids present in the livers of fetal lambs contained a very much lower proportion of phosphatidyl choline with compensatory increases in the proportions of phosphatidyl ethanolamine and spingomyelin (Downing, 1964). Similar observations have been made on the phospholipid composition of livers from newborn lambs (Noble *et al.*, 1971c).

During the early neonatal period, considerable changes occur in the relative proportions of the major phospholipids present in the liver, in particular the phosphatidyl choline and phosphatidyl ethanolamine fractions, towards achieving a composition similar to that displayed by the adult. Apart from the detailed analyses that have been reported for the fatty acid compositions of the individual phospholipid components of the livers of newborn lambs (Noble *et al.*, 1971b,c), fatty acid compositional data for the livers of fetal and newborn ruminants appear to be confined to overall fatty acid compositions of the total lipid of newborn calf liver (Ter Meulen *et al.*, 1970) and the total phospholipid fractions of livers from calves (Poukka, 1966) and fetal or newborn lambs (Scott *et al.*, 1967). It was clear that the fatty acid compositions of the liver phospholipids of fetal and newborn ruminants were characterized by extremely low proportions of linoleic acid and relatively high proportions of the eicosatrienoic acid, 20:3(n-9),* giving rise to the high triene : tetraene [20:3(n-9) acid : arachidonic acid] ratios which, in nonruminant species, have been equated with essential fatty acid deficiency (Holman, 1971). Within 8–10 days after birth, the proportion of linoleic acid had increased markedly, and that of 20:3(n-9) acid had decreased with a concomitant reduction in the triene : tetraene ratio (Noble *et al.*, 1971b, 1972). A similar picture, i.e., an increase during the first 8 days of life in the proportions of linoleic acid accompanied by a reduction in the proportions of 20:3(n-9) acid was seen in the fatty acid compositions of the other major lipid classes in the livers of lambs during the early neonatal period (Noble *et al.*, 1971a). The poor essential fatty acid status of the newborn ruminant and its possible effect on the longer-term viability of the animal are discussed later (see Section 7.3).

2.2. Adipose Tissue

Since it constitutes the main lipid storage site, adipose tissue contains large amounts of fatty acids, almost exclusively esterified to glycerol, which can be rapidly mobilized and subsequently utilized by the young ruminant animal. The composition of ruminant adipose tissue, especially from sheep and cattle, has been extensively studied (Christie, 1978). Subcutaneous (Garton and Duncan, 1969a) and perirenal (Downing, 1964; Noble *et al.*, 1971a) adipose tissue from newborn lambs, which contained over 90% triglycerides, had fatty acid compositions that were essentially similar; some 65% of the total fatty acids present were *cis*-monoenoic components, largely oleic acid with some palmitoleic acid, with the remainder of the fatty acids consisting of palmitic and stearic acids in approximately equal proportions.

Virtually none of the components such as odd-chain, branched-chain, *trans*-monoenoic acids or the essential fatty acids which are derived exclusively from exogenous sources, were found in the adipose tissue triglycerides of the newborn lamb. The triglycerides from the subcutaneous (Garton and Duncan, 1969b) and perirenal (Ter Meulen *et al.*, 1970) adipose tissues of newborn calves contained higher proportions of palmitic and palmitoleic acids and less oleic acid than did the triglycerides from the corresponding ovine tissues. The only other

*The recommendations of the IUPAC-IUB Commission on Biochemical Nomenclature for the identification of double bond position in unsaturated fatty acids are given in *Biochem. J.* **105**:902 (1967).

lipid classes present in significant amounts in the perirenal adipose tissue, and probably in other adipose tissue sites, of the newborn lamb were unesterified fatty acids and phospholipids (Noble *et al.*, 1971a). The unesterified fatty acids contained lower proportions of palmitic and stearic acids than the triglycerides and correspondingly higher proportions of oleic acid; in the phospholipids, the proportions of palmitic and oleic acids were lower than in the triglycerides because of the presence of significant amounts of the C_{20} metabolites of linoleic and oleic acids.

In the lamb during the early neonatal period, there were no significant changes in the relative proportions of the major lipid fractions within the adipose tissues (Noble *et al.*, 1971a). However, under normal dietary conditions during this period, there are considerable changes in fatty acid composition, in particular, an accumulation of significant amounts of stearic and linoleic acids and concomitant reductions in the proportions of palmitic and oleic acids.

2.3. Body Fluids

2.3.1. Blood

(a) Plasma. The plasma of newborn ruminants contains considerably less lipid than that of the adult animal (Shannon and Lascelles, 1966; Leat, 1967), the concentration ranging from 60 to 140 mg lipid per 100 ml plasma compared with adult levels of 160 to 490 mg per 100 ml plasma. Whereas in the adult ruminant cholesteryl esters and phospholipids were the principal lipid components and were accompanied by very small amounts of triglycerides, free cholesterol, and unesterified fatty acids, the plasma of the ruminant at birth contained higher proportions of triglycerides and unesterified fatty acids and lower amounts of cholesteryl esters (Leat, 1967; Noble *et al.*, 1971d). Within hours after birth, the concentrations of the unesterified fatty acids increased considerably (Noble *et al.*, 1971d), an action by the newborn animal to meet the energy demands of the new environment (see Section 7.2.); following the onset of suckling, the plasma concentrations of the cholesteryl esters, triglycerides, free cholesterol, and phospholipids also increased rapidly (Noble *et al.*, 1971d).

Within 24 hr after birth, then, considerable increases had occurred in the overall plasma lipid concentration of the ruminant animal. During the early neonatal period, it may therefore be speculated that there is a shift in the distribution of the major plasma lipoprotein fractions away from an HDL-dominated pattern similar to that exhibited by the adult (Nelson, 1973; Raphael *et al.*, 1973) toward one containing considerably increased proportions of VLDL and chylomicron fractions.

Although some slight species variation did exist, there was an overall similarity in the fatty acid compositions of the major plasma lipid fractions of the newborn calf, kid, and lamb (Leat, 1967). The bulk of the fatty acids in all the major plasma lipid fractions were comprised of palmitic, palmitoleic, and oleic acids. In contrast to the adult animal, the cholesteryl ester and phospholipid fractions contained small amounts only of C_{18} polyunsaturated fatty acids and their longer-chain metabolites at birth, but their levels increased rapidly once suckling had begun (Leat, 1966; Noble *et al.*, 1971d).

(b) Cellular Constituents. In contrast to the numerous studies on the lipid and fatty acid compositions of erythrocytes and other cellular blood constituents from adult ruminants, information for fetal, newborn and early neonatal animals is very sparse. It would appear, however, that the proportions of the major phospholipid fractions, i.e., spingomyelin, phosphatidyl ethanolamine, and phosphatidyl choline, which together comprise the majority of the lipid present within the erythrocyte, were very similar to those of the adult sheep (De Gier and Van Deenen, 1964; Leat, 1967). Spingomyelin has therefore been shown to account for some 60%, phosphatidyl ethanolamine, 30%, and phosphatidyl choline, 10% of the total phospholipids present. During the first few days after birth, there was a tendency for the proportion of sphingomyelin to increase even further at the expense of phosphatidyl choline (De Gier and Van Deenen, 1964). The erythrocyte lipids of the newborn ruminant were characterized by a strikingly high content of C_{18} monoenoic acids and a low content of polyunsaturated fatty acids, in particular arachidonic acid (De Gier and Van Deenen, 1964). From our observations, nervonic acid accounted for more than 50% of the fatty acids present in the sphingomyelin fraction. The fatty acid pattern of the total lipids from the lamb erythrocytes was found to be practically identical to that for the normal adult sheep.

In the neonatal calf, the thymus gland, an accumulation of lymphoid tissue, has been used to provide information on the lipid composition of the circulating lymphocytes (Rose and Frenster, 1965). About 50% of the total lipid present in the calf thymus consisted of triglycerides, the remainder being mainly phospholipids (31%) and free cholesterol (6%); only minor amounts of cholesteryl esters, methyl esters, free fatty acids, and partial glycerides were present. Analyses of the phospholipids from whole calf thymus (Rose and Frenster, 1965) revealed that the phospholipid fraction consisted almost wholly of phosphatidyl choline (57%), phosphatidyl ethanolamine (34%), and sphingomyelin (6%), with only small amounts of phosphatidyl inositol, phosphatidyl serine, and cardiolipin. Some of these minor phospholipids, together with lysophosphatides which were not detected in the whole tissue, were present in significant amounts in the subcellular organelles (Jarasch *et al.*, 1973). An interesting feature of the fatty acids of the calf thymus, which were comprised mainly of palmitic, stearic, and oleic acids, was the significant proportion of linoleic acid in both the phosphatidyl choline and phosphatidyl ethanolamine fractions and of arachidonic acid in the phosphatidyl ethanolamine fraction. In addition, the triglyceride fraction contained about 3% of 20:1(n-9) acid, a metabolite of oleic acid not normally present in ruminant tissues.

2.3.2. Lymph

In spite of the importance of the lymphatic system in the absorption and transport of lipids in ruminants (Noble, 1978), the lipid composition of the lymph in neonatal animals has largely been neglected, probably because of the difficulties involved in the surgical procedures required for cannulation. When the lipids in the thoracic duct lymph of newborn and early neonatal calves have been analyzed (Shannon and Lascelles, 1966), it has been shown that during the first 24 hr after birth the amount of lipid transported increased about threefold, the concentration of lipids rising from very low levels at birth to almost those of the adult. Concomitant changes in the relative proportions of the component lipid classes also occurred. During the first day of suckling, the proportion of triglyceride in the lipid increased from about 60% to 81%, and the proportion of phospholipid decreased from 28% to 15%. Over this period, significant reductions in the levels of the minor components, i.e., cholesteryl esters, free cholesterol, and unesterified fatty acids, also occurred. In general, the lipid composition of the lymph of the suckling calf then resembled that of the adult animal. Clearly the fatty acid compositions of the major lymph lipids, in particular that of the triglyceride fraction, would reflect the fatty acid composition of the maternal milk.

2.3.3. Bile

The composition of adult ruminant bile has been studied extensively; by comparison, only limited data are available on the composition of bile from fetal and neonatal animals. The overall lipid composition of ruminant bile remains fairly constant with age, with the phospholipids comprising by far the largest component and accompanied by only small amounts of other lipid fractions (Adams and Heath, 1963). However, marked differences in the proportions of the individual phospholipid classes between the bile from neonatal lambs and adult sheep have been noted (Adams and Heath, 1963). Whereas the phospholipids of the bile from the lambs were comprised of about 93% phosphatidyl choline, with small amounts of phosphatidyl ethanolamine and sphingomyelin and practically no *lyso*-phosphatidyl choline, the phospholipids from the bile of the adult sheep contained almost equal proportions of phosphatidyl choline and *lyso*-phosphatidyl choline; the proportions of phosphatidyl ethanolamine and sphingomyelin remained essentially unchanged.

Analyses of the fatty acid compositions of the phospholipids from the bile of the lambs and adult sheep indicated that those of the lamb contained much higher proportions of linoleic and linolenic acids and their associated C_{20} metabolites than those of the adult sheep, an observation that appears to be in complete contradiction to the polyunsaturated fatty acid pattern of all other newborn lamb tissues (Adams and Heath, 1963). Bile acid composition changes significantly as the young ruminant develops (Peric-Golia and Socic, 1968). Although at birth and during the first week after birth taurocholic acid predominated within the bile acids, its proportion soon diminished, and there was a compensatory increase in the proportion of taurochenodeoxycholic acid. Further changes then occurred, the most notable of these being a gradual decline in the proportion of taurochenodeoxycholic acid and increases in the proportions of taurodeoxycholic and glycocholic acids. Analyses of the bile from stillborn sheep fetuses (Sheriha *et al.*, 1968) have shown that, in comparison to adult sheep bile, there were only low levels of both cholic and deoxycholic acids but high proportions of lithocholic and chenodeoxycholic acids, both of which are virtually absent from the adult animal. The relationship between the changes in bile acid composition and the extensive changes that occur in the digestive system of the ruminant animal from the neonatal period up to adulthood has been fully discussed (Noble, 1978).

2.4. Heart, Kidney, Lung, and Brain

2.4.1. Heart

The lipid composition of heart tissue during the perinatal period has been quite extensively studied in the lamb (Masters, 1964). The lipids of newborn heart tissue were composed mainly of phospholipids (62%) with approximately equal proportions of triglycerides and unesterified fatty acids (15–17%) and a minor amount (<5%) of cholesterol. Analyses of heart tissue from the fetal lamb have shown that the phospholipids contained large amounts of phosphatidyl choline and phosphatidyl ethanolamine together with their plasmalogen analogues (Scott *et al.*, 1967). Whereas in the adult, choline plasmalogen alone accounted for about 20% of the total phospholipids, in fetal heart tissue, phosphatidal ethanolamine was the predominant plasmalogen comprising about 15% of the phospholipids.

Results obtained for the compositions of the phospholipids present in the heart tissue of fetal and neonatal calves (De Rooij and Hooghwinkel, 1967) were in general similar to those obtained for the lamb. As in the adult animal, cardiolipin accounted for a small but significant proportion (4–6%) of the phospholipids present in the heart tissue of fetal lambs (Scott *et al.*, 1967). The fatty acids of the phospholipids of ovine fetal heart displayed extremely low concentrations of linoleic acid and high proportions of oleic acid, thereby contrasting with the adult where linoleic acid was predominant. In the neutral lipids also, ovine fetal heart lipid contained very high proportions of oleic acid and significantly lower proportions of stearic acid than the adult. More recent analytical data (Ter Meulen *et al.*, 1970; Payne, 1978) have shown that the fatty acid compositions of the phospholipids from the hearts of neonatal lambs and calves were essentially similar to those of the fetal tissues.

2.4.2. Kidney

The major lipid component of fetal and neonatal ovine kidney tissue is the phospholipid fraction that may account for up to 73% of the total lipid present (Masters, 1964). Except for the virtual absence of cardiolipin and choline plasmalogen and an appreciably higher level of sphingomyelin, which is thought to play a role in ion transport, the proportions of individual phospholipid fractions (Scott *et al.*, 1967) were similar to those of heart tissue. The fatty acid compositions of the neutral and phospholipid fractions of kidney tissue from fetal lambs were also similar to those of the corresponding fractions of the heart and liver. Some 30–50% of the total fatty acids were accounted for by oleic acid, whereas, as in most other fetal sheep tissues, a somewhat higher proportion of arachidonic acid appeared to be maintained at the expense of a low level of linoleic acid (Scott *et al.*, 1967). The fatty acid compositions of the lipids from kidney tissue of newborn calves were essentially similar to those of the ovine tissue (Ter Meulen *et al.*, 1970).

2.4.3. Lung

Although no data appear to be available for the lipid composition of lung tissue *per se*, a limited study has been made on the lipids of the lung surfactant of fetal lambs at various stages of gestation (Chida *et al.*, 1966; Fujiwara *et al.*, 1968). Of the total lung surfactant, phospholipids were found to account for 30–60%, and total neutral lipids 3–11%, both increasing as term approached (Chida *et al.*, 1966). Phosphatidyl choline comprised the major phospholipid and was found to contain some 60–94% of saturated fatty acids, their proportion increasing with gestational age. Disaturated species of phosphatidyl choline were later shown to be the principal components of near-term fetal and newborn lamb lung surfactant (Fujiwara *et al.*, 1968). Analyses of the fatty acid compositions of the total lung lipids from perinatal lambs and calves have confirmed an extremely high level of palmitic acid which may be attributable to increased secretion of surfactant by the alveolar cells just before birth (Ter Meulen *et al.*, 1970; Payne, 1978).

2.4.4. Brain

Fairly extensive analyses have been made of the total neutral lipids and phospholipids of brain tissue from fetal lambs (Scott *et al.*, 1967) and of the phospholipids from the brain tissue of fetal calves and newborn lambs (Payne, 1978). The phospholipid composition of brain from the fetal lamb was broadly similar to that of the heart but with an increased proportion of phosphatidyl choline at the expense of the plasmalogen fraction. The fatty acid compositions of the phospholipids from brain of fetal and neonatal sheep displayed higher proportions of palmitic acid and lower proportions of oleic acid than fetal or neonatal calf brain (Ter Meulen *et al.*, 1970; Payne, 1978). Both fetal and newborn sheep brain lipids have been shown to contain large proportions of hydroxyl-substituted fatty acids (Downing, 1964; Scott *et al.*, 1967).

2.5. Rumen, Abomasum, and Intestine

The main lipid components present in the rumen and abomasal tissues of fetal lambs were triglycerides (37 and 32%, respectively) and phospholipids (about 41% in each tissue) (Body *et al.*, 1970). Significant amounts of hydrocarbons, unesterified fatty acids, partial glycerides, and free cholesterol, together with minor amounts of cholesteryl esters, were also present. After birth, the proportion of triglycerides increased, and that of the phospholipids decreased in both tissues. Differences in the relative proportions of individual phospholipid components were noted between the rumen and abomasal tissues of the fetal lamb (Body *et al.*, 1970); in particular, the proportions of phosphatidic acid and *lyso* derivatives were higher, and that of phosphatidyl ethanolamine lower in rumen tissue than in the abomasum. Both tissues contained a high but variable (14–21%) proportion of sphingomyelin.

The overall phospholipid composition did not appear to change in either tissue after birth. The fatty acid compositions of the triglycerides and phospholipids in the fetal ovine rumen and abomasum were similar and contained mainly palmitic, stearic, and oleic acids except for the higher proportions of lauric acid and lower proportions of *trans* fatty acids present in the rumen tissue (Body *et al.*, 1970; Body and Shorland, 1974). During the first month after birth, both tissues accumulated significant amounts of linoleic acid and *trans*-monoenoic fatty acids which were accompanied by decreased proportions of lauric, palmitic, and oleic acids. In contrast to results obtained with newborn calves (Ter Meulen *et al.*, 1970), analyses of various segments of the intestinal tract of fetal and newborn lambs (Payne, 1978) have revealed a virtually uniform fatty acid composition which was similar to that of the liver. The phospholipids from the intestine of the newborn lamb characteristically contained a much lower proportion of linoleic acid and higher proportion of 20:3(n-9) acid than the intestine of the calf.

2.6. Skin Surface and Sebaceous Glands

Except in the cases of some specialized areas such as the tongue, the mammary teat canal, and the hoof, data on the lipid composition of ruminant skin tissue do not appear to be available. Analysis of neonatal ruminant skin lipids has virtually been confined to the secretions of the sebaceous glands and the lipids of the skin surface. Although they have a common origin, their lipid compositions may differ significantly, especially in the sheep, because of autoxidation and microbial modification as a result of prolonged exposure on the skin surface. The skin surface lipids of the adult ox are characterized by high proportions (20%) of triglycerides containing unusually high levels of linoleic acid (Noble *et al.*, 1974).

In a study of the changes in the fatty acid composition of the skin surface triglycerides of the calf from birth to 6 weeks of age (Noble *et al.*, 1975b), it has been found that although saturated fatty acids comprised the major proportion of the fatty acids of the lipids at birth, their proportion rapidly decreased, and there was a rise in the proportion of oleic acid. It was some 3 to 4 weeks after birth before the proportion of linoleic acid in the triglycerides attained adult levels. The fatty acids of wool wax from newborn sheep have been found to contain very high levels (45%) of hydroxylated fatty acids, almost half of which were branched species of the C_{16} family (Downing, 1964). Of the remaining unhydroxylated fatty acids, 62% were found to be branched-chain acids, a large proportion of which were of chain length C_{20} and above.

3. Transfer of Lipids across the Placenta

3.1. Phospholipids

In common with other mammalian species, the plasma of the newborn ruminant contains considerably lower concentrations of phospholipids than that of the maternal plasma (Noble *et al.*, 1971d, 1975b); the compositions of the phospholipids in the fetal and newborn plasmas also differ considerably from that of the maternal plasma. As a result of these differences, it has been concluded that some form of placental "barrier" must exist between the fetal phospholipids and the maternal supply (Zee, 1967). On the basis of similarities in the proportions of the major phospholipid components within whole fetal lamb and maternal lipid extracts (Body and Shorland, 1974) and also rumen and abomasal tissue from fetal lamb and adult sheep (Body *et al.*, 1970), it has however been concluded that phospholipids are able to pass freely across the placenta. On the other hand, analysis of the phospholipid compositions of specific tissues from fetal and newborn lambs compared with maternal and adult sheep fails to substantiate such conclusions, as marked differences were found between the relative distributions of the major phospholipid fractions (Scott *et al.*, 1967; Noble *et al.*, 1970b, 1971c). These latter findings were in fact entirely consistent with similar investigations on nonruminant species in which considerable differences have been noted between the phospholipid compositions of the major tissues of the fetus and its mother (Biezenski *et al.*, 1963; Weinhold and Villee, 1965).

Considerable differences between the fatty acid compositions of the phospholipids in the tissues of the fetal ruminant compared with those of the maternal and adult animal have also been shown. In particular, the phospholipids of the fetal tissues contained only low proportions of the C_{18} polyunsaturated fatty acids but appreciable concentrations of 20:3(n-9) acid, contrasting with the maternal and adult animal in which the proportions of the C_{18} polyunsaturated fatty acids were high, and those of 20:3(n-9) acid extremely low (Shorland *et al.*, 1966; Scott *et al.*, 1967; Body *et al.*, 1970; Body and Shorland, 1974). However, in spite of the lack of linoleic acid, the tissue phospholipids of the fetal ruminant displayed high levels of arachidonic acid, resulting in arachidonic acid/linoleic acid ratios far greater than those in the tissue phospholipids of the maternal animal. Similar distinctive differences in the distribution of the C_{18} and C_{20} polyunsaturated fatty acids have also been noted between the plasmas of fetal and adult or maternal ruminants (Noble *et al.*, 1978a) and between the plasma and tissues of newborn and adult animals (Leat, 1966; Noble *et al.*, 1971a–d, 1975b; Palmquist *et al.*, 1977; Payne, 1978).

In many instances, the difference between the distribution of the polyunsaturated fatty acids could be equated with differences in the percentage distribution of the major diacyl phospholipids (Noble *et al.*, 1970b). Clearly, these observations on the C_{18} and C_{20} polyunsaturated fatty acid contents of the maternal and fetal tissue phospholipids would not be consistent with the view that intact phospholipids pass through the placenta. However, the almost exclusive presence of odd-numbered saturated fatty acids, branched-chain fatty acids, polyunsaturated fatty acids, and *trans* unsaturated fatty acids within the phospholipids of the fetal tissues, all of which have to be considered as exogenous in origin, could still be taken as lending support to the hypothesis that phospholipids or their *lyso*-derivatives were capable of being transported across the placenta from mother to fetus. In the absence of any passage of intact phospholipid molecules across the placenta, any similarity in fatty acid composition between the phospholipids of the fetal and maternal tissues would have to arise through a synthesis of fetal phospholipids using an identical selection of fatty acids to that of the mother.

More direct evidence, in particular, through the use of radiotracer techniques, on the origin of the fetal phospholipids appears to be confined to nonruminant species. From experiments involving the intravenous injection of [^{32}P]-labeled phospholipids into the circulation of pregnant rabbits, extensive uptake of label by the maternal liver and placenta was not accompanied by an indication that any unchanged phospholipids were transported across the placenta from the mother to the fetus (Popjak and Beeckmans, 1950a).

The possibility that absorption of the maternal phospholipids by the placenta may lead to the appearance of some degraded phospholipid moieties in the fetus has been suggested (Popjak, 1954), but *in-vitro* incubations of human placenta have failed to provide any evidence for the ability of the placenta to degrade naturally occurring exogenous and endogenous phospholipids into smaller fragments (Robertson and Sprecher, 1967). However, there is evidence that during pregnancy a considerable decline in the concentration of *lyso*-phosphatidyl choline occurs within the maternal plasma (Svanborg and Vikrot, 1965), whereas in experiments where rats and mice were injected intravenously with ^{14}C- and ^{32}P-labeled *lyso*-phosphatidyl choline, extensive uptake of the compounds into the placental tissues took place followed by rapid reacylation to form phosphatidyl choline (Eisenberg *et al.*, 1967). With the detection of significant quantities of ^{14}C-labeled triglycerides and phospholipids in the fetal tissues, it was clear that the *lyso*-phosphatidyl choline injected into the maternal circulation had been able to provide substantial quantities of fatty acid for incorporation into the fetal lipids.

Similar work with rabbits (Biezenski, 1970; Biezenski *et al.*, 1971) was also unable to show any direct placental transfer of intact phospholipids from mother to fetus but did show an abundant transfer into the fetus of phospholipid fragments derived from the maternal phospholipids. That the placenta has an important role to play in the supply of phospholipids to the fetus in both ruminant and nonruminant species is now apparent (see Section 4).

3.2. Cholesterol and Cholesteryl Esters

In the absence of any investigations with ruminants, the weight of evidence from nonruminant species would indicate that the contribution to the fetus through the transfer across the placenta of cholesterol circulating within the mother is quite considerable. By the use of radioactively labeled cholesterol fed to rats and guinea pigs, it has been estimated that some 11 to 22% of the cholesterol in some of the major fetal tissues, including the liver, heart, kidney, brain, muscle, and lung, was derived from the maternal plasma (Goldwater and Stetten, 1947; Connor and Lin, 1967). In humans, measurements of the venous–arterial differences in the umbilical cord blood have also indicated a direct supply of cholesterol to the fetus from maternal sources; thus, consistent positive values for umbilical venous–arterial cholesterol concentration differences, together with a positive correlation between maternal vein and umbilical vein–artery blood cholesterol concentrations, have been obtained (Spellacy *et al.*, 1974).

Direct evidence for the transfer of cholesteryl esters across the placenta from the maternal plasma into the fetus is lacking, but from the abundance of indirect evidence available, direct transfer of maternal cholesteryl esters across the placenta does not appear to take place. In nonruminant species, therefore, distinct fatty acid compositional differences have been noted between the cholesteryl esters of the maternal circulation and those of the fetus. The existence of distinction "fetal" and "maternal" fatty acid patterns have been established in a number of nonruminant species under a wide range of maternal dietary regimes (Muldrey *et al.*, 1961; Lopez-Santolino *et al.*, 1965). Similar indirect evidence available from analysis of plasma lipid compositions of fetal or newborn and maternal plasmas of ruminants would also indicate that the direct transfer of cholesteryl esters across the placenta is a most unlikely process. There are, for instance, extremely large differences in the relative proportions of the C_{18} polyunsaturated fatty acids between the maternal and fetal plasmas. Whereas in the cow, linoleic acid may account for up to 80% of the total fatty acids present in the plasma cholesteryl esters, in the plasma of the newborn calf, linoleic acid accounts for only about 6% of the total fatty acids present (Noble *et al.*, 1975b; Steele *et al.*, 1971). Similarly, in the sheep, the plasma cholesteryl esters of the ewe contain some 35% linoleic acid compared to only about 2–3% in the newborn lamb (Noble *et al.*, 1971d,e; Noble and Moore, 1974).

3.3. Triglycerides

It has been established in a number of species, including various ruminants, that large differences exist between the fatty acid compositions of the maternal and fetal or newborn plasma triglycerides. Fetal lamb plasma is characterized by a very much higher level of oleic acid in the plasma triglycerides than is present in the ewe, and there are compensatory decreases in the proportions of a wide range of straight- and branched-chain fatty acids and polyunsaturated fatty acids (Shorland *et al.*, 1966). These differences in the fatty acid compositions were taken to indicate that direct transfer of maternal triglycerides across the placenta of the ewe into the fetus did not occur. Distinctive fatty acid compositional differences between the plasma triglycerides of the ewe and cow compared to the newborn lamb and calf have resulted in similar conclusions being drawn (Noble *et al.*, 1971d,e; Steele *et al.*, 1971; Noble and Moore, 1974, Noble *et al.*, 1975b).

Confirmation of the inability of the maternal plasma triglycerides to be directly transferred across the placenta to the fetus has been obtained from a series of experiments using radioactive substrates. Following the injection into pregnant guinea pigs of chylomicrons containing ^{14}C-labeled triglycerides that had been harvested from the lymph of donor animals, no significant accumulation of label in the tissue of the fetus could be detected (McBride and Korn, 1964). Similarly, following the injection of albumin complexed with 1-[^{14}C]palmitic acid into pregnant rabbits, the relationship between the appearance of activity in the triglycerides of the maternal and fetal plasmas was such as to preclude the possibility of direct transfer of triglycerides across the placenta (Van Duyne *et al.*, 1962). In the sheep also, evidence has recently been obtained substantiating the inability of maternal triglycerides to directly transfer across the placenta into the fetal plasma (Shand and Noble, 1979a). Following the injection of ^{14}C-labeled tripalmitin into the circulation of the pregnant ewe, no significant appearance of radioactivity in the plasma triglycerides of the fetus had occurred up to 20 min later. However, the possibility exists that under certain circumstances rapid uptake of the maternal plasma triglycerides by peripheral tissues followed by subsequent release of fatty acids back into the circulation could lead to a rapid and significant appearance of maternal plasma triglyceride fatty acids in the fetal bloodstream.

3.4. Unesterified Fatty Acids

It has been shown in many mammalian species, including ruminants, that the total concentration of the unesterified fatty acids in the fetal plasma is considerably less than that of the mother (Van Duyne and Havel, 1959; Hershfield and Nemeth, 1968; Alexander *et al.*, 1969a; Noble *et al.*, 1971d,e, 1978c; Comline and Silver, 1972; Elphick and Hull, 1977a). The possibility that the placenta was not freely permeable to the passage of unesterified fatty acids was therefore suggested. On the other hand, injections of radioactive fatty acids into the maternal circulation of a variety of nonruminant species had invariably led to a rapid appearance of radioactivity in the plasma of the fetus (Van Duyne and Havel, 1959; Van Duyne *et al.*, 1962; McBride and Korn, 1964; Elphick *et al.*, 1975; Elphick and Hull, 1977a), suggesting an active movement of unesterified fatty acids across the placenta. In some instances, the accumulation of significant amounts of radioactivity in the fetal plasma could be detected within 30 sec of the introduction of the labeled fatty acids into the maternal circulation (Elphick and Hull, 1977a). It soon became apparent that, contrary to initial conclusions, the maternal plasma unesterified fatty acids were able to cross the placenta and enter and leave the placenta in an unesterified form before incorporation into the fetal tissues (Hershfield and Nemeth, 1968; Portman *et al.*, 1969).

From evidence involving experiments in which radioactive fatty acids were injected into the pregnant animal, it has been established that in ruminant species as well, the unesterified fatty acids of the maternal circulation were able to pass across the placenta in an unesterified form (Van Duyne *et al.*, 1960; Noble *et al.*, 1978c). These conclusions have been further substantiated by the differences that have been observed between the concentrations of the unesterified fatty acids in the arterial and venous umbilical blood supplies (Persson and Tunell, 1971; Elphick *et al.*, 1975; Elphick and Hull, 1977a).

Consistent with the view that the unesterified fatty acids constitute the only maternal plasma fatty acid-containing fraction that can pass through the placenta is the evidence that has been obtained from comparative fatty acid compositional studies of maternal, fetal, and neonatal plasmas (Leat, 1966; Hershfield and Nemeth, 1968). In nonruminant species, appreciable concentrations of linoleic acid were found in the plasma of fetal and newborn animals in contrast to ruminant species, in which the concentrations of this polyunsaturated fatty acid were extremely low (Leat, 1966; Noble *et al.*, 1971d; 1975b).

This observation may readily be explained by differences in maternal plasma lipid composition. Although the maternal plasmas of both ruminant and nonruminant species contained high concentrations of linoleic acid, in ruminants it was almost wholly associated with the cholesteryl ester and phospholipid fractions, the concentration in the unesterified fatty acids being extremely low (Garton and Duncan, 1964; Noble *et al.*, 1971e). The partition of linoleic acid between the major plasma lipids of nonruminant species, on the other hand, was far less evident, with the result that the unesterified fatty acids contained appreciable concentrations of linoleic acid (Leat, 1966).

It is clear, however, that in both ruminant and nonruminant species, distinct differences may exist among the rates of passage of individual fatty acids across the placenta into the fetal circulation. Although in the rabbit both palmitic and linoleic acids displayed similar abilities to cross the placenta, there appeared to be a relatively higher accumulation of arachidonic acid in the fetal circulation (Elphick and Hull, 1977a). The possibility that this could be accounted for by either a preferential transfer from the maternal plasma into the fetus or through the placental synthesis from linoleic acid was suggested. In the guinea pig, linoleic acid would appear to possess a far greater ability to cross the placenta than palmitic acid (Hershfield and Nemeth, 1968). Recent investigations with sheep, on the other hand, have shown an obvious discrimination against the passage of linoleic acid relative to palmitic acid across the placenta into the fetal circulation (Noble *et al.*, 1978c). As in the rabbit, the suggestion was made that this may also be associated with the synthesis of arachidonic acid by the placenta (see Section 4).

Passage of the unesterified fatty acids is not confined to the direction of mother to fetus. Injection of labeled fatty acids into the fetal circulation of several species, including ruminants, has shown that unesterified fatty acids may also pass from the fetus back across the placenta into the mother (Kayden *et al.*, 1969; Portman *et al.*, 1969; Elphick *et al.*, 1975; Noble *et al.*, 1979). Although under the conditions of the experiments the net flow of the acids was in the direction of mother to fetus, i.e., down the existing concentration gradient, constant fatty acid exchange between the two circulations was obviously occurring. As a result, it has been suggested that under certain physiological conditions where maternal unesterified fatty acid concentrations are severely reduced, a net flow of fatty acids from the fetus back into the mother is possible (Hull, 1975). Although the main determinant of fatty acid flow between mother and fetus is clearly the concentration gradient between the two circulations, many other factors have also been implicated, not the least being the relative binding affinities of the various fatty acids for plasma albumin (Hershfield and Nemeth, 1968).

3.5. Short-Chain Fatty Acids and Ketone Bodies

A comparison of the total carbon requirement of the fetus and its waste products with the rate of passage of major nutrients from the mother has indicated a large discrepancy between maternal supply and fetal requirements (Bassett and Wallace, 1966). The deficit amounted to some 15% of the total fetal carbon requirement. The importance of acetate and other short-chain fatty acids to the metabolism of the parent ruminant is well established (Vernon, 1980). That maternally derived short-chain fatty acids, in particular, acetic acid, play a similar role in the metabolism of the fetus is now apparent. Early conclusions (Alexander *et al.*, 1967) that acetate transfer from mother to fetus was minimal have been shown to be incorrect. From correlative data between maternal and fetal plasma acetate concentrations under a variety of physiological conditions, together with observations of placental acetate contribution, it is apparent that maternally derived short-chain fatty acids, in particular, acetic acid, may contribute considerably to the nutrient supply of the fetus in both sheep and cows (Char and Creasy, 1976; Comline and Silver, 1976). The possibility that under certain conditions, in particular, starvation, maternal plasma β-hydroxybutyrate and acetoacetate may contribute significantly to the fetal metabolism has also been investigated (Katz and Bergmann, 1969; Morriss *et al.*, 1974). Although it could be demonstrated that both metabolites make some contribution to fetal metabolism, in each case, this contribution accounted for no more than 2–3% of the total oxidizable substrate required by the fetus even under conditions of starvation and, therefore, could not be considered as a major metabolic fuel for the fetus.

4. Lipid Metabolism in the Placenta

With the passage across the placenta of all of the nutrients required to nourish the fetus, the placenta has every opportunity to play a considerable role as a modifying influence and reservoir, however temporary, in the lipid supply to the fetus. The presence of many enzymic processes in addition to those purely necessary for nutrient transfer clearly indicates the involvement of the placenta in other diverse functions (Hagerman, 1964). Apart from the direct provision of lipid to the placenta through transport from the maternal blood supply, the lipids of the placenta may arise through lipid synthesis either from the incorporation of already absorbed lipid fractions or from nonlipid precursors. In the ruminant, as in nonruminant species (Portman *et al.*, 1969), transfer of labeled fatty acids from mother to fetus has resulted in considerable accumulation of activity within the placenta lipids, in particular, the phospholipids which comprise by far the largest fraction present (R. C. Noble and J. H. Shand, unpublished observation).

The possibility also exists that the ruminant placenta may also be able to take up and incorporate maternal lipids originating from the chylomicrons or very-low-density lipoprotein fractions of the plasma; certainly in rat, rabbit, and man, there is evidence that this occurs (Hummel, *et al.*, 1976b; Elphick and Hull, 1977b; Elphick *et al.*, 1978), whereas the identifiable presence of an active liproprotein lipase within the placenta would also indicate the probability of such a process (Mallov and Alousi, 1965; Elphick and Hull, 1977b). In spite of the existence of the required enzyme systems within the placenta, the extent of lipid synthesis from nonlipid precursors appears to be extremely low (Diamant *et al.*, 1975; Char and Creasy, 1976; Hummel *et al.*, 1976a). On the other hand, lipid synthesis and turnover involving preformed lipid fragments of maternal origin is very extensive in the placenta (Mallov and Alousi, 1965; Robertson and Sprecher, 1967; Eisenberg *et al.*, 1967).

Evidence has now been obtained that the placenta possesses a considerable ability to desaturate maternally derived fatty acids before releasing them for uptake by the fetus. The presence of very much higher concentrations of arachidonic acid within the fetal plasma than among the unesterified fatty acids circulating in the maternal plasma has been demonstrated in a number of ruminant and nonruminant species (Portman *et al.*, 1969; Phanteliadis and Troll, 1976; Elphick and Hull, 1977a; Noble *et al.*, 1978a,c; Shand and Noble, 1979b).

In the placenta also, the proportion of arachidonic acid has been shown to be very much higher than that of the maternal plasma unesterified fatty acids (Robertson *et al.*, 1968; Phanteliadis and Troll, 1976; Pascaud *et al.*, 1977; Troll *et al.*, 1977). In the ruminant, the differences between the distributions of linoleic and arachidonic acid in the plasma lipids of the mother and fetus are particularly apparent: in the sheep, for instance, the arachidonic/linoleic acid ratio in the plasma of the mother during gestation was only 0.21 compared with a ratio of 2.72 in the fetal plasma (Noble *et al.*, 1978a,c; Shand and Noble, 1979b). Similarly, the placenta also showed a high arachidonic/linoleic acid ratio (Shand and Noble, 1979b). In both ruminant and nonruminant species, the ability to account for the higher proportions of arachidonic acid within the placenta and fetal plasma through either a preferential uptake and transfer of the acid from the maternal plasma or synthesis within the fetal tissues has been ruled out.

In the sheep, as in other species, the placenta appears to be the most likely source of the increased levels of arachidonic acid (Phanteliadis and Troll, 1976; Pascaud *et al.*, 1977; Elphick and Hull, 1977a; Noble *et al.*, 1978a,c). Direct evidence to substantiate this conclusion for ruminants has recently been obtained with the identification in sheep of high levels of both $\Delta 6$ and $\Delta 9$ desaturase activities in the placenta (Shand and Noble, 1979b) which far exceeded the levels that had previously been found in other tissues from both adult and early neonatal animals (Wahle, 1974; Shand *et al.*, 1978). Without doubt, then, the placenta plays a considerable role in the provision of essential fatty acids to the developing fetus, a role that may be of particular importance in the case of the ruminant animal (see Section 7.3).

5. Maternal Contribution to Fetal Lipid Requirements

Estimates of the maternal contribution to fetal lipid requirements appear to be open to considerable variation. However, in spite of several observations to the contrary, it is now generally agreed that in many nonruminant species the maternal contribution of lipid to the fetal requirements is extensive, and in spite of a fetal ability to synthesize considerable amounts of lipid, sufficient fatty acids may cross the placenta to meet the demands of the fetus (Hershfield and Nemeth, 1968; Portman *et al.*, 1969; Persson and Tunell, 1971; Hummel *et al.*, 1975; Elphick *et al.*, 1978).

Transfer of fatty acids across the placenta of ruminants is clearly much more limited than in nonruminant species. In the sheep, it has been concluded that only a small proportion of the fetal fatty acids had been derived from maternal sources (Van Duyne *et al.*, 1960; Elphick and Hull, 1978). From comparisons of the specific activities of the unesterified fatty acids from the maternal and fetal plasmas following maternal injection of ^{14}C-labeled palmitic acid, it was concluded that the placenta of the sheep was not freely permeable to the passage of fatty acids (Van Duyne *et al.*, 1960). More recently (Noble *et al.*, 1978c), similar experiments in sheep involving the maternal injection of ^{14}C-labeled palmitic and linoleic acids have shown a discrimination against the passage of linoleic acid across the placenta.

A far more extreme situation was indicated to exist from experiments in which measurements were made of venous–arterial concentration differences of the unesterified fatty acids in the umbilical circulation of the fetal lamb, since no significant passage of unesterified fatty acids across the maternal uterine circulation could be demonstrated (James *et al.*, 1971). From the evidence so far available, it would therefore appear that in ruminant animals, maternal unesterified fatty acids may provide a considerably lower proportion of the total fetal lipid requirement than in nonruminants.

On entering the fetal circulation, the maternally derived fatty acids undergo rapid metabolism by the fetal tissues. In both ruminant and nonruminant species, it has been demonstrated that the appearance of labeled fatty acids within the fetal circulation was followed by a rapid accumulation of activity in the liver and adipose tissues followed by the reappearance of activity in the plasma in esterified lipid form (Van Duyne *et al.*, 1960; Hershfield and Nemeth, 1968; Portman *et al.*, 1969). In the sheep, it has been shown that the turnover rates of the fetal lipids far exceed those of the mother (Van Duyne *et al.*, 1960); metabolism by the fetus of any phospholipids that have been indirectly derived from the maternal plasma is also extremely rapid. As in the adult, there are considerable variations in the ultimate distributions and rates of metabolism of different fatty acids within the fetus, especially between saturated and unsaturated fatty acid species (Portman *et al.*, 1969). In ruminant species, this may be of particular significance in fetal and neonatal animals where the availability of essential fatty acids for metabolic purposes is particularly limited (see Section 7.3.).

From the previous discussion, it is clear that the role of maternal fatty acid transfer in satisfying the lipid requirements of the developing fetus is dependent on several major factors. For instance, the ability of the maternal circulation to supply the fatty acids in sufficient quantity is of prime importance. The levels of circulatory lipids, including the unesterified fatty acids, are known to undergo considerable increases in concentration during pregnancy in nonruminant species (Burt, 1960; Svanborg and Vikrot, 1965; Edson *et al.*, 1975; Vernon, 1980), consequently affecting the fatty acid supply to the fetus. Although the concentration of unesterified fatty acids in the maternal plasma of ruminant species also shows an increase, it appears to be mainly confined to the very latter stage of pregnancy (Noble *et al.*, 1971e; Comline and Silver, 1972).

Clearly, any change in the physiological state of the mother that effects an alteration in the concentration of the plasma unesterified fatty acids can be expected to result in changes in the fatty acid supply to the fetus. The state of maternal nutrition could be expected to affect fetal fatty acid supply considerably. In nonruminant species, the large rise in the plasma concentration of unesterified fatty acids that occurs within the mother during starvation has been shown to lead to a considerable increase in the passage of fatty acids across the placenta and, somewhat paradoxically, to increased levels of lipid in the liver and adipose tissues of the fetus (Edson *et al.*, 1975; Hull, 1975). Other mediators of lipolytic fatty acid release into the maternal circulation, e.g., increased neural activity resulting from fear, excitement, or stress, changes in hormonal activity, would be expected to increase fetal lipid uptake.

Although the ability of starvation, physiological stress, and the administration of several lipolytic hormones to cause increases in the concentrations of the maternal plasma unesterified fatty acids in ruminants is well established (Noble *et al.*, 1969b; Comline and Silver, 1972; Thompson and Clough, 1972), only small effects on fetal plasma concentration and fetal uptake of fatty acids have been noted (Battaglia and Meschia, 1973). Conversely, under conditions in which there is a reduction in the maternal plasma unesterified fatty acid concentration, fetal fatty acid levels and uptake would be expected to be reduced accordingly. It is also to be expected that the pattern of fatty acids taken up by the fetus would be directly related to the composition of the maternal plasma unesterified fatty acids and also to a large extent to the fatty acid composition of the maternal adipose tissue stores. Accordingly, changes in the fatty acid compositions of the maternal diet have, in both ruminant and nonruminant species, been accompanied by related changes in the plasma and tissue fatty acid profiles of the fetus and newborn (Hull, 1975; Noble *et al.*, 1978b).

To allow for adequate maternal contribution to fetal lipid supply, there must also be sufficient ability for the placenta to transfer the fatty acids. In this respect, it is clear from previous discussion (see Section 4) that the placenta plays a considerable role in both the qualitative and quantitative aspects of fatty acid transfer between mother and fetus. The ultimate regulator in the quantitative significance of maternal fatty acid transfer is the extent to which the fetal tissues are able to take up and metabolize the fatty acids. Over and above the lipid required for structural purposes, lipid utilization by the fetus would require that the tissues, in particular the adipose tissue, possess the capacity to store the transported fat. In both ruminant and nonruminant species, extensive development of adipose tissue stores occurs during the last third of gestation in the fetus. During this period, therefore, the ability to synthesize and take up lipid is high (see Section 7). Compared to maternal depots, the fetus is well able to maintain these fat stores as, under normal circumstances, the fetus is unlikely to be subjected to extreme stresses similar to those experienced by the mother. In ruminant species in which maternal fatty acid transfer is very much lower than in nonruminant species, the fetus has to rely to a greater extent on fatty acid synthesis to build up any body stores (Hershfield and Nemeth, 1968).

6. Lipid Digestion in the Newborn

6.1. The Diet

6.1.1. Lipid Composition of Ruminant Milk

In view of the importance of milk both as a food for the newborn animal and as a commercial product for human consumption, its lipid composition has been extensively studied (Morrison, 1970; Patton and Jensen, 1975; Christie, 1978, 1979). This brief discussion will therefore be confined to the lipid composition of colostrum, i.e., the secretion of the mammary gland prior to and immediately after parturition, and of the milk secreted during the first week post-partum, these being of primary importance in the nutrition of the neonatal animal.

The lipids contained in bovine colostrum and early postparturient milk are comprised almost totally (>95%) of triglycerides (Morrison, 1970), with the minor components consisting mainly of unesterified fatty acids, monoglycerides, diglycerides, free sterols, and phospholipids. Extensive data are available on the detailed changes that occur in the overall lipid composition of bovine milk during the lactational cycle (Rook, 1961a,b; Storry, 1970; Jenness, 1974), and it is generally agreed that the relative proportion of fat in milk is at a maximum during the early stages of lactation. More precisely, fat levels are highest in the milk secreted about 24 hr post-partum and not in colostrum itself.

The fatty acid compositions of the lipids of bovine colostrum and of the early postparturient milk have been extensively studied (Senft and Klobasa, 1970), and, depending on time post-partum, the major fatty acids were found to be myristic acid (10–13%), palmitic acid (26–36%), stearic acid (5–9%), oleic acid (20–24%), and short- and medium-chain acids from C_4 to C_{12} (15–20%). Although present, only minor amounts of linoleic and linolenic acids were detected. In general, colostral lipids contained lower proportions of butyric, caproic, capric, lauric, stearic, and oleic acids and higher proportions of caprylic, myristic, palmitic, linoleic, and linolenic acids than the lipids of milk secreted subsequently. The possible implication of differential rates of fat secretion on the one hand and of the synthesis of fatty acids by the mammary gland or their mobilization from adipose tissue on the other, in changes in composition of these fatty acids has been suggested (Decaen *et al.*, 1970). The relative proportions of C_4 to C_{14} fatty acids in ovine colostrum (Noble *et al.*, 1978b) were essentially similar to those of bovine colostral lipids. However, there were marked differences in the proportions of fatty acids from C_{16} onwards. In particular, ovine colostral lipids contained a lower proportion of palmitic acid, higher levels of stearic and oleic acids, and even lower levels of linoleic acid than were present in bovine colostrum.

6.1.2. Dietary Effects on the Fatty Acid Composition of Milk Lipids

Compositional changes in the diet of the early neonatal animal are reflected in rapid compositional changes in plasma and tissue compositions. Thus, fatty acids specifically associated with the mothers' milk, for example, linoleic acid and *trans*-monoenoic species, may readily be detected in the adipose tissue of lambs within 2 days of the onset of suckling (Noble *et al.*, 1971a).

Compositional changes in the diet during the early postpartum stages of development are therefore extremely pertinent to any lipid compositional studies of early neonatal tissues. In recent years, for example, a considerable amount of work has been devoted to improving the essential fatty acid status of the newborn animal through dietary means (see Section 7.3.3). Detailed accounts are available for the effects on milk fat composition of maternal dietary supplementation with fatty acids of chain length C_{12}–C_{18} (Christie, 1979).

In general, increased maternal dietary intake of the acids from C_{12} to C_{16} has led to corresponding increases in the proportions of the fatty acids in the milk fat. The effect of dietary supplementation with the C_{18} fatty acids on milk fat composition is far more complex because of both the biohydrogenation processes in the rumen and extensive manipulation in the mammary gland and other tissues. Supplementation of the diet of lactating cows with cottonseed oil (5–10% of the concentrate ration) (Steele and Moore, 1968a,b) or stearic acid itself (Steele and Moore, 1968d) has led to significant increases in proportions of stearic and oleic acids in the milk fat at the expense of a decrease in the proportions of the medium-chain length fatty acids. In view of the lack of any effect on rumen volatile fatty acid concentration (Steele and Moore, 1968b–d), it was concluded that the reduction in the proportions of medium-chain fatty acids was mediated through inhibition of *de novo* fatty acid synthesis by the additional dietary intake of stearic acid (Steele and Moore, 1968d; Noble *et al.*, 1969a).

In the absence of any change in the concentration of oleic acid in the maternal plasma triglycerides, the accumulation of oleic acid had clearly occurred through the action of the highly active Δ9-desaturase in the mammary gland (Moore and Christie, 1979). Where there has been dietary supplementation with oleic or linoleic acids, the effects on milk fat composition are complicated by the biohydrogenation processes in the rumen. Supplementation of the diet of cows with oleic acid at levels of 5 and 10% of the concentrate ration increased the proportions of both stearic and oleic acids in the milk fat at the lower level of feeding but increased the proportion of oleic acid and decreased that of stearic acid at the higher level of feeding (Steele and Moore, 1968c).

Under these dietary regimes, it was obvious that the effects on milk fat composition were the net products of both rumen and mammary tissue processes. Similar results were obtained when the diet of cows was supplemented with tallow (Storry *et al.*, 1973). Where cows have been fed tallow and other fats containing high proportions of stearic and oleic acids "protected" from intraruminal effects by a formaldehyde-treated casein coat, results have been somewhat contradictory. Thus, in some experiments, feeding the "protected" supplement merely had an effect on milk fat composition similar to the feeding of an "unprotected" supplement (Dunkley *et al.*, 1977), whereas in others, no effect was observed on fatty acid synthesis in the mammary gland (Wrenn *et al.*, 1977).

The effects of feeding vegetable oils containing high levels of polyunsaturated fatty acids on milk fat composition have been the subject of extensive experimentation (Christie, 1979). Virtually all of the results from this type of work have shown that increasing the dietary levels of polyunsaturated fatty acids gives rise to increased proportions of stearic and oleic acids and decreased proportions of medium-chain fatty acids in the milk fat. Because of biohydrogenation in the rumen, there were virtually no changes in the proportions of linoleic or linolenic acids in the milk fat.

In recent years, a great deal of work has been done on raising the levels of the C_{18} polyunsaturated fatty acids in milk fat by feeding cows dietary supplements in which the polyunsaturated fatty acids were "protected" from rumen biohydrogenation by a formaldehyde-treated casein coat (Scott *et al.*, 1971). Considerably increased levels of C_{18} polyunsaturated fatty acids (up to 35% of the total fatty acids present) in the milk fat were achieved and were accompanied by overall increased milk fat yields. The proportions of stearic and oleic acids were also increased slightly, whereas those of the medium-chain fatty acids were diminished. Feeding similar dietary supplements to sheep during the last 8 weeks of pregnancy has resulted in marked increases in the proportions of linoleic acid in the colostrum and early milk of the ewes (Noble *et al.*, 1978b). Apart from a compensatory decrease in the proportions of oleic acid, no effects were observed on the proportions of any other fatty acids.

6.2. Development of the Gastrointestional Tract

The principal feature of the gastrointestinal system of the adult ruminant animal is the characteristically specialized polygastric arrangement of the early part of the tract (Harfoot, 1978a). It is sufficient to say that four compartments exist between the esophagus and the small intestine; these are, in descending order, the reticulum, the rumen, the omasum, and the abomasum (or true stomach), the whole being dominated by the rumen which, in the case of the adult sheep and cow, may constitute up to 64–69% of the total weight of the four compartments. Through the presence of a vast microbial population within the rumen, extensive fermentation and other dietary changes are able to be achieved that enable the efficient utilization and assimilation of the mainly roughage diet (Harfoot, 1978b). Unlike the simple-stomached animal in which the processes of digestion and assimilation of dietary fats do not begin until the small intestine is reached, in the ruminant, events occurring within the anterior part of the gastro-intestional tract have a profound effect on the lipid absorption that occurs in the small intestine (Noble, 1978). In terms of digestion, the metabolic processes that occur in the rumen are by far the most important, since not only does fermentation of the normally indigestible plant material occur there (resulting in the formation of readily assimilable short-chain fatty acids), but extensive hydrolysis and chemical alteration (in particular hydrogenation of unsaturated fatty acid components) of the plant lipids also occur (Harfoot, 1978b) with profound effects on subsequent tissue lipid metabolism and composition (Christie, 1978).

In the ruminant at birth, only the abomasum is functional (Phillipson, 1977); although the three compartments of the forestomach are present within the embryo and at birth, they are relatively small and virtually nonfunctional. Thus, whereas in the adult animal the abomasum may account for only about 8% of the total stomach volume, in the calf at birth, the abomasum accounts for more than 70% of the total stomach volume and far exceeds in weight the rumen and the reticulum together (Radostits and Bell, 1970). However, within a very short time after birth, the forestomachs, in particular the rumen, undergo an extremely rapid development with the result that in the 4-week-old calf, the rumen and reticulum constitute some 65% of the total stomach volume and may be considered to have reached adult proportions by some 8–12 weeks after birth (Godfrey, 1961). Adult proportions for each of the compartments in the lamb are reached at about 8 weeks of age (Wardrop and Coombe, 1960, 1961).

The absolute size and volume of each of the compartments of course continue to increase for some considerable time after this. In the lamb, three distinct phases of forestomach development have been recognized (Wardrop and Coombe, 1960; Church *et al.*, 1962). The period from birth to 3 weeks is considered to be a nonruminant phase during which, with the dependence of the young animal on its mother's milk, no solid food is consumed, and digestion may be considered to be similar to that of a neonatal monogastric animal. The period of 3 to 8 weeks after birth is considered to be a transition phase during which there is an increase in the intake of solid food and less reliance on the mother's milk. During this period, there is a considerable increase in the microbial population of the rumen. After about the eighth week, intake of solid food predominates, and the forestomachs are considered to be adult in function.

The physical changes that take place in the stomach compartments are accompanied by all the required histological changes necessary to achieve efficient utilization of the metabolic products resulting from the microbial attack on the diet. With the first appearance of solid food in the diet, these changes are particularly rapid (Wardrop, 1961; Steven and Marshall, 1970). Volatile fatty acids can be detected in the rumen compartment within a very short time after birth, and their production rapidly increases from about the first week after birth onwards, so that by the sixth to eighth week after birth, their concentration in the rumen contents is similar to that found in the adult (Lengemann and Allen, 1955; Wardrop and Coombe, 1961; Poe *et al.*, 1971).

The increased concentrations of the short-chain fatty acids in the rumen contents are accompanied by an increasing ability of the rumen mucosa to absorb and metabolize short-chain fatty acids and by concomitant increases in the concentrations of short-chain fatty acids within the circulation (Reid, 1953). Apart from the physical capacity of the forestomachs, it is usually considered that by a little over 3 weeks after birth, the forestomachs of the ruminant animal are for all purposes adult in function.

The development of the ruminant forestomachs after 3 weeks of age is very much dependent on the inclusion in the diet of solid food, in particular the presence of fibrous material. This contrasts with the first 3 weeks after birth during which the diet is of little importance in controlling the development of the forestomachs (Wardrop and Coombe, 1960; Church *et al.*, 1962). When solid food material is deliberately withheld from inclusion in the mainly milk diet during the neonatal period, all aspects of forestomach development are arrested, but development of the abomasum remains unaffected (Warner *et al.*, 1956; Wardrop, 1960). Subsequent inclusion in the diet of fibrous material rapidly rectifies the situation (Lengemann and Allen, 1959). Although under normal circumstances there is a natural inclination for the young ruminant animal to begin to take in solid food at about 3–4 weeks of age, the ability to control forestomach development through dietary means has proven to be a useful tool in animal husbandry. Prime mediatory factors in the functional development of the ruminant forestomachs are clearly the short-chain fatty acids produced within the rumen from fermentation of the increasing amounts of roughage, in particular the production of butyric acid which, of all the short-chain fatty acids, has been shown to be a particularly effective stimulant to development (Sander *et al.*, 1959; Tamate *et al.*, 1962).

During the early neonatal period, the milk taken in by the suckling ruminant bypasses the forestomachs and is delivered directly to the well-developed abomasum through the closure of the esophageal or reticular groove (Harfoot, 1978a). Any entry of milk into the rumen compartment during the period immediately after birth has a detrimental effect on the digestive abilities of the animal and has to be avoided. Although the ability of the esophageal groove to respond diminishes with age, the reflex can be maintained artificially with a liquid diet for up to 1 year of age (Phillipson, 1977). Prior to feeding, the contents of the abomasum are composed of a clear, slightly viscous fluid, of pH

1–2, containing only small clots of milk. With feeding, the milk and saliva entering the abomasum are immediately clotted through the action of rennin and, to a lesser extent, pepsin (Hill *et al.*, 1970), and the milk lipids undergo slow digestion (see Section 6.3). With entry of the milk and saliva into the abomasum, there is an immediate rise in the pH of the contents to above 6 which then decreases gradually over the next 5–6 hr back to prefeeding levels because of the secretion of acid by the stomach (Porter, 1969). It is at about neutral pH that the clotting action of the rennin is particularly potent.

The pH changes in the abomasum are concomitantly reflected more distally in the small intestine. One of the major metabolic functions of clot formation in the abomasum is to delay the passage of dietary material into the small intestine in order to achieve a steady flow of nutrients through the remainder of the intestinal tract, in particular the small intestine. Interference with clot formation therefore leads to extensive digestive upsets (Roy, 1974). The differential flow of nutrients into the small intestine that is achieved through clot formation in the abomasum is also of vital importance in the efficient utilization of dietary lipids. As a result of the slow digestion of the clot within the abomasum, about one-half of the abomasal contents have passed into the duodenum within the first 2 hr after feeding. It is, however, some 3–4 hr after feeding before predominantly fatty material begins to pass from the abomasum into the small intestine with considerable consequential changes in the appearance and nature of the digesta (Porter, 1969; Roy, 1974). The flow of parameters for the digesta may be considerably affected by alterations to the diet (Poe *et al.*, 1971).

6.3. Lipid Digestion in the Abomasum

Immediately after feeding in the young ruminant, most of the milk fat becomes entrapped within the casein coagulate resulting from the clotting mechanism within the abomasum. The amount of milk fat that finds its way directly into the small intestine is almost negligible. Extensive hydrolysis of the fat thus entrapped within the abomasum occurs almost immediately because of the action of the pregastric esterase or oral lipase that has become mixed with the milk during its passage through the oral cavity following secretion of this enzyme from glands situated at the base of the tongue, the glotto-epiglottis region of the oral cavity, and the pharyngeal end of the esophagus (Ramsey *et al.*, 1956; Olivercrona *et al.*, 1973). Hydrolytic activity within the abomasum from lipases of gastric origin or through pancreatic enzymes that have arisen through regurgitation can be discounted (Otterby *et al.*, 1964b; Toothill *et al.*, 1976). The presence of milk lipases can also be discounted as a source of any significant lipase activity within the abomasum (Christie, 1979). The role of pregastric esterase as an enzyme responsible for a major part of the digestion of the milk fat in the early neonatal animal is now well established (Gooden and Lascelles, 1973).

The ability of the pregastric esterase to hydrolyze short-chain fatty acids from water-insoluble triglycerides over a relatively wide pH range has been thoroughly documented. This esterase does not appear to be active against ester linkages of other lipid classes (Brockerhoff and Jensen, 1974). Optimum working pH values do exist. In the calf, optimum hydrolytic release has been shown to occur at pH 5.3, 6.2, and 7.5 (Harper and Gould, 1955), whereas in the lamb and kid, activity peaks occurred at pH 4.8 and 5.5 (Richardson and Nelson, 1967). However, extensive hydrolytic activity can be said to occur over the pH range 4.5–7.0; hydrolysis is still evident at pH 3.0–4.0, but below pH 3.0, activity is considerably reduced (Siewart and Otterby, 1970).

Initial investigations soon defined a selective ability of the enzyme to release butyric acid from a wide range of ruminant milk triglycerides (Harper, 1955), but as a result of subsequent work, the fatty acid specificity has been extended to include both short- and medium-chain fatty acids up to and including capric acid (Jensen and Sampugna, 1964; Hamilton and Raven, 1973). In ruminant milk fats, short- and medium-chain fatty acids account for a very high proportion of the total fatty acids present in the triglycerides (see Section 6.1.1.).

The ability of pregastric esterase to release fatty acids of chain length greater than lauric acid is almost negligible. It has also been shown that the hydrolytic attack by the pregastric esterase is specifically directed to the *sn*-3 ester linkage of the triglycerides with a relative inability to degrade further any of the diglyceride products resulting from the hydrolytic event (Edwards-Webb and Thompson, 1977). The major products of pregastric esterase action on the milk fat triglycerides in the abomasum are, therefore, a mixture of short- and medium-chain fatty acids and diglycerides.

The ability of the pregastric esterase to affect digestion in the abomasum is clearly considerably influenced by lipid compositional changes in the diet. As opposed to diets having a high content of short- and medium-chain fatty acids, milk-replacer diets containing relatively high proportions of long-chain fatty acids have been fed to calves and have led to a considerable reduction in pregastric esterase action with resultant effects on subsequent lipid composition and digestion within the small intestine (Siewert and Otterby, 1971). The changes in pH that occur in the abomasum with time after feeding also influence the ability of the pregastric esterase to hydrolyze the milk fat triglycerides. In the calf, therefore, it has been shown that for at least 2 hr after feeding, the pH of the abomasal contents is sufficiently near the optimum working range for pregastric esterase action to effect extensive hydrolysis (Siewert and Otterby, 1970). However, by 3 hr after feeding, the pH has fallen to a sufficiently low value (3.4) to considerably reduce the ability of the pregastric esterase to act on the milk fat triglycerides.

Pregastric esterase action within the abomasum is known to be controlled by many independent factors. Age itself does not appear to be a major factor in the control of pregastric esterase action, as secretion has been detected in aminals well into maturity (Leidy *et al.*, 1975). Output of pregastric esterase is minimally affected by changes in dietary composition (Leidy *et al.*, 1975). However, the mode of feeding is a considerable influence on pregastric esterase output and action. Whereas suckling is a large stimulus to pregastric esterase secretion, *ad libitum* feeding in a manner not involving suckling results in a considerable inhibition of pregastric esterase output (Wise *et al.*, 1976). The effect of any insufficiency in pregastric esterase output does not appear to be confined to consequences on the lipolysis and digestive functions described above. Several indirect benefits to the young animal arising from pregastric esterase action have been described. Thus, beneficial effects on the physical characteristics of the abomasal contents have been noted to be associated with pregastric esterase secretion (Ramsey and Young, 1961a), whereas there is strong evidence that the hydrolysis and fatty acid release brought about through the action of pregastric esterase are associated with an extensive inhibition of potentially harmful microorganisms found in the upper part of the gastrointestinal tract of the young ruminant animal (Wise *et al.*, 1968).

6.4. Lipid Digestion in the Small Intestine

Following their extensive metabolism is the rumen, the lipids of the digesta in the adult ruminant remain virtually unaltered during their passage through the omasum and abomasum and are, therefore, delivered into the duodenum mainly as unesterified fatty acids accompanied by a small amount of phospholipids. The ability of the adult ruminant to adapt to these specialized conditions and to achieve an overall efficient lipid absorption in the small intestine has been the subject of a previous review (Noble, 1978). In contrast to the adult animal, where fat accounts for no more than 3-10% of the total dry matter of leaf tissue, in the young suckling ruminant, fat accounts for some 70% or more of the total energy of the diet.

In spite of the action of the pregastric esterase in the abomasum, hydrolysis of the dietary lipid prior to entry into the small intestine is far from complete. Various estimates of the extent of hydrolysis that has occurred within the abomasum have been made ranging from about 20% of the total ester linkages (Otterby *et al.*, 1964a) up to 65-70% of the total lipid offered in a milk diet (Gooden, 1973). It is, however, agreed that the time spent by the dietary lipid in the abomasum, enabling extensive triglyceride hydrolysis to occur and permitting a slow and modulated release of the lipid material into the small intestine, is an essential function for lipid digestion in the young ruminant.

In the duodenum, the digesta are augmented by the addition of the bile and pancreatic secretions which, unlike the situation in the monogastric animal, are discharged together into the duodenum through a common duct (Harfoot, 1978a). As in the adult (Noble, 1978), both bile and pancreatic juice play major roles in the digestion and absorption of lipid in the young ruminant receiving a milk diet (Heath and Morris, 1963; Gooden and Lascelles, 1973).

As measured by the changes in the flow and concentration of lipid in the lymph, deprivation of bile and pancreatic secretions in lambs led to significantly decreased lipid absorption (Heath and Morris, 1963). Similarly, in young lambs deprived of bile and pancreatic juice, the appearance in the lymph of ^{14}C-labeled lipids following their introduction into the small intestine was considerably reduced. Insufficient lipase activity has been shown to be the major causative factor in the malabsorption of lipids in the young calf deprived of pancreatic secretion (Gooden and Lascelles, 1973). Lack of pancreatic lipase activity was indicated by high triglyceride/unesterified fatty acid ratios in the duodenum. The lipolysis that had occurred was largely the result of previous pregastric esterase action within the abomasum, which in the young ruminant deprived of pancreatic secretion is usually sufficient to maintain a certain minimum level of lipid absorption.

Any possibility of pregastric esterase action continuing within the duodenum is eliminated once the dietary material becomes mixed with the bile secretion, since the presence of taurine conjugates are known to be inhibitory to pregastric esterase activity (Ramsey and Young, 1961b). Although in ruminants, secretion of taurine conjugates remains far in excess of glycine conjugates throughout life, exceedingly high taurine-to-glycine ratios are found at birth and do not alter significantly until about 1–2 months of age (Noble, 1978). In the adult ruminant, where highly acidic conditions prevail over such a large proportion of the jejunum, it has been suggested that the high taurine-to-glycine ratios have a particular role to play in lipid solubilization (Noble, 1978). In the ruminant immediately after birth, the concentration of lipase in the pancreas and the rate of secretion of pancreatic lipase into the small intestine are very low (Huber *et al.*, 1961; Gooden, 1973), but considerable increases occur within a few days after birth. Thus, by the eighth day after birth, the concentration of lipase within the pancreas of the calf is some three times higher than the levels found at 1 day of age (Huber *et al.*, 1961), whereas pancreatic fluid secretion, accompanied by proportional increases in lipase and phospholipase outputs, has been observed to increase in calves up to 37 days after birth (Gooden, 1973; Ternouth and Buttle, 1973; Ternouth *et al.*, 1976).

Whereas in the adult animal, control of pancreatic output is mediated mainly through stimuli released from the small intestine (Noble, 1978), in the young ruminant, the concentrations of the enzymes in the pancreatic secretion and their rate of output are controlled to a greater extent through stimuli emanating directly from general dietary influences, e.g., nature of the diet (Ternouth and Buttle, 1973; Ternouth *et al.*, 1974, 1976). Clearly, the hydrolytic actions of the pregastric esterase in the abomasum and the pancreatic lipase in the small intestine are ideally suited to act together to achieve efficient lipid absorption during a period when fat intake is so high. Thus, the initial fatty acid release directed specifically towards the short-chain fatty acids through the action of pregastric esterase is followed up by the hydrolytic action of the pancreatic lipase which has an extensive ability to degrade diglycerides and is virtually unprejudiced by any fatty acid specificity. The action of pregastric esterase may be particularly important during the period immediately after birth when a high lipid intake from the diet is combined with low pancreatic secretion.

6.5. Lipid Absorption

The balance between the hydrolytic actions of the pregastric esterase and pancreatic lipase and their ability to achieve high levels of fat digestion in the young ruminant may be easily upset, resulting in deleterious effects on the health and viability of the young animal. Primary among the many causative factors of lipid digestive upsets is the inability to achieve the correct pattern of abomasal clotting in order to allow sufficient lipid hydrolysis to occur and to permit the slow release of lipid into the small intestine. In general, the ability of ruminant animals to digest dietary fats is much higher than in nonruminants (Noble, 1978). Although the fatty acid composition of the total lipids present in the lymph resembles that of the jejunal digesta, some selective absorption and incorporation into the lymph lipids do occur. For the young ruminant, as in the adult, efficiency of uptake and incorporation of fatty acids into the lymph lipids is known to decrease with increasing fatty acid chain length and increase with the degree of unsaturation of any given chain length (Walker and Stokes, 1970; Noble, 1978). The utilization of any fatty acid is also affected by the position of esterification on the triglyceride molecule. Without doubt, a major factor affecting lipid utilization by the young ruminant is the size of the fat globule (Roy *et al.*, 1961; Roy, 1974). Therefore, where artificial diets are being fed, correct emulsification and homogenization of the dietary lipid components are vitally important. Clearly, the nonlipid constituents of the diet, e.g., the stability of the casein matrix of the abomasal clot, also play a considerable role in the maintenance of efficient lipid absorption.

Absorption of lipid from the lumen of the small intestine is rapidly followed by resynthesis within the mucosal cells. Two major pathways are known to exist for the triglyceride synthesis, the monoglyceride pathway and the α-glycerophosphate pathway. In the young suckling ruminant animal, where, as in nonruminant species, lipid digestion results in a considerable accumulation of monoglyceride products in the intestinal contents, the monoglyceride pathway is extremely active within the mucosal cells and accounts for a considerable proportion of triglyceride formation (Cunningham and Leat, 1969). The presence of the monoglyceride system has also been shown in the intestinal mucosal cells of the fetal ruminant (Cunningham and Leat, 1969). In contrast, the monoglyceride pathway is of little biological significance in the adult ruminant where, because of a predominance of unesterified fatty acids, monogycerides hardly exist within the contents of the small intestine (Noble, 1978). Under these circumstances, synthesis by the monoglyceride pathway is almost completely suppressed, and triglyceride formation is almost wholly by the α-glycerophosphate system (Bickerstaffe and Annison, 1969; Noble, 1978). However, should the need arise, the ability to use the monoglyceride pathway has been retained by the adult ruminant from the earlier neonatal period when the presence of such a mechanism was more necessary (Cunningham and Leat, 1969). Following resynthesis in the intestinal mucosa, the lipid components are then packaged into lipoprotein particles, in particular, the chylomicron and very-low-density lipoprotein (VLDL) fractions (see Section 2.3.1a), which enter the lacteals, and, via the lymphatic system, are finally delivered into the plasma for distribution to the tissues (Noble, 1978).

7. Metabolism of Lipids by the Fetus and Newborn

7.1. General Aspects

Although, compared to nonruminant species, comparatively little information exists on the synthesis of lipid by the fetal ruminant, it is apparent from the limited amount of evidence available that, in common with nonruminant species, lipid synthesis within the fetus is extremely active. In several nonruminant species, it has been shown by use of a variety of labeled substrates that the synthesis of fatty acids and various major lipid fractions *de novo* within the fetal tissues is extremely active (Popjak, 1947; Popjak and Beeckmans, 1950b). Whereas the extrahepatic tissues appeared to be capable of satisfying virtually all their requirements for fatty acids through synthesis in the liver, only a proportion of the fatty acids present were derived from synthesis within the fetus (Fain and Scow, 1966).

Similarly, in the ruminant, it has been shown that the fetal tissues display lipid synthetic capabilities using both lipid and nonlipid precursors. Thus, ^{32}P-labeled orthophosphate has been shown to be readily incorporated into slices of ovine fetal heart (Scott *et al.*, 1967), whereas following the infusion of either [^{14}C]glucose or fructose into the umbilical vein of the fetal lamb *in utero,* ^{14}C from either substrate was found to be rapidly incorporated into the triglycerides of the adipose tissue.

As measured by the rates of incorporation of labeled substrates into various tissue lipids, clear differences exist among the rates of lipid synthesis at various stages of gestation. In both ruminant and nonruminant species, it would generally appear that the rate of lipid synthesis within fetal tissues is considerably higher than in corresponding tissues of the adult, but this decreases rapidly with the approach of term (Villee and Hagermann, 1958; Roux, 1966; Iliffe *et al.*, 1973). As a result, studies with lambs have shown that the rate of adipose tissue lipogenesis was barely detectable at birth, but after a lag phase of 2–3 days, it had increased over tenfold by 10 days of age (Vernon, 1975). However, during the early part of the lamb's development, the diet provides the main source of fatty acids for decomposition, and fatty acid synthesis *de novo* contributes only about 2–3% of the total fatty acids esterified in adipose tissue.

Although the ruminant fetus possesses the ability to synthesize lipid from both acetate and glucose, the use of glucose for fatty acid synthesis far exceeds that of acetate (Ballard *et al.*, 1969). However, with the development of a functional rumen soon after birth, the capacity to utilize glucose for lipogenic purposes rapidly diminishes. Measurements of the activity levels of the enzymes that control the synthesis of fatty acids within the animal body have been able to show marked differences between the tissues of the adult and those of the fetus (Iliffe *et al.*, 1973).

The concentrations of the unesterified fatty acids in the plasma of ruminants at birth are extremely low. However, it has been shown that within a very short time after birth their concentrations in the plasma undergo considerable increases. In the newborn lamb, for instance, increased plasma unesterified fatty acid concentrations could be detected within 1 hr of birth, and by 6 hr after birth, the concentrations of the unesterified fatty acids had increased some tenfold (Van Duyne and Havel, 1959; Van Duyne *et al.*, 1960). By some 2–4 hr after birth, the unesterified fatty acids accounted for over 30% of the total fatty acids present (Noble *et al.*, 1971d).

From both metabolic and fatty acid compositional studies (Van Duyne *et al.*, 1960; Thompson and Clough, 1972; Battaglia and Meschia, 1973), it is now apparent that these increases in the plasma unesterified fatty acid concentrations of the newborn within such a short time after birth result from a rapid mobilization of the adipose tissue lipid stores in order to meet the energy needs of the animal. Clearly, the increased levels of the plasma unesterified fatty acids must arise from specific responses of the newborn to the various stimuli associated with the sudden exposure to an extrauterine existence, in particular stimuli resulting from exposure to a lower environmental temperature or loss of heat through evaporation which are known to have a large effect on plasma unesterified fatty acid concentrations (Thompson and Clough, 1972).

The possibility that the rapid mobilization of the depot lipids after birth may be the result of a sudden increase in sympathetic nervous activity and catecholamine release associated with muscular activity has also been suggested (Van Duyne *et al.*, 1960; Comline and Silver, 1972). In this respect, it is interesting to note that no increase in the concentration of unesterified fatty acids was observed in the plasma of a number of lethargic lambs that survived for only a short time after birth. With the onset of suckling and the intake of extremely high levels of fat associated with the colostrum and early milk of the mother, considerable increases take place in the plasma concentrations of other major lipid fractions in the newborn. Thus, in the newborn lamb, kid, and calf, the concentrations of the triglycerides, phospholipids, cholesteryl esters, and free cholesterol in the plasma have all been shown to undergo large increases once suckling has begun (Leat, 1967), whereas at birth, in the lamb, the concentration of total plasma fatty acids was similar to that of the adult, i.e., 80–90 mg/100 ml plasma (Noble *et al.*, 1971d).

With the approach of weaning and a diminished reliance on milk intake, reductions in the plasma lipid concentrations of the young ruminant are observed (Leat, 1967). The compositional differences that have been observed to exist between the fatty acids of the plasma lipids of the fetal and newborn ruminant can readily be correlated with their predominantly endogenous origin in the case of the fetus and their predominantly exogenous source in the case of suckling newborn. Therefore, whereas during fetal life the major plasma lipids of the lamb were observed to contain relatively high levels of palmitic and palmitoleic acids, in the suckling animal, the proportions of these acids were very much lower, and the levels of oleic acid were considerably increased (Noble *et al.*, 1971d, 1978c). The fetal plasma also contained only small concentrations of C_{18} polyunsaturated fatty acids but relatively high levels of C_{20} and C_{22} polyunsaturated fatty acids; after birth, however, the concentrations of the C_{18} components increased markedly, and there were reductions in the concentrations of C_{20} and C_{22} acids.

In the lamb, these changes in the relative concentrations of the polyunsaturated fatty acids in the plasma have been correlated with changes in enzyme activities associated with their metabolism (Noble *et al.*, 1975a). In newborn lambs maintained on a low-fat milk diet, the plasma fatty acid composition remained similar to that of the fetus. Similar results have been obtained with the newborn calf (Noble *et al.*, 1975b).

Considerable differences exist in ruminants between the lipid composition of the fetal tissues and those of the adult (Christie, 1978). Within a very short time after birth, extensive compositional changes in the tissue lipids have occurred (see Section 2). As in the case of the plasma, the changes that occur in the proportions of the various tissue lipids between the fetal and neonatal animal are accompanied by extensive alterations in fatty acid compositions. In particular, there are changes in the proportions of C_{18}, C_{20}, and C_{22} polyunsaturated fatty acids, and the fatty acid patterns of the neonatal tissues rapidly reflect the increasing metabolic importance of fatty acids of exogeneous rather than endogenous origin (Noble *et al.*, 1970b, 1971b,c). In the livers of suckling lambs, therefore, significant concentrations of *trans*-octadecenoic acids have been detected within a short time after birth, whereas in newborn ruminants that have been prevented from suckling, tissue lipid compositions have remained similar to those observed in the fetus or at birth. From correlative data between the changes in the relative proportions of major lipid fractions of fetal and neonatal ruminant tissues, together with comparative data from nonruminant fetal and neonatal tissues, it has been possible to implicate extensive changes in the relative importance of several major pathways associated with lipid synthesis during the early neonatal period (Noble *et al.*, 1970b, 1971c).

7.2. Brown Adipose Tissue and Thermogenesis in the Newborn Ruminant

Brown adipose tissue is now generally recognized as the major site for "nonshivering" thermogenesis in newborn mammals, including ruminants (Smith and Horwitz, 1969; Hahn and Novak, 1975). Whereas in the past, cytological features visible under the light microscope were considered to be sufficient to distinguish between brown and white adipose tissue (Jenkinson *et al.*, 1968; Thompson and Jenkinson, 1969), subsequent investigations have shown the necessity for the application of far more stringent morphological criteria (Smith and Horwitz, 1969). Through the use of the electron microscope then, the previous confusion over the presence of brown adipose tissue in newborn calves, lambs, and kids has largely been resolved with the identification of multilocular brown adipose tissue cells containing similar ultrastructural features where previously, under the light microscope, the presence of large intracellular lipid droplets had given the appearance of unilocular white adipose tissue (Thompson and Jenkinson, 1970; Alexander *et al.*, 1975; Cannon *et al.*, 1977). Where samples of adipose tissue from a wide variety of sites in both newborn calves and lambs have been studied under the electron microscope (Gemmell *et al.*, 1972; Alexander *et al.*, 1975), all but the subcutaneous tissue, which in any case is very sparse except in the fatter breeds of sheep, possessed the characteristic morphology of brown adipose tissue. In spite of this, however, it was estimated that the brown adipose tissue comprised only some 1.5–2% of the total body weight of newborn ruminants (Alexander, 1975), an amount that is considerably less than that found in the newborn of most smaller mammals.

7.2.1. Function of Brown Adipose Tissue

The initial evidence for the thermogenic role of brown adipose tissue in the newborn ruminant was obtained indirectly through its metabolic response to injection of norepinephrine which mimics the effect of cold exposure in the neonatal calf and lamb (Thompson and Jenkinson, 1969; Alexander *et al.*, 1975). This response to norepinephrine diminished at the same rate as the adipose tissue lost the morphological characteristics of brown adipose tissue. Indirect evidence implicating brown adipose tissue as the site of nonshivering thermogenesis in the newborn ruminant was also obtained from the dramatic increase observed to occur in the blood flow to all the major brown adipose tissue sites in the lamb during cold stress (Alexander *et al.*, 1973a). More recently, it has been shown that both norepinephrine infusion and cold exposure stimulated heat production in the brown adipose tissue of newborn calves (Ter Meulen *et al.*, 1976; Thompson and Bell, 1976). Various methods have been used to calculate oxygen consumption in the brown adipose tissue of cold-stressed newborn lambs (Alexander and Williams, 1970; Alexander *et al.*, 1973a; Alexander and

Bell, 1975b), and through these studies, it has been confirmed that the increased oxygen consumption alone could account for the nonshivering thermogenic response of the tissue.

The biochemistry of thermogenesis in brown adipose tissue has been extensively studied in hibernators and newborn or cold-adapted animals (Nicholls, 1976; Foster and Frydman, 1978), and it is generally accepted that the thermogenic response to cold exposure is mediated through the sympathetic nervous stimulation of adenylate cyclase in brown adipose tissue leading to an increase in lipolysis caused by raised levels of cyclic AMP. The rate of oxidation of the resultant long-chain fatty acids must then determine the capacity of the tissue for thermogenesis, although measurements of the heat-producing capacity of brown adipose tissue have indicted that it may exceed the capacity for ATP hydrolysis. This implies that in brown adipose tissue, during its thermogenically active state, respiration must be at least partially uncoupled from phosphorylation, and it has therefore been suggested that the control of energy dissipation in the tissue through gradual uncoupling of oxidative phosphorylation can be explained in terms of Mitchell's chemiosmotic model (Flatmark and Pedersen, 1975). Experimental support for this suggestion has come from the demonstration that in brown adipose tissue mitochondria, the proton conductance mechanism in the inner membrane can be short-circuited to allow respiration to proceed without the stoichiometric phosphorylation of ADP (Nicholls, 1976).

The mitochondrial biochemistry of brown adipose tissue of newborn ruminants has not been widely studied. A single experiment (Cannon *et al.*, 1977) performed on the brown adipose tissue of newborn lambs showed that, apart from the response to NAD-linked substrates and a high ability to synthesize ATP (probably the result of the presence in ruminant brown adipose tissue of an active ATPase inhibitor protein), its mitochondrial biochemistry resembled in most respects that of the brown adipose tissue of other animals. Brown adipose tissue from newborn ruminants also possesses a nonthermogenic function not observed in the tissue from the newborn of nonruminant species. Starvation in a thermoneutral environment very effectively depleted the lipid content of newborn lamb perirenal brown adipose tissue (Alexander *et al.*, 1969b; Alexander and Bell, 1975a), whereas in the rabbit, starvation from birth for several days resulted in virtually no change in the triglyceride content of the brown adipose tissue (Hardman *et al.*, 1969; Schenk *et al.*, 1974). This may, however. merely be a reflection of the relative maturity of the lamb at birth, since the brown adipose tissue of the rabbit can assume this emergency nonthermogenic function after about a week of life (Hahn and Novak, 1975); it could also be an adaptation in the newborn ruminant to the almost complete lack of white adipose tissue.

7.2.2. Development of Brown Adipose Tissue

Brown and white adipose tissues begin to develop in the fetal lamb and calf just before midgestation (Ter Meulen *et al.*, 1972; Alexander, 1975), although preadipose cells from which the two tissues arise could be identified during the first trimester. After midgestation, however, there was a rapid differentiation of the two tissues in the fetal lamb, and by 80–90 days, the perirenal fat had acquired many of the morphological characteristics of brown adipose tissue (Gemmell and Alexander, 1978). By 115 days, the weight of the perirenal tissue had reached a maximum, falling slightly near term. In contrast, fetal lamb white adipose tissue began growth much later and regressed from about day 115 until term (Alexander, 1978). Although no systematic study appears to have been done on the effect of gestational age on the rate of lipogenesis in fetal brown adipose tissue, it has been found (R. G. Vernon, unpublished observations) that at 130 days, the rate of long-chain fatty acid synthesis in the perirenal adipose tissue was much greater than in the adult sheep. Very little is known about the prenatal development of thermogenic activity in brown adipose tissue, although studies on fetal rabbits (Hull, 1976) and lambs (Alexander *et al.*, 1973b) have suggested that functional maturity does not occur until just before or during parturition.

In the lamb and calf, brown adipose tissue lost its morphological characteristics during the first 2 weeks after parturition (Thompson and Jenkinson, 1969; Gemmell *et al.*, 1972). Ultrastructural (Gemmell *et al.*, 1972) and biochemical (Vernon, 1975, 1977) studies on brown adipose tissue from lambs have suggested that this is accompanied by a conversion process to white adipose tissue cells. Involution of the brown adipose tissue was accompanied by a concomitant loss of thermogenic response in lambs (Thompson and Jenkinson, 1969; Alexander *et al.*, 1970) and calves (Alexander *et al.*, 1975; Thompson and Bell, 1976), this process occurring particularly rapidly in calves. The factors controlling the morphological and functional development of brown adipose tissue before and after birth are ill understood. Nevertheless, hormonal, environmental, and nutritional factors have all been implicated in the development of the tissue in ruminants and other species (Alexander *et al.*, 1970; Sack *et al.*, 1976; Alexander, 1978).

7.3. Metabolism of Polyunsaturated Fatty Acids in the Newborn

7.3.1. Essential Fatty Acid Deficiency

It is now some 50 years since the observations of Burr and Burr (1929) implicated linoleic acid as an essential dietary constitutent. It was shown that a new deficiency disease could be produced by the absence of dietary fat and that the polyunsaturated fatty acids, in particular linoleic acid, were the constituents of the fat necessary to prevent the appearance of these deficiency symptoms. Subsequent investigations have confirmed that in spite of the many possible interconversions that may occur among the polyunsaturated fatty acids, linoleic acid cannot be synthesized within the animal body (Holman, 1971). Thus, the term essential fatty acid has been adopted to describe its indispensible presence in the diet. The undesirable consequences of a deficiency of linoleic acid in the diet on maintenance, growth, and many physiological processes have been demonstrated in a wide variety of animals, and a considerable amount of information has now been obtained on the metabolism, function, and dietary requirements for linoleic acid or its metabolites by various animals (Holman, 1971). However, investigations into the metabolism of the essential fatty acids have almost wholly been confined to nonruminant species; the metabolism of the essential fatty acids by the ruminant animal has received little attention.

The polyunsaturated fatty acids of normal mammalian tissue may be divided into two separate groups, linoleic and linolenic acids and their respective metabolic derivatives. Neither linoleic nor linolenic acid can be synthesized *de novo* in animal tissue, but each of the acids can serve as the initial precursor for the biosynthesis of independent families of C_{20} and C_{22} polyunsaturated fatty acids through a series of chain elongations and desaturations as follows:

(a) Linoleic (n-6) Series.

$$18{:}2 \rightarrow 18{:}3 \rightarrow 20{:}3 \rightarrow \underset{\text{arachidonic acid}}{20{:}4} \rightarrow 22{:}4 \rightarrow 22{:}5 \quad (1)$$

(b) Linolenic (n-3) Series.

$$18{:}3 \rightarrow 18{:}4 \rightarrow 20{:}4 \rightarrow 20{:}5 \rightarrow 22{:}5 \rightarrow 22{:}6 \quad (2)$$

These chain elongation and desaturase systems present in the microsomal fractions of animal tissues have a high substrate specificity for linoleic and linolenic acid but only very low substrate specificities for monoenoic fatty acids such as palmitoleic and oleic acid (Mead, 1971). However, under conditions in which the dietary supply of linoleic acid is severely limited, the systems will then accept oleic acid as a substrate and give rise to a third series of polyunsaturated fatty acids with the resultant accumulation in the tissue of the trienoic acid, 20:3(n-9) (Mead and Slaton, 1956).

(c) Oleic (n-9) Series.

$$18{:}1 \rightarrow 18{:}2 \rightarrow 20{:}2 \rightarrow 20{:}3 \rightarrow 22{:}3 \quad (3)$$

In nonruminant species, the accumulation of 20:3(n-9) acid has been positively correlated with a deficiency of dietary linoleic acid and the appearance of deficiency symptoms (Mohrhauer and Holman, 1963). It is now accepted that a measure of the essential fatty acid status of an animal can be assessed from the relationship between the tissue concentrations of 20:3(n-9) acid and arachidonic acid. This ratio is known as the triene : tetraene ratio, and investigations with nonruminant animals under varying degrees of essential fatty acid deficiency suggest that a triene : tetraene ratio greater than 0.4 in various tissues, including the plasma, is indicative of a deficiency of linoleic acid (Holman, 1971, 1973). Such values for this ratio were observed in the tissues only when linoleic acid constituted 1 to 2% or less of the total dietary energy.

7.3.2. Changes in the Polyunsaturated Fatty Acid Composition of Neonatal Ruminant Tissues

Investigations of the metabolism of polyunsaturated fatty acids in the young ruminant animal have revealed many interesting features that deserve critical comparison with the findings that have been reported from nonruminant species. The metabolism of the polyunsaturated fatty acids by the adult ruminant presents no less an intriguing picture than that of its newborn and has a distinct bearing on the polyunsaturated fatty acid metabolism of the young ruminant (Noble, 1978). The investigations into the polyunsaturated fatty acid metabolism of the newborn ruminant may be summarized as follows.

1. In the ruminant, very low levels of essential fatty acids are found in the plasma and tissue lipids of the fetus and the newborn (Leat, 1966; Noble *et al.*, 1971a–d, 1975b). The concentrations found are very much lower than those in the plasma and tissues of fetuses and newborn of nonruminant species and are considerably lower than the concentrations found in adult ruminants. For instance, the total carcass of the newborn lamb weighing 5.2 kg contained only 0.3 g of linoleic acid and 1.3 g of arachidonic acid (Noble *et al.*, 1972). Much higher proportions of essential fatty acids have been found, for example, in the tissues of the newborn rat (Sklan *et al.*, 1972). Compared with adult sheep, in which linoleic acid may constitute up to 40% of the total fatty acids circulating in the plasma, the concentration of linoleic acid did not exceed 2–3% of the total fatty acids present in any of the major plasma lipid fractions of the newborn (Noble *et al.*, 1971d). At birth, the tissue lipids of the lamb also contain very high levels of palmitoleic acid, the accumulation of which in tissues of nonruminant species has also been reported to occur when there is a dietary deficiency of essential fatty acids.

2. In spite of the lack of C_{18} polyunsaturated fatty acids, the plasma and tissues of the fetal and newborn ruminants contained appreciably high proportions of C_{20} and C_{22} polyunsaturated fatty acids, in particular 20:3(n-9) acid (Leat, 1966; Noble *et al.*, 1971a–d, 1975b, 1978a,c; Palmquist *et al.*, 1977; Payne, 1978; Shand *et al.*, 1978). As a result, the ruminant displays exceedingly high triene : tetraene ratios at birth which greatly exceed the critical essential fatty acid deficiency value of 0.4 designated for nonruminant species. For instance, in the plasma lipids of the newborn lamb and kid, respective triene : tetraene ratios of 1.7 and 2.8 were found; similarly high values have been observed in the tissues of newborn and fetal lambs and newborn calves.

3. During the first 3–4 days after birth, there are large increases in the concentrations of linoleic acid in the plasma and tissues of the ruminant, de-

creases in the proportions of 20:3(n-9) acid, and large reductions in the triene : tetraene ratios (Leat, 1966; Noble *et al.*, 1971a–d, 1975b; Palmquist *et al.*, 1977; Payne, 1978). In lambs reared on ewes' milk, for instance, the concentrations of linoleic acid by the fourth day after birth had risen to 22% and 18%, respectively, in the cholesteryl ester and phospholipid fractions of the plasma. Although large increases in the linoleic acid contents of the tissue phospholipids and cholesteryl esters also occurred, they were slightly less pronounced than those in the corresponding plasma lipid fractions. During the first 48 hr of life, the concentration of 20:3(n-9) acid in the plasma phospholipids of lambs decreased from 5.2 to 0.25% and from 1.4% to trace amounts in the plasma cholesteryl esters. In the liver lipids of lambs, the concentration of 20:3(n-9) acid decreased less rapidly than it did in the plasma. It was calculated that during the first 2 days of life, the triene : tetraene ratio decreased from 1.02 to 0.15 in the total plasma fatty acids and from 0.88 to 0.22 in in the total liver fatty acids. In the plasma, the changes in the polyunsaturated fatty acid content during the period immediately after birth have also been directly correlated with significant changes in the activity level of specific enzymes associated with their metabolism (Noble *et al.*, 1975a).

Thus, on biochemical criteria based on the tissue fatty acid compositions of nonruminant species, the newborn ruminant may be classified as being deficient in essential fatty acids. This feature of ruminant lipid metabolism may be explained by the relationship between the maternal lipid metabolism and that of the developing fetus. The only possible source of C_{18} polyunsaturated fatty acids and their derivatives available to the fetus is the maternal circulation, and it has been demonstrated that of the plasma lipid fractions, only the unesterified fatty acids contribute significant quantities of fatty acids to the fetus (see Section 3). Not only are the concentrations of the unesterified fatty acids in the maternal plasma of the ruminant very low, but the distribution of polyunsaturated fatty acids between the major plasma lipid fractions is such that no polyunsaturated fatty acids are associated with the unesterified fatty acids (Garton and Duncan, 1964; Noble *et al.*, 1971e). This contrasts with nonruminant species in which the partition of polyunsaturated fatty acids among the various maternal plasma lipid fractions is very much less extreme, and, as a result, appreciable concentrations of essential fatty acids are found in the tissues of the newborn (Leat, 1966). Therefore, with regard to accumulating any significant concentrations of polyunsaturated fatty acids in the tissues during the gestation period, the ruminant is presented with a far more difficult problem than its nonruminant counterpart. After birth, however, large increases occur in the concentrations of the essential fatty acids in the tissues of newborn ruminants in spite of the fact that during this period the diet normally consists entirely of milk.

The suggestion that one of the functions of colostrum in the ruminant is to provide increased levels of linoleic acid during the critical period immediately after birth (Leat, 1964) has not been substantiated by later investigations. The milk fat of ruminants is known to contain only very small concentrations of the C_{18} polyunsaturated fatty acids, whereas detailed analyses of ruminant colostrum and milk secreted during early lactation have shown that the concentration of linoleic acid rarely exceeds 1% of the total fatty acids present and may provide only about 0.4–0.6% of the total calories available in the diet (Noble *et al.*, 1970a).

Therefore, rapid and pronounced improvements in the essential fatty acid status of young ruminants are achieved when they consume a diet that, if given to a nonruminant animal, would result in the appearance of external symptoms of essential fatty acid deficiency. For example, in contrast to the newborn calf, when human infants were given a diet of cows' milk from birth, there was a progressive increase in the plasma triene : tetraene ratio from 0.38 on the day of birth up to 1.0 on the 40th day after birth (Holman, 1973). When newborn calves were fed a diet in which the linoleic acid provided 0.1, 0.32, and 1.00% of the total calories, the linoleic and arachidonic acid concentrations of the erythrocytes and plasma increased rapidly. In contrast, even when rats were given the diet in which linoleic acid provided 0.32% of the total calories, there were marked decreases in the concentrations of essential fatty acids and increases in the triene : tetraene ratios (Sklan *et al.*, 1972).

With the knowledge that the young ruminant animal possesses no stores of essential fatty acids for mobilization during the immediate period after birth (Noble *et al.*, 1972), interpretation of the above observations would indicate that the young ruminant may be extremely efficient, indeed, far more so than its nonruminant counterpart, in utilizing and conserving linoleic acid obtained from the diet during the initial period immediately after birth. It would also indicate that the level of linoleic acid in the diet needed to satisfy the minimum requirement for essential fatty acids in the young ruminant is far less than that needed by nonruminant species.

Results from comparative balance studies on lambs during the first 30 days after birth would substantiate these conclusions (Noble *et al.*, 1972). Furthermore, attempts to induce essential fatty acid deficiency in the young ruminant appear to have met with mixed success. Calves that received a low-fat semisynthetic milk diet have exhibited various symptoms that are normally associated with essential fatty acid deficiency (Cunningham and Loosli, 1954), and recovery could be effected by the inclusion in the diet of linoleic acid, suggesting that linoleic acid is also an essential dietary requirement for cattle. However, similar work in which calves received a fat-free milk diet showed that the various metabolic lesions produced could be cured by the addition to the diet of hydrogenated coconut oil (Lambert *et al.*, 1954) and therefore were not necessarily associated with essential fatty acid deficiency. That the newborn ruminant is sensitive to diets of extremely low essential fatty acid content is apparent, for when lambs have been fed from birth to 8 days on a diet in which linoleic acid only provided 0.001% of the total energy, the triene : tetraene ratio in the plasma and tissues increased to values well above those observed at birth (Noble *et al.*, 1971b,d).

7.3.3. Improving the Essential Fatty Acid Status of the Newborn

It appears to be beyond dispute that at birth the essential fatty acid status of the ruminant animal is poor and that there is an immediate requirement for essential fatty acids during the period after birth, a requirement that, in the case of the lamb, does not appear to be satisfied until some 10–20 days after birth (Noble *et al.*, 1972). Although little is known about the specific role and fate of dietary polyunsaturated fatty acids in the tissues of the ruminant, the importance of the polyunsaturated fatty acids in the control and regulation of various apsects of the metabolism in biological systems of nonruminant species including their roles in the maintenance of membrane structure and function, water balance, and resistance to infectious diseases is well known (Holman, 1971).

Whether or not there are circumstances under which the reduced essential fatty acid status of the ruminant during the early neonatal period is in any way detrimental to the animal is open to debate. Certainly, there are many circumstances, such as conditions of high infection and other environmental stresses, the occurrence of multiple births, and lack of vigor at birth, under which an improved bodily level of essential fatty acids immediately after birth might be of benefit. For instance, it is known that exposure to a high temperature drastically reduces the polyunsaturated fatty acid levels in tissues (Bobek and Ginter, 1966) and resistance of the body to invasion by bacteria or other exogeneous agents (Pan, 1970), and evidence is now available to suggest that in the case of the newborn ruminant the stresses that may accompany the poor essential fatty acid status at birth may be considerably heightened under such adverse environmental conditions (O'Kelly, 1968; Noble *et al.*, 1973; Noble and Moore, 1974).

Certain difficulties exist with regard to improving the essential fatty acid status of the newborn ruminant. The use of dietary means to improve the essential fatty acid status during prenancy is severely limited on two scores. First, there is the extensive hydrogenation of dietary fatty acids that occurs in the rumen of the mother and prevents the passage of any significant quantities of polyunsaturated fatty acids into the absorptive areas of the small intestine (Harfoot, 1978b). Second, any polyunsaturated fatty acids that do escape rumen biohydrogenation and are absorbed into the maternal circulation rapidly become incorporated into the plasma cholesteryl ester and phospholipid fractions, thereby rendering them virtually unavailable to the developing fetus (see Sections 3.1 and 3.2).

The possibility of increasing the polyunsaturated fatty acid content of the colostrum and milk during early lactation by normal dietary means is also limited by rumen biohydrogenation and by the selective uptake by the mammary gland of plasma lipid fractions that contain only very low proportions of polyunsaturated fatty acids (Barry *et al.*, 1963). The recent inclusion in the maternal diet of polyunsaturated fatty acids "protected" from rumen biohydrogenation by a protein/formaldehyde coating obviously presents a method of overcoming the problems of biohydrogenation and thereby enables the proportions of polyunsaturated fatty acids in the maternal plasma lipid fractions, including the unesterified fatty acids and low-density lipoproteins, to be considerably increased (Noble *et al.*, 1977). As a result, in recent experiments, when pregnant sheep have been fed a diet containing "protected" linoleic acid during the latter part of pregnancy, the considerable increases in the linoleic acid contents of the maternal plasma unesterified fatty acids have resulted in significant increases in the linoleic and arachidonic acid contents of the plasma and tissues of the fetal and newborn lamb together with significant reductions in the triene : tetraene ratios (Noble *et al.*, 1978b; Shand *et al.*, 1978). Furthermore, the elevated concentrations of linoleic acid in the low-density lipoprotein triglycerides of the maternal plasma resulted in increased concentrations of linoleic acid in the colostrum and early milk secreted to well above the 1% of the total caloric intake suggested to be the minimum dietary level needed to satisfy the essential fatty acid requirement in nonruminant species.

The paradox in the case of the polyunsaturated fatty acid metabolism of the ruminant animal is therefore the conflict that exists between the ability of the mother to overcome her own problem of polyunsaturated fatty acid availability and the pressures that this creates with regard to the supply of polyunsaturated fatty acids to the fetus, which at best is extremely precarious.

References

Adams, E. P., and Heath, T. J., 1963, The phospholipids of ruminant bile, *Biochim. Biophys. Acta* **70**:688.

Alexander, D. P., Britton, H. G., and Nixon, D. A., 1967, Acetate metabolism in the isolated sheep fetus, *J. Physiol. (Lond.)* **190**:295.

Alexander, D. P., Britton, H. G., Cohen, N. M., and Nixon, D. A., 1969a, Plasma concentrations of insulin, glucose, free fatty acids and ketone bodies in the fetal and new born sheep and the response to a glucose load before and after birth, *Biol. Neonate* **14**:178.

Alexander, D. P., Britton, H. G., and Nixon, D. A., 1969b, Comparison of the metabolism of the human fetus and fetal sheep: Problems of measurement: Effects of glucose administraton, in: *Perinatal Medicine* (P. J. Huntingford, K. A. Huler, and E. Saling, eds.), pp. 183–187, Academic Press, New York.

Alexander, G., 1975, Body temperature control in mammalian young, *Br. Med. Bull.* **31**:62.

Alexander, G., 1978, Quantitative development of adipose tissue in fetal sheep, *Aust. J. Biol. Sci.* **31**:489.

Alexander, G., and Bell, A. W., 1975a, Maximum thermogenic response to cold in relation to the proportion of brown adipose tissue and skeletal muscle in the body and to other parameters in young lambs, *Biol. Neonate* **26**:182.

Alexander, G., and Bell, A. W., 1975b, Quantity and calucalted oxygen consumption during summit metabolism of brown adipose tissue in new born lambs, *Biol. Neonate* **26**:214.

Alexander, G., and Williams, D. J., 1970, Cardiovascular function in young lambs during summit metabolism, *J. Physiol. (Lond.)* **208**:65.

Alexander, G., Bell, A. W., and Williams, D., 1970, Metabolic response of lambs to cold, effects of prolonged treatment with thyroxine and of acclimation to low temperatures, *Biol. Neonate* **15**:198.

Alexander, G., Bell, A. W. and Hales, J. R. S., 1973a, Effects of cold exposure on tissue blood flow in the new born lamb, *J. Physiol. (Lond.)* **234**:65.

Alexander, G., Nicol, D., and Thorburn, G. D., 1973b, Thermogenesis in prematurely delivered lambs, in: *Fetal and Neonatal Physiology* (K. S. Comline, K. W. Cross, G. S. Dawes, and P. W. Nathanielsz, eds.), pp. 410–417, Cambridge University Press, London.

Alexander, G., Bennett, J. W. and Gemmell, R. T., 1975, Brown adipose tissue in the new born calf (*Bos taurus*), *J. Physiol. (Lond.)* **244**:223.

Ballard, F. J., Hanson, R. W., and Kronfield, D. S., 1969, Gluconeogenesis and lipogenesis in tissue from ruminant and non-ruminant animals, *Fed. Proc.* **28**:218.

Barry, J. M., Bartley, W., Linzell, J. L., and Robinson, D. S., 1963, The uptake from the blood of triglyceride fatty acids of chylomicra and low density lipoproteins by the mammary gland of the goat, *Biochem. J.* **89**:6.

Bassett, J. M., and Wallace, A. L. C., 1966, Short-term effects of ovine growth hormone on plasma glucose, free fatty acids and ketones in sheep, *Metabolism* **15**:933.

Battaglia, F. C., and Meschia, G., 1973, Fetal metabolism and substrate utilisation, in: *Fetal and Neonatal Physiology* (R. S. Comline, K. W. Cross, G. S. Dawes, and P. W. Nathanielsz, eds.), pp. 382–397, Cambridge University Press, London.

Bickerstaffe, R., and Annison, E. F., 1969, Triglyceride synthesis by the small-intestinal epithelium of the pig, sheep and chicken, *Biochem J.* **111**:419.

Biezenski, J. J., 1970, Role of placenta in fetal lipid metabolism ii. Phospholipid transfer in early rabbit gestation, *Am. J. Obstet. Gynecol.* **108**:638.

Biezenski, J. J., Spaet, T. H., and Gordon, A. L., 1963, Phospholipid patterns in subcellular fractions of adult and immature rat organs, *Biochim. Biophys. Acta* **70**:75.

Biezenski, J. J., Carrozza, J., and Li, J., 1971, Role of placenta in fetal lipid metabolism. iii. Formation of rabbit plasma phospholipids, *Biochim. Biophys. Acta* **239**:92.

Bobek, P., and Ginter, E., 1966, Metabolism of lipids in rats exposed to heat under conditions of a normal and high fat–high cholesterol diet, *J. Nutr.* **89**:373.

Body, D. R., and Shorland, F. B., 1974, The fatty acid composition of the main phospholipid fractions of the rumen and abomasum tissues of fetal and adult sheep, *J. Sci. Food Agric.* **25**:197.

Body, D. R., Shorland, F. B., and Czochanska, Z., 1970, Changes in the composition of the rumen and abomasum lipids of sheep from birth to maturity, *J. Sci. Food Agric.* **21**:220.

Brockerhoff, H., and Jensen, R. G., 1974, *Lipolytic Enzymes,* Academic Press, London.

Burr, G. O., and Burr, M. M., 1929, A new deficiency disease produced by the rigid exclusion of fat from the diet, *J. Biol. Chem.* **82**:345.

Burt, R. L., 1960, Plasma nonesterified fatty acids in normal pregnancy and the puerperium, *Obstet. Gynecol.* **15**:460.

Cannon, B., Romert, L., Sundin, U., and Barnard, T., 1977, Morphology and biochemical properties of perirenal adipose tissue from lamb (*Ovis aries*). A comparison with brown adipose tissue, *Comp. Biochem. Physiol.* **56B**:87.

Char, V. C., and Creasy, R. K., 1976, Acetate as a metabolic substrate in the fetal lamb, *Am. J. Physiol.* **230**:357.

Chida, N., Adams, F. H., Nozaki, M., and Norman, A., 1966, Changes in lamb-lung lipids during gestation, *Proc. Soc. Exp. Biol. Med.* **122**:60.

Christie, W. W., 1978, The composition, structure and function of lipids in the tissues of ruminant animals, *Prog. Lipid Res.* **17**:111.

Christie, W. W., 1979, The effects of diet and other factors on the lipid composition of ruminant tissues and milk, *Prog. Lipid Res.* **17**:245.
Church, D. C., Jessup, G. L., and Bogart, R., 1962, Stomach development in the suckling lamb, *Am. J. Vet. Res.* **23**:220.
Comline, R. S., and Silver, M., 1972, The composition of fetal and maternal blood during parturition in the ewe, *J. Physiol. (Lond.)* **222**:233.
Comline, R. S., and Silver, M., 1976, Some aspects of fetal and uteroplacental metabolism in cows with indwelling umbilical and uterine vascular catheters, *J. Physiol. (Lond.)* **260**:571.
Connor, W. E., and Lin, D. S., 1967, Placental transfer of cholesterol-4-^{14}C into rabbit and guinea pig fetus, *J. Lipid Res.* **8**:558.
Cunningham, H. M., and Leat, W. W. F., 1969, Lipid synthesis by the monoglyceride and α-glycerophosphate pathways in sheep intestine, *Can. J. Biochem.* **47**:1013.
Cunningham, H. M., and Loosli, J. K., 1954, The effect of fat-free diets on young dairy calves with observations on metabolic fecal fat and digestion coefficients for lard and hydrogenated coconut oil, *J. Dairy Sci.* **37**:453.
Diamant, Y. Z., Mayorek, N., Neuman, S., and Shafrir, E., 1975, Enzymes of glucose and fatty acid metabolism in early and term human placenta, *Am. J. Obstet. Gynecol.* **121**:58.
Decaen, C., Adda, J., Lefaivre, R., Marquis, B., and Hoden, A., 1970, Changes in secretion of milk fat during lactation in cows, *Ann. Biol. Anim. Biochim. Biophys.* **10**:659.
De Gier, J., and Van Deenen, L. L. M., 1964, A dietary investigation on the variations in phospholipid characteristics of red cell membranes, *Biochim. Biophys. Acta* **84**:294.
De Rooij, R. E., and Hooghwinkel, G. J. M., 1967, Method for the simultaneous determination of phospholipids and their plasmalogens. The role of plasmalogens in the embryonic development of the bovine brain, *Acta Physiol. Pharmacol. Neerl.* **14**:410.
Downing, D. T., 1964, Branched chain fatty acids in the lipids of the newly born lamb, *J. Lipid Res.* **5**:210.
Dunkley, W. L., Smith, N. E., and Franke, A. A., 1977, Effects of feeding protected tallow on composition of milk and milk fat, *J. Dairy Sci.* **60**:1863.
Edson, J. L., Hudson, D. G., and Hull, D., 1975, Evidence for increased fatty acid transfer across the placenta during a maternal fast in rabbits, *Biol. Neonate* **27**:50.
Edwards-Webb, J. D., and Thompson, S. Y., 1977, Studies on lipid digestion in the preruminant calf. 2. A comparison of the products of lipolysis of milk fat by salivary and pancreatic lipases *in vitro*, *Br. J. Nutr.* **37**:431.
Eisenberg, S., Stein, Y., and Stein, O., 1967, The role of placenta in lysolecithin metabolism in rats and mice, *Biochim. Biophys. Acta* **137**:115.
Elphick, M. C., and Hull, D., 1977a, The transfer of free fatty acids across the rabbit placenta, *J. Physiol. (Lond.)* **264**:751.
Elphick, M. C., and Hull, D., 1977b, Rabbit placental clearing-factor lipase and transfer to the fetus of fatty acids derived from triglycerides injected into the mother, *J. Physiol. (Lond.)* **273**:475.
Elphick, M. C., and Hull, D., 1978, The transfer of fatty acids across the sheep placenta, *J. Physiol. (Lond.)* **276**:56.
Elphick, M. C., Hudson, D. G., and Hull, D., 1975, Transfer of fatty acids across the rabbit placenta, *J. Physiol. (Lond.)* **252**:29.
Elphick, M. C., Filshie, C. M., and Hull, D., 1978, The passage of fat emulsion across the human placenta, *Br. J. Obstet. Gynaecol.* **85**:610.
Fain, J. N., and Scow, R. O., 1966, Fatty acid synthesis *in vivo* in maternal and fetal tissues in the rat, *Am. J. Physiol.* **210**:19.
Flatmark, T., and Pedersen, J. I., 1975, Brown adipose tissue mitochondria, *Biochim. Biophys. Acta* **416**:53.
Foster, D. O., and Frydman, M. L., 1978, Comparison of microspheres and $^{86}Rb^+$ as tracers of the distribution of cardiac output in rats indicates invalidity of $^{86}Rb^+$ based measurements, *Can. J. Physiol. Pharmacol.* **56**:97.

Fujiwara, T., Adams, F. H., El-Salaway, A., and Sipos, S., 1968, 'Alveolar' and whole lung phospholipids in new born lambs, *Proc. Soc. Exp. Biol. Med.* **127**:962.

Garton, G. A., and Duncan, W. R. H., 1964, Blood lipids. 5. The lipids of sheep plasma, *Biochem. J.* **92**:472.

Garton, G. A., and Duncan, W. R. H., 1969a, Composition of adipose tissue triglycerides of neonatal and year-old lambs, *J. Sci. Food Agric.* **20**:39.

Garton, G. A., and Duncan, W. R. H., 1969b, Effect of diet and rumen development on the composition of adipose tissue triglycerides in the calf, *Br. J. Nutr.* **23**:421.

Gemmell, R. T., and Alexander, G., 1978, Ultrastructural development of adipose tissue in fetal sheep, *Aust. J. Biol. Sci.* **31**:505.

Gemmell, R. T., Bell, A. W., and Alexander, G., 1972, Morphology of adipose tissue cells in lambs at birth and during subsequent transition of brown to white adipose tissue in cold and warm conditions, *Am. J. Anat.* **133**:143.

Godfrey, N. W., 1961, The functional development of the calf, *J. Agric. Sci.* **57**:173.

Goldwater, W. H., and Stetten, D., 1947, Studies in fetal metabolism, *J. Biol. Chem.* **169**:723.

Gooden, J. M., 1973, The importance of lipolytic enzymes in milk-fed and ruminating calves, *Aust. J. Biol. Sci.* **26**:1189.

Gooden, J. M., and Lascelles, A. K., 1973, Relative importance of pancreatic lipase and pregastric esterase on lipid absorption in calves 1–2 weeks of age, *Aust. J. Biol. Sci.* **26**:625.

Hagerman, D. D., 1964, Enzymatic capabilities of the placenta, *Fed. Proc.* **23**:785.

Hahn, P., and Novak, M., 1975, Development of brown and white adipose tissue, *J. Lipid Res.* **16**:79.

Hamilton, R. K., and Raven, A. M., 1973, The influence of chemical modification on the hydrolysis of fats by pancreatic lipase and pregastric esterase with reference to fat digestion in the pre-ruminant calf, *J. Sci. Food Agric.* **24**:257.

Hardman, M. J., Hey, E. N., and Hull, D., 1969, The effect of prolonged cold exposure on heat production in new born rabbits, *J. Physiol. (Lond.)* **205**:39.

Harfoot, C. G., 1978a, Anatomy, physiology and microbiology of the ruminant digestive tract, *Prog. Lipid Res.* **17**:1.

Harfoot, C. G., 1978b, Lipid metabolism in the rumen, *Prog. Lipid Res.* **17**:21.

Harper, W. J., 1955, Apparent selective liberation of butyric acid from milk fat by the action of various lipase systems, *J. Dairy Sci.* **38**:1391.

Harper, W. J., and Gould, I. A., 1955, Lipase systems used in the manufacture of italian cheese. 1. General characteristics, *J. Dairy Sci.* **38**:87.

Heath, T. J., and Morris, B., 1963, The role of bile and pancreatic juice in the absorption of fat in ewes and lambs, *Br. J. Nutr.* **17**:465.

Hershfield, M. S., and Nemeth, A. M., 1968, Placental transport of free palmitic and linoleic acids in the guinea pig, *J. Lipid Res.* **9**:460.

Hill, K. J., Noakes, D. E., and Lowe, R. A., 1970, Gastric digestive physiology of the calf and piglet, in: *Physiology of Digestion and Metabolism in the Ruminant* (A. T. Phillipson, ed.), pp. 166–179, Oriel Press, Newcastle-upon-Tyne.

Holman, R. T., 1971, Essential fatty acid deficiency, in: *Progress in the Chemistry of Fats and Other Lipids,* Vol. 9 (R. T. Holman, ed.), pp. 275–348, Pergamon Press, Oxford.

Holman, R. T., 1973, Essential fatty acid deficiency in humans, in: *Dietary Lipids and Postnatal Development* (C. Galli, G. Jacine, and A. Pecile, eds.), pp. 127–143, Raven Press, New York.

Huber, J. T., Jacobson, N. L., Allen, R. S., and Hartman, P. A., 1961, Digestive enzyme activities in the young calf, *J. Dairy Sci.* **44**:1494.

Hull, D., 1975, Storage and supply of fatty acids before and after birth, *Br. Med. Bull.* **31**:32.

Hull, D., 1976, The function of brown adipose tissue in the new born, *Biochem. Soc. Trans.* **4**:226.

Hummel, L., Schirrmeister, W., and Wagner, H., 1975, Quantitative evaluation of the maternal-fetal transfer of free fatty acids in the rat, *Biol. Neonate* **26**:263.

Hummel, L., Zimmermann, T., Schirrmeister, W., and Wagner, H., 1976a, Synthesis, turnover and compartment analysis of the free fatty acids in the placenta of rats, *Acta Biol. Med. Ger.* **35**:1311.

Hummel, L., Schwartze, A., Schirrmeister, W., and Wagner, H., 1976b, Maternal plasma triglycerides as a source of fetal acids, *Acta Biol. Med. Ger.* **35**:1635.

Iliffe, J., Knight, B. L., and Myant, N. B., 1973, Fatty acid synthesis in the brown fat and liver of fetal and newborn rabbits, *Biochem. J.* **134**:341.

James, E., Meschia, G., and Battaglia, F. C., 1971. A-V differences of free fatty acids and glycerol in the ovine umbilical circulation, *Proc. Soc. Exp. Biol. Med.* **138**:823.

Jarasch, E. D., Reilly, C. E., Comes, P., Kartenbeck, J., and Franke, W. W., 1973, Isolation and characterisation of nuclear membranes from calf and rat thymus, *Hoppe Seylers Z. Physiol. Chem.* **354**:974.

Jenkinson, D. McE., Noble, R. C., and Thompson, G. E., 1968, Adipose tissue and heat production in the new born ox (*Bos taurus*), *J. Physiol. (Lond.)* **195**:639.

Jenness, R., 1974, Composition of milk, in: *Lactation,* vol. III (B. L. Larson and V. R. Smith, eds.), pp. 3–107, Academic Press, New York.

Jensen, R. G., and Sampugna, J., 1964, Lipolysis of synthetic and milk triglycerides by pregastric esterase, *J. Dairy Sci.* **47**:664.

Katz, M. L., and Bergmann, E. N., 1969, Hepatic and portal metabolism of glucose, free fatty acids and ketone bodies in the sheep, *Am. J. Physiol.* **216**:953.

Kayden, H. J., Dancis, J., and Money, W. L., 1969, Transfer of lipids across the guinea pig placenta, *Am. J. Obstet. Gynecol.* **104**:564.

Lambert, M. R., Jacobson, N. L., Allen, R. S., and Zaletel, J. H., 1954, Lipid deficiency in the calf, *J. Nutr.* **52**:259.

Leat, W. M. F., 1964, Plasma fatty acids of new born and neonatal ruminants, *Biochem. J.* **93**:22P.

Leat, W. M. F., 1966, Fatty acid composition of the plasma lipids of newborn and maternal ruminants, *Biochem. J.* **98**:598.

Leat, W. M. F., 1967, Plasma lipids of new born and adult ruminants and of lambs from birth to weaning, *J. Agric. Sci.* **69**:241.

Leidy, R. R., Russell, R. W., and Wise, G. H., 1975, Pregastric esterase in milk sham fed to adult jersey steers, *J. Dairy Sci.* **58**:563.

Lengemann, F. W., and Allen, N. N., 1955, The development of rumen function in the diary calf. i. Some characteristics of the rumen contents of cattle of various ages, *J. Dairy Sci.* **38**:651.

Lengemann, F. W., and Allen, N. N., 1959, Development of rumen function in the dairy calf. ii. Effect of diet upon characteristics of the rumen flora and fauna of young calves, *J. Dairy Sci.* **42**:1171.

Lopez-Santolino, A., Miller, O. N., and Muldrey, J. E., 1965, Effect of prenatal diet on serum cholesteryl ester fatty acids in new born and adult rats, *Proc. Soc. Exp. Biol. Med.* **118**:829.

Mallov, S., and Alousi, A. A., 1965, Lipoprotein lipase activity of rat and human placenta, *Proc. Soc. Exp. Biol. Med.* **119**:301.

Masters, C. J., 1964, The fatty acid components of ovine extrahepatic tissues after rumen development, *Aust. J. Biol. Sci.* **17**:200.

McBride, O. W., and Korn, E. D., 1964, Uptake of free fatty acids and chylomicron glycerides by guinea pig mammary gland in pregnancy and lactation, *J. Lipid Res.* **5**:453.

Mead, J. F., 1971, The metabolism of the polyunsaturated fatty acids, in: *Progress in the Chemistry of Fats and Other Lipids,* Vol. 9 (R. T. Holman, ed.) pp. 159–192, Pergamon Press, Oxford.

Mead, J. F., and Slaton, W. M., 1956, Metabolism of essential fatty acids. iii. Isolation of 5,8,11-eicosatrienoic acid from fat-deficient rats, *J. Biol. Chem.* **219**:705.

Mohrhauer, H., and Holman, R. T., 1963, The effect of dose level of essential fatty acids upon fatty acid composition of the rat liver, *J. Lipid Res.* **4**:151.

Moore, J. H., and Christie, W. W., 1979, Lipid metabolism in the mammary gland of ruminant animals, *Prog. Lipid Res.* **17**:347.

Morrison, W. R., 1970, Milk lipids, in: *Topics in Lipid Chemistry* (F. D. Gunstone, ed.), pp. 51-106, Logos Press, London.

Morriss, F. H., Boyd, R. D. H., Makowski, E. L., Meschia, G., and Battaglia, F. C., 1974, Umbilical V-A differences of acetoacetate and β-hydroxybutyrate in fed and starved ewes, *Proc. Soc. Exp. Biol. Med.* **145**:879.

Muldrey, J. E., Hamilton, J. G., Wells, J. A., Swartwout, J. R., and Miller, O. N., 1961, Differences in the cholesteryl ester fatty acid pattern in human maternal and cord serum, *Fed. Proc.* **20**:277.

Nelson, G. J., 1973, The lipid composition of plasma lipoprotein density classes of sheep *Ovis aries*, *Comp. Biochem. Physiol.* **46B**:81.

Nicholls, D. G., 1976, The bioenergetics of brown adipose tissue mitochondria, *FEBS Lett.* **61**:103.

Noble, R. C., 1978, Digestion, absorption and transport of lipids in ruminant animals, *Prog. Lipid Res.* **17**:55.

Noble, R. C., and Moore, J. H., 1974, Heat exposure and the fatty acid composition of the plasma of the young lamb, *Res. Vet. Sci.* **17**:204.

Noble, R. C., Steele, W., and Moore, J. H., 1969a, The effects of dietary palmitic and stearic acids on milk fat composition in the cow, *J. Dairy Res.* **36**:375.

Noble, R. C., Thompson, G. E., and Moore, J. H., 1969b, The effects of the intravenous administration of noradrenalin on the composition of the plasma lipids of sheep, *Res. Vet. Sci.* **10**:555.

Noble, R. C., Steele, W., and Moore, J. H., 1970a, The composition of ewe's milk fat during early and late lactation, *J. Dairy Res.* **37**:297.

Noble, R. C., Steele, W., and Moore, J. H., 1970b, Diet and the diacyl phospholipids in the livers of lambs, *Biochim. Biophys. Acta* **218**:359.

Noble, R. C., Christie, W. W., and Moore, J. H., 1971a, Diet and the lipid composition of adipose tissue in the young lamb, *J. Sci. Food Agric.* **22**:616.

Noble, R. C., Steele, W., and Moore, J. H., 1971b, Fatty acid composition of liver lipids of young lambs, *Br. J. Nutr.* **26**:97.

Noble, R. C., Steele, W., and Moore, J. H., 1971c, Postnatal changes in the phospholipid composition of livers from young lambs, *Lipids* **6**:926.

Noble, R. C., Steele, W., and Moore, J. H., 1971d, Diet and the fatty acids in the plasma of lambs during the first eight days after birth, *Lipids* **6**:26.

Noble, R. C., Steele, W., and Moore, J. H., 1971e, The plasma lipids of the ewe during pregnancy and lactation, *Res. Vet. Sci.* **12**:47.

Noble, R. C., Steele, W., and Moore, J. H., 1972, The metabolism of linoleic acid by the young lamb, *Br. J. Nutr.* **27**:503.

Noble, R. C., O'Kelly, J. C., and Moore, J. H., 1973, Observations on changes in lipid composition and lecithin-cholesterol acyl transferase reaction of bovine plasma induced by heat exposure, *Lipids* **8**:216.

Noble, R. C., Crouchman, M. L., and Moore, J. H., 1974, The presence of linoleic acid in the skin surface lipids of the ox, *Res. Vet. Sci.* **17**:372.

Noble, R. C., Crouchman, M. L., and Moore, J. H., 1975a, Plasma cholesterol ester formation in the neonatal lamb, *Biol. Neonate* **26**:117.

Noble, R. C., Crouchman, M. L., Jenkinson, D. McE., and Moore, J. H., 1975b, Relationship between lipids in plasma and skin secretions of neonatal calf with particular reference to linoleic acid, *Lipids* **10**:128.

Noble, R. C., Vernon, R. G., Christie, W. W., Moore, J. H., and Evans, A. J., 1977, The effect of dietary fats on the plasma lipid composition of sheep, *Lipids* **12**:423.

Noble, R. C., Shand, J. H., and Moore, J. H., 1978a, Implications for the role of the the placenta in synthesis of arachidonic acid [20:4(n-6)] in the sheep, *Int. Res. Commun. Syst. J. Med. Sci.* **6**:235.

Noble, R. C., Shand, J. H., Drummond, J. T., and Moore, J. H., 1978b, "Protected" polyunsaturated fatty acid in the diet of the ewe and the essential fatty acid status of the neonatal lamb, *J. Nutr.* **108**:1868.

Noble, R. C., Shand, J. H., Bell, A. W., Thompson, G. E., and Moore, J. H., 1978c, The transfer of free palmitic and linoleic acids across the ovine placenta, *Lipids* **13**:610.

Noble, R. C., Shand, J. H., and Bell, A. W., 1979, Fetal to maternal transfer of palmitic and linoleic acids across the sheep placenta, *Biol. Neonate* **36**:113.

O'Kelly, J. C., 1968, Comparative studies of lipid metabolism in Zebu and British cattle in a tropical environment, *Aust. J. Biol. Sci.* **21**:1013.

Olivercrona, T., Hernell, O., Egelrud, T., Billstrom, A., Helander, H., Samuelson, G., and Frederikzon, B., 1973, Studies on the gastric lipolysis of milk lipids in suckling rats and in human infants, in: *Dietary Lipids and Postnatal Development* (C. Galli, G. Jacini, and A. Pecile, eds.), pp. 77–89, Raven Press, New York.

Otterby, D. E., Ramsey, H. A., and Wise, G. H., 1964a, Lipolysis of milk fat by pregastric esterase in the abomasum of the calf, *J. Dairy Sci.* **47**:993.

Otterby, D. E., Ramsey, H. A., and Wise, G. H., 1964b, Source of lipolytic enzymes in the abomasum of the calf, *J. Dairy Sci.* **47**:997.

Palmquist, D. L., McClure, K. E., and Parker, C. F., 1977, Effect of protected or polyunsaturated fat fed to pregnant and lactating ewes on milk composition, lamb plasma fatty acids and growth, *J. Anim. Sci.* **45**:1152.

Pan, Y. S., 1970, Breed and seasonal differences in quantities of lipids on skin surface and hair in cattle, *J. Agric. Sci.* **75**:41.

Pascaud, M., Rougier, A., and Delhaye, N., 1977, Assimilation of ^{14}C-linoleic acid by the rat fetus, *Nutr. Metab.* **21**:310.

Patton, S., and Jensen, R. G., 1975, Lipid metabolism and membrane functions of the mammary gland, in: *Progress in the Chemistry of Fats and Other Lipids,* Vol. 14 (R. T. Holman, ed.), pp. 163–277, Pergamon Press, Oxford.

Payne, E., 1978, Fatty acid composition of tissue phospholipids of the fetal calf and neonatal lamb, deer, calf and piglet as compared with the cow, sheep, deer and pig, *Br. J. Nutr.* **39**:45.

Peric-Golia, L., and Socic, H., 1968, Biliary bile acids and cholesterol in developing sheep, *Am. J. Physiol.* **215**:1284.

Persson, B., and Tunell, R., 1971, Influence of environmental temperature and acidosis on lipid mobilisation in the human infant during the first two hours after birth, *Acta Paediatr. Scand.* **60**:385.

Phanteliadis, C., and Troll, U., 1976, Fatty acid pattern of the serum lipids (normal values in mothers, placentae and infants), *Z. Ernaehrungswiss.* **15**:305.

Phillipson, A. T., 1977, Ruminant digestion, in: *Dukes' Physiology of Domestic Animals,* Ninth Edition (M. J. Swenson, ed.), pp. 250–286, Cornell University Press, Ithaca.

Poe, S. E., Ely, D. G., Mitchell, G. E., Glimp, H. A., and Deweese, W. P., 1971, Rumen development in lambs ii. Rumen metabolite changes, *J. Anim. Sci.* **32**:989.

Popjak, G., 1947, Synthesis of phospholipids in the fetus, *Nature* **160**:841.

Popjak, G., 1954, The origin of fetal lipides, *Cold Spring Harbor Symp. Quant. Biol.* **19**:200.

Popjak, G., and Beeckmans, M. L., 1950a, Are phospholipins transmitted through the placenta? *Biochem. J.* **46**:99.

Popjak, G., and Beeckmans, M. L., 1950b, Synthesis of cholestrol and fatty acids in fetuses and in mammary glands of pregnant rabbits, *Biochem. J.* **46**:547.

Porter, J. W. G., 1969, Digestion in the pre-ruminant animal, *Proc. Nutr. Soc.* **28**:115.

Portman, O. W., Behrman, R. E., and Soltys, P., 1969, Transfer of free fatty acids across the primate placenta, *Am. J. Physiol.* **216**:143.

Poukka, R., 1966, Tissue lipids in calves suffering from muscular dystrophy, *Br. J. Nutr.* **20**:245.

Radostits, O. M., and Bell, J. M., 1970, Nutrition of the pre-ruminant dairy calf with special reference to the digestion and absorption of nutrients: A review, *Can. J. Anim. Sci.* **50**:405.

Ramsey, H. A., and Young, J. W., 1961a, Role of pregastric esterase in the abomasal hydrolysis of milk fat in the young calf, *J. Dairy Sci.* **44**:2227.

Ramsey, H. A., and Young, J. W., 1961b, Substrate specificity of pregastric esterase from the calf, *J. Dairy Sci.* **44**:2304.

Ramsey, H. A., Wise, G. H., and Tove, S. B., 1956, Esterolytic activity of certain alimentary and related tissues from cattle in different age groups, *J. Dairy Sci.* **39**:1312.

Raphael, B. C., Dimick, P. S., and Pappione, D. L., 1973, Lipid characterisation of bovine serum lipoproteins throughout gestation and lactation, *J. Dairy Sci.* **56**:1025.

Reid, R. L., 1953, Studies on the carbohydrate metabolism of sheep. vi. Interrelationships between changes in the distribution and levels of glucose and in the levels of volatile fatty acid in the blood of lambs, *Aust. J. Agric. Res.* **4**:213.

Richardson, G. H., and Nelson, J. H., 1967, Assay and characterisation of pregastric esterase, *J. Dairy Sci.* **50**:1061.

Robertson, A., and Sprecher, H., 1967, Human placental lipid metabolism. iii. Synthesis and hydrolysis of phospholipids, *Lipids* **2**:403.

Robertson, A., Sprecher, H., and Wilcox, J., 1968, Free fatty acid patterns of human maternal plasma, perfused placenta and umbilical cord plasma, *Nature* **217**:378.

Rook, J. A. F., 1961a, Variations in the chemical composition of the milk of the cow—Part 1, *Dairy Sci. Abstr.* **23**:251.

Rook, J. A. F., 1961b, Variations in the chemical composition of the milk of the cow—Part II, *Dairy Sci. Abstr.* **23**:303.

Rose, H. G., and Frenster, J. H., 1965, Composition and metabolism of lipids within repressed and active chromatin of interphase lymphocytes, *Biochim. Biophys. Acta* **106**:577.

Roux, J. F., 1966, Lipid metabolism in the fetal and neonatal rabbit, *Metabolism* **15**:856.

Roy, J. H. B., 1974, Feeding the new born: Comparative problems in animals and man, *Proc. Nutr. Soc.* **33**:79.

Roy, J. H. B., Shillam, K. W. G., Thompson, S. Y., and Dawson, D. A., 1961, The effect of emulsification of a milk-substitute diet by mechanical homogenisation and by the addition of soya bean lecithin on plasma lipid and vitamin A levels and on the growth rate of the new born calf, *Br. J. Nutr.* **15**:541.

Sack, J., Beaudry, M., Oh, W., and Fisher, D. A., 1976, Umbilical cord cutting triggers hyper-triiodothyroninemia and non-shivering thermogenesis in the new born lamb, *Pediatr. Res.* **10**:169.

Sander, E. G., Warner, R. G., Harrison, H. N., and Loosli, J. K., 1959, The stimulatory effect of sodium butyrate and sodium propionate on the development of rumen mucosa in the young calf, *J. Dairy Sci.* **42**:1600.

Schenk, H., Heim, T., Wagner, H., Winkler, L., Varga, F., and Goetze, E., 1974, *In vivo* metabolism of ^{14}C-labeled palmitic acid in serum and in brown and white adipose tissue of well fed and starved new born rabbits. 2. Tracer kinetic analysis of the incorporation rate of serum free fatty acids into the main lipid fractions of brown and white adipose tissues, *Biol. Neonate* **24**:256.

Scott, T. W., Setchell, B. P., and Bassett, J. M., 1967, Characterisation and metabolism of ovine fetal lipids, *Biochem. J.* **104**:1040.

Scott, T. W., Cook, L. J., and Mills S. C., 1971, Protection of dietary polyunsaturated fatty acids against microbial hydrogenation in ruminants, *J. Am. Oil Chem. Soc.* **48**:358.

Senft, B., and Klobasa, F., 1970, Studies on the composition of cow's colostrum, *Milchwissenschaft,* **25**:391.

Shand, J. H., and Noble, R. C., 1979a, The role of maternal triglycerides in the supply of lipids to the ovine fetus, *Res. Vet. Sci.* **26**:117.

Shand, J. H., and Noble, R. C., 1979b, Δ9- and Δ6-desaturase activities of the ovine placenta and their role in the supply of fatty acids to the fetus, *Biol. Neonate* **36**:298.

Shand, J. H., Noble, R. C., and Moore, J. H., 1978, Dietary influences on fatty acid metabolism in the liver of the neonatal lamb, *Biol. Neonate* **34**:217.

Shannon, A. D., and Lascelles, A. K., 1966, Changes in the concentration of lipids and some other constituents in the blood plasma of calves from birth to six months of age, *Aust. J. Biol. Sci.* **19**:831.

Sheriha, G. M., Waller, G. R., Chan, T., and Tillman, A. D., 1968, Composition of bile acids in ruminants, *Lipids* **3**:72.

Shorland, F. B., Body, D. R., and Gass, J. P., 1966, The foetal and maternal lipids of Romney sheep. ii. The fatty acid composition of the lipids from the total tissues, *Biochim. Biophys. Acta* **125**:217.

Siewert, K. L., and Otterby, D. E., 1970, Effects of *in vivo* and *in vitro* acid environments on activity of pregastric esterase, *J. Dairy Sci.* **53**:571.

Siewert, K. L., and Otterby, D. E., 1971, Effect of fat source on relative activity of pregastric esterase and on glyceride composition of intestinal contents, *J. Dairy Sci.* **54**:258.

Sklan, D., Volcani, R., and Budowski, P., 1972, Effects of diets low in fat or essential fatty acids on the fatty acid composition of blood lipids of calves, *Br. J. Nutr.* **27**:365.

Smith, R. E., and Horwitz, B. A., 1969, Brown fat and thermogenesis, *Physiol. Rev.* **49**:330.

Spellacy, W. N., Asbacher, L. V., Harris, G. K., and Buhi, W. C., 1974, Total cholesterol content in maternal and umbilical vessels in term pregnancies, *Obstet. Gynecol.* **44**:661.

Steele, W., and Moore, J. H., 1968a, The effects of dietary tallow and cottonseed oil on milk fat secretion in the cow, *J. Dairy Res.* **35**:223.

Steele, W., and Moore, J. H., 1968b, Further studies on the effects of dietary cottonseed oil on milk fat secretion in the cow, *J. Dairy Res.* **35**:343.

Steele, W., and Moore, J. H., 1968c, The effects of mono-unsaturated and saturated fatty acids in the diet on milk fat secretion in the cow, *J. Dairy Res.* **35**:353.

Steele, W., and Moore, J. H., 1968d, The effects of a series of saturated fatty acids in the diet on milk fat secretion in the cow, *J. Dairy Res.* **35**:361.

Steele, W., Noble, R. C., and Moore, J. H., 1971, The relationship between plasma lipid composition and milk fat secretion in cows given diets containing soybean oil, *J. Dairy Res.* **38**:57.

Steven, D. H., and Marshall, A. B., 1970, Organisation of the rumen epithelium, in: *Physiology of Digestion and Metabolism in the Ruminant,* (A. T. Phillipson, ed.), pp. 80–100, Oriel Press, Newcastle-upon-Tyne.

Storry, J. E., 1970, Ruminant metabolism in relation to the synthesis and secretion of milk, *J. Dairy Res.* **37**:139.

Storry, J. E., Hall, A. J., and Johnson, V. W., 1973, The effects of increasing amounts of dietary tallow on milk fat secretion in the cow, *J. Dairy Res.* **40**:293.

Svanborg, A., and Vikrot, O., 1965, Plasma lipid fractions, including individual phospholipids, at various stages of pregnancy, *Acta Med. Scand.* **178**:615.

Tamate, H., McGilliard, A. .D, Jacobson, N. L., and Getty, R., 1962, Effect of various dietaries on the anatomical development of the stomach in the calf, *J. Dairy Sci.* **45**:408.

Ter Meulen, U., Molnar, S., and Neumann, H., 1970, Fatty acid composition of tissue lipids of the new born calf, *Z. Tierphysiol.* **26**:206.

Ter Meulen, U., Mueller, D., and Molnar, S., 1972, Development of the brown adipose kidney tissue during the fetal growing period in calves, *Z. Tierphysiol.* **29**:85.

Ter Meulen, U., Duchanowo, H., Hakles, H., Nordbeck, H., Preusse, C. J., and Molnar, S., 1976, Investigations of the non-shivering thermogenesis of new born calves under cold stress, *Z. Tierphysiol.* **36**:283.

Ternouth, J. H., and Buttle, H. L., 1973, Concurrent studies on the flow of digesta in the duodenum and of exocrine pancreatic secretion of calves. i. The collection of the exocrine pancreatic secretion from a duodenal cannula, *Br. J. Nutr.* **29**:387.

Ternouth, J. H., Roy, J. H. B., and Siddons, R. C., 1974, Concurrent studies of the flow of digesta in the duodenum and of exocrine pancreatic secretion of calves. 2. The effects of addition of fat to skim milk and of 'severe' pre-heating treatment of spray-dried skim milk powder, *Br. J. Nutr.* **31**:13.

Ternouth, J. H., Roy, J. H. B., and Shotton, S. M., 1976, Concurrent studies of the flow of digesta in the duodenum and of exocrine pancreatic secretion of calves. 4. The effect of age, *Br. J. Nutr.* **36**:523.

Thompson, G. E., and Bell, A. W., 1976, Heat production in the new born ox during noradrenaline infusion, *Biol. Neonate* **28**:375.

Thompson, G. E., and Clough, D. P., 1972, The effect of cold exposure on plasma lipids of the new born and adult ox, *J. Exp. Physiol.* **57**:192.

Thompson, G. E., and Jenkinson, D. McE., 1969, Non-shivering thermogenesis in the new born lamb, *Can J. Physiol. Pharmacol.* **47**:249.

Thompson, G. E., and Jenkinson, D. McE., 1970, Adipose tissue in the new born goat, *Res. Vet. Sci.* **11**:102.

Toothill, J., Thompson, S. Y., and Edwards-Webb, J. D., 1976, Studies on lipid digestion in the preruminant calf. The source of lipolytic activity in the abomasum, *Br. J. Nutr.* **26**:439.

Troll, U., Phanteliadis, C., and Hexel, R., 1977, Comparison of the serum fatty acid patterns of mothers and their new born infants, *Monatsschr. Kinderheilkd.* **125**:28.

Van Duyne, C. M., and Havel, R. J., 1959, Plasma unesterified fatty acid concentration in fetal and neonatal life, *Proc. Soc. Exp. Biol. Med.* **102**:599.

Van Duyne, C. M., Parker, H. R., Havel, R. J., and Holm, L. W., 1960, Free fatty acid metabolism in fetal and new born sheep, *Am. J. Physiol.* **199**:987.

Van Duyne, C. M., Havel, R. J., and Felts, J. M., 1962, Placental transfer of palmatic acid-1-C^{14} in rabbits, *Am. J. Obstet. Gynecol.* **84**:1069.

Vernon, R. G., 1975, Effect of dietary safflower oil upon lipogenesis in neonatal lamb, *Lipids* **10**:284.

Vernon, R. G., 1977, Development of perirenal adipose tissue in the neonatal lamb: Effects of dietary safflower oil, *Biol. Neonate* **32**:15.

Vernon, R. G., 1980, Lipid metabolism in the adipose tissue of ruminant animals, *Prog. Lipid Res.* **19**:23.

Villee, C. A., and Hagermann, D. A. D., 1958, Effect of oxygen deprivation on the metabolism of fetal and adult tissues, *Am. J. Physiol.* **194**:457.

Wahle, K. J., 1974, Desaturation of long chain fatty acids by tissue preparations of the sheep, rat and chicken, *Comp. Biochem. Physiol.* **48B**:87.

Walker, D. M., and Stokes, G. B., 1970, The nutritive value of fat in the diet of the milk-fed lamb. i. The apparent and corrected digestibilities of different dietary fats and their constituent fatty acids, *Br. J. Nutr.* **24**:425.

Wardrop, I. D., 1960, The postnatal growth of the visceral organs of the lamb, ii. The effect of diet on growth rate, with particular reference to the parts of the alimentary tract, *J. Agric. Sci.* **55**:127.

Wardrop, I. D., 1961, Some preliminary observations on the histological development of the fore-stomachs of the lamb. i. Histological changes due to age in the period from forty-six days of fetal life to seventy-seven days of postnatal life, *J. Agric. Sci.* **57**:335.

Wardrop, I. D., and Coombe, J. B., 1960, The postnatal growth of the visceral organs of the lamb. i. The growth of the visceral organs of the grazing lamb from birth to sixteen weeks of age, *J. Agric. Sci.* **54**:140.

Wardrop, I. D., and Coombe, J. B., 1961, The development of rumen function in the lamb, *Aust. J. Agric. Res.* **12**:661.

Warner, R. G., Flatt, W. P., and Loosli, J. K., 1956, Dietary factors influencing the development of the ruminant stomach. *J. Agric. Food Chem.* **4**:788.

Weinhold, P. A., and Villee, C. A., 1965, Phospholipid metabolism in the liver and lung of rats during development, *Biochim. Biophys. Acta* **106**:540.

Wise, G. H., Miller, P. G., Anderson, G. W., and Jones, J. C., 1968, Changes in milk products sham fed to calves, iii. Effects of concentration of fat, *J. Dairy Sci.* **51**:1077.

Wise, G. H., Miller, P. G., Anderson, G. W., and Linnerud, A. C., 1976, Changes in milk products sham fed to calves. iv. Suckling from a nurse cow versus consuming from either a nipple feeder or an open pail, *J. Dairy Sci.* **59**:97.

Wrenn, T. R., Bitman, J., Weyant, J. R., Wood, D. L., Wiggers, K. D., and Edmondson, L. F., 1977, Milk and tissue lipid composition after feeding cows protected polyunsaturated fat for two years, *J. Dairy Sci.* **60**:521.

Zee, P., 1967, Lipid metabolism in the new born. i. Phospholipids in cord and maternal sera, *Pediatrics* **39**:82.

Index

www.ingramcontent.com/pod-product-compliance
Ingram Content Group UK Ltd.
Pitfield, Milton Keynes, MK11 3LW, UK
UKHW041830200726
13854UKWH00002BA/979

* 9 7 8 1 4 6 1 3 9 9 3 5 3 *